Springer Series in Solid-State Sciences

Volume 189

The Springer Series in Solid-State Sciences consists of fundamental scientific books prepared by leading researchers in the field. They strive to communicate, in a systematic and comprehensive way, the basic principles as well as new developments in theoretical and experimental solid-state physics.

More information about this series at http://www.springer.com/series/682

Sanju Gupta · Avadh Saxena
Editors

The Role of Topology in Materials

Editors
Sanju Gupta
Department of Physics and Astronomy
Western Kentucky University
Bowling Green, KY
USA

Avadh Saxena
Theoretical Division
Los Alamos National Laboratory
Los Alamos, NM
USA

ISSN 0171-1873 ISSN 2197-4179 (electronic)
Springer Series in Solid-State Sciences
ISBN 978-3-030-09529-1 ISBN 978-3-319-76596-9 (eBook)
https://doi.org/10.1007/978-3-319-76596-9

Printed on acid-free paper

This Springer imprint is published by the registered company Springer International Publishing AG part of Springer Nature
The registered company address is: Gewerbestrasse 11, 6330 Cham, Switzerland

Foreword

Topology studies the properties of geometrical objects that remain unaltered when subjected to transformations known as deformations in the materials context and homeomorphisms with regard to mathematical objects. It is no wonder that the concepts and methods of topology permeate modern mathematics and physics. Over the past few decades, it has also clearly emerged that topological notions have become pervasive in materials science. New synthesis, characterization, modeling, and simulation techniques are extending the frontiers of topological materials. For this reason, there has been a need to bring these new developments and excitement in a cohesive book to the broader materials community. Both co-editors are well qualified in this interdisciplinary field as they have diligently worked at the interface of materials science and topology. They have earnestly promoted this field by organizing international symposia and workshops on this topic. Similarly, all the chapter authors bring forth their passion and technical rigor in the book. This book seamlessly bridges a gap between how materials science is traditionally viewed and how topological concepts can be gainfully employed to understand complex materials in an accessible way. This book thus emphatically promotes the emerging paradigm of geometry/topology → properties → applications in the context of soft and hard condensed matter, biomaterials, and a suite of novel topological materials including topological insulators, low-dimensional carbons, Dirac- and Weyl semimetals. There is an emphasis on the role of topological defects such as vortices, skyrmions, and disclinations which are discussed in the materials that harbor them. The book will be accessible to a broad cohort of researchers ranging from graduate students to beginners and advanced practitioners alike.

Sapporo, Japan

Satoshi Tanda
Professor, Department of Applied Physics,
Center of Topological Science and Technology
Hokkaido University

Preface

Role of Topology in Materials Science is a rapidly evolving area of research with significant cross-fertilization between topological concepts and materials science. Topology is ubiquitous in physical systems yet most materials are not usually viewed from this perspective. Beyond the usual structural and geometric descriptions, it is shown how topological considerations lead to the emergence of novel phases of matter and how to understand their properties from this new vantage point. One of the major objectives of the book is to establish why topology is important for understanding modern and future functional materials, to identify common topology inspired synthetic routes, to develop novel experimental characterization techniques, to evolve theoretical and simulation strategies as well as topological data analysis, related predictive modelling and phenomenology in topological materials. The book is likely to be broadly accessible and caters to junior researchers, graduate students as well as experts in physics and applied mathematics besides materials science and engineering.

The book contains eleven chapters, which represent partly a review with a broad perspective and partly original research aimed at identifying open issues in topology of materials. The first two chapters describe the basic notions of topology in a broad spectrum of materials and invoke the concepts of genus, Euler characteristic, network topology and homotopy classes in the context of low-dimensional carbons and magnetic materials and interlinking topology and geometry. Both real space (*e.g.* Möbius ribbons) and momentum space (*e.g.* Dirac materials) topological materials are explored using the topological concepts.

Chapters 3–6 illustrate the importance of topology in condensed matter systems. Chapter 3 demonstrates the emergence of curvature when non-hexagonal (e.g. pentagon, heptagon) carbon rings are embedded into the nanoscale graphene sheets. In particular, emphasis is placed on the geometry-property correlations that are inherent in fullerene polymers and single walled carbon nanocoils. These structures show curvature or torsion arising from the local change in the carbon network topology. Chapter 4 focuses on artificial spin ice as a model system to study topology by design in magnetic nanomaterials such as permalloy. The role of geometric frustration is accentuated here in addition to the concepts of magnetic monopoles, Dirac strings and 'magnetricity'.

In Chap. 5, a magnetic topological excitation called a skyrmion is studied in confined geometries. A skyrmion is a magnetic whirl or a spin configuration quite distinct from a magnetic vortex and a domain wall. Due to its topological stability and an extremely low current needed to move it, potential applications in low-power spintronics and memory devices are discussed. The importance of the Dzyloshinskii-Moriya interaction in stabilizing skyrmions and the role of confinement (using stripes) in creating individual skyrmions are emphasized. Chapter 6 provides an in-depth understanding of topological phases in quantum matter by virtue of electronic structure calculations and angle-resolved photoemission spectroscopy. Broadly described as Dirac materials (i.e. with linear electronic dispersion), they include both two- and three-dimensional topological insulators, topological superconductors, topological crystalline insulators as well as Weyl and Dirac semi-metals, and are delineated via the Fermi surface evolution and relevant symmetries. Finally, the effects of electron correlations in creating exotic topological phases, e.g. topological Kondo insulators, are discussed.

The next two chapters delve into what we may refer to as biological-mathematical applications of topology. Chapter 7 deals with nanomaterials which may form as wire network graphs from triply periodic surfaces including gyroids. The analysis entails symmetric groups, singularity theory and representation theory. The esoteric analysis, however, leads to very useful results in terms of understanding band structure and Dirac points in materials discussed in Chap. 4 such as Weyl and Dirac semi-metals. In the presence of a magnetic field the analysis indicates that the spectrum of the gyroid wire network can be characterized as the three-dimensional analogue of the Hofstadter's butterfly. In chapter 8, knot theory is invoked to explain a variety of entangled biomolecules such as proteins. Such proteins form knots, links, slipknots and lassos and their knotting fingerprints are described by relevant databases, e.g. KnotProt. The key idea here is that entanglement is a global property of a protein chain. Besides folding of the proteins their functions are portrayed via knotting, which is not accidental but must play certain biological role.

The subsequent two chapters elaborate on the important role of topology in soft matter. Chapter 9 nicely captures a slew of rich phenomena in liquid crystals: nematic, smectic and cholesteric mesophases. Within the homotopy theory of defects such topological entities as Schlieren textures, disclination loops, hedgehogs, skyrmions, torons and Hopf textures are identified and classified. The impact of geometric constraints on topological properties is clearly illustrated. The topological characterization of knotted defect lines and construction of the director field with arbitrary knotted disclination lines are illustrated. Chapter 10 explores the topologically complex morphologies of block copolymers and associated phase diagrams. Using both experiments and simulations, it is shown that molecular architecture (i.e. chain topology) is a defining variable that influences the resulting self-assembly morphologies. In addition, the interfacial dimensionality changes as the molecular topology is altered and the structural response is a combination of certain motifs. Similarly, colloids and related soft matter also exhibit topology dictated structures and properties.

Finally, Chap. 11 describes the role of topology in the emergence of structural colors and iridescence observed in arthropods including beetles, weevils and butterflies. Based on electron microscopy and small angle X-ray scattering analysis, various biophotonic nanostructures are explained in terms of gyroid-like minimal surfaces. Interestingly, gyroids and other triply periodic minimal surfaces are invoked in Chap. 7 as well, albeit in the context of wire network graphs to explain Dirac points in certain lattices.

These eleven chapters discuss a multitude of open questions and set the stage for future research in this highly multidisciplinary and evolving field. They also provide a much needed integration between the two broad subject areas, materials science and topology, which is expected to usher into further insights and a better understanding of materials from a topological perspective. However, much remains to be improved and learned such as the detailed topological metrology and computational predictability. Besides seasoned researchers the book will also serve as a valuable resource for graduate students in materials science and engineering, condensed matter materials physics, applied mathematics, physical chemistry, biophysics and other related disciplines.

Bowling Green, Kentucky, USA Sanju Gupta
Los Alamos, New Mexico, USA Avadh Saxena

Contents

Part I Introductory

Contributors

Gareth P. Alexander Department of Physics and Centre for Complexity Science, University of Warwick, Coventry, UK

Arun Bansil Department of Physics, Northeastern University, Boston, MA, USA

Rossen Dandoloff Laboratoire de Physique Théorique et Modélisation, Université de Cergy-Pontoise, Cergy-Pontoise, France

Haifeng Du The Anhui Province Key Laboratory of Condensed Matter Physics at Extreme Conditions, and High Magnetic Field Laboratory, Chinese Academy of Science (CAS), Hefei, Anhui, China; School of Physics and Materials Science, Anhui University, Hefei, Anhui, China

Sanju Gupta Department of Physics and Astronomy, Western Kentucky University, Bowling Green, KY, USA

Ralph M. Kaufmann Department of Mathematics, Purdue University, West Lafayette, IN, USA

J. J. K. Kirkensgaard Niels Bohr Institute, University of Copenhagen, Copenhagen, Denmark

Hsin Lin Institute of Physics, Academia Sinica, Taipei, Taiwan

Cristiano Nisoli Theoretical Division, Los Alamos National Laboratory, Los Alamos, USA

Jun Onoe Department of Energy Science and Engineering, Nagoya University, Chikusa-ku, Nagoya, Japan

Vinodkumar Saranathan Division of Science, Yale-NUS College, Singapore, Singapore; NUS Nanoscience and Nanotechnology Initiative, National University of Singapore, Singapore, Singapore; Department of Biological Science, National University of Singapore, Singapore, Singapore; Lee Kong Chian Natural History Museum, National University of Singapore, Singapore, Singapore

Avadh Saxena Theoretical Division, Los Alamos National Laboratory, Los Alamos, NM, USA

Hiroyuki Shima Department of Environmental Sciences, University of Yamanashi, Kofu, Yamanashi, Japan

Joanna I. Sulkowska Centre of New Technologies, University of Warsaw, Warsaw, Poland

Piotr Sułkowski Faculty of Physics, University of Warsaw, Warsaw, Poland; Walter Burke Institute for Theoretical Physics, California Institute of Technology, Pasadena, CA, USA

Mingliang Tian The Anhui Province Key Laboratory of Condensed Matter Physics at Extreme Conditions, and High Magnetic Field Laboratory, Chinese Academy of Science (CAS), Hefei, Anhui, China; School of Physics and Materials Science, Anhui University, Hefei, Anhui, China

Wei-Feng Tsai School of Physics, Sun Yat-Sen University, Guangzhou, China

Birgit Wehefritz-Kaufmann Department of Mathematics and Department of Physics and Astronomy, Purdue University, West Lafayette, IN, USA

Part I
Introductory

Chapter 1
Importance of Topology in Materials Science

Sanju Gupta and Avadh Saxena

Abstract We underscore the substantial need for understanding a wide range of multifunctional materials through the notions of topology-geometry interrelationships such as genus, Euler characteristic and network connectivity. After introducing the basic concepts of topology we first illustrate these notions on nanocarbon allotropes as a case study. Next, we consider the growing class of emergent topological materials that encompass both real-space and k-space topological materials including Dirac materials, topological insulators, Weyl semimetals as well as soft and polymeric matter, supramacromolecular assemblies and biophotonic materials. Finally, we emphasize and evaluate metrics to quantify topology in order to study and classify materials properties relevant for wide ranging modern and future technologies.

1.1 Introduction

The recent blooming of topological notions in condensed matter physics, synthetic materials chemistry, supramacromolecular chemistry, materials science and biophysics has given impetus to the development of new and the revision of many old concepts in the physical world [1, 2]. These sub-disciplines are being greatly benefitted by invoking topological concepts to understand novel, complex and emergent states of matter such as quantum Hall systems, topological insulators, Dirac materials and Weyl semimetals, to name just a few of these new classes of materials. The 2016 Nobel Prizes in Physics and Chemistry are a direct testament to this observation [1]. If the mainstream materials science can tap into the full power of

S. Gupta (✉)
Department of Physics and Astronomy, Western Kentucky University, Bowling Green, KY 42101, USA
e-mail: sanju.gupta@wku.edu

A. Saxena (✉)
Theoretical Division, Los Alamos National Laboratory, Los Alamos, NM 87545, USA
e-mail: avadh@lanl.gov

S. Gupta and A. Saxena (eds.), *The Role of Topology in Materials*,
Springer Series in Solid-State Sciences 189,
https://doi.org/10.1007/978-3-319-76596-9_1

topology, that would certainly open avenues for novel materials synthesis, property characterization as well as applications [2, 3].

Our aim in this chapter is to answer the key question: why is topology important for understanding materials? Topology refers to the fact that certain materials properties remain invariant under continuous deformation such as stretching, bending or twisting (but without cutting or puncturing at any place in the material systems). It also means that nearby points remain neighbors during deformation. In this sense a sphere and ellipsoid are topologically equivalent; so are a cone and disk [2].

First and foremost, we can classify the global topology of a material in terms of its characteristic genus (or handlebars or holes), number of open boundaries and local connectivity. Different topological phases of matter can be distinguished by a topological invariant (usually an integer such as the genus, Chern number, winding number, etc.) [1]. While a carbon nanoring and $NbSe_3$ Möbius strip represent real-space topological materials, recent interest in Dirac materials and topological insulators refers to topology in the k- or momentum space usually in terms of the electronic band structure. Quantum oscillations such as the de Haas-van Alphen (dHvA) and Shubnikov-de Haas (SdH) oscillations are a direct consequence of the topology of the Fermi surface of a crystal in a magnetic field. Based on these and many other illustrative examples mentioned below, we aim to show that the intersection of topology and materials science is both physically insightful and aesthetically appealing.

1.2 Essentials of Topology

1.2.1 Genus and Euler Characteristics

First, we introduce the basics of topology. Global topology of a material is described by a parameter called genus (g), which is an integer characterizing number of holes [2]. For instance it is zero for a sphere but equal to one for a torus. The genus is related to another parameter (of a surface) called the Euler characteristic, defined by $\chi = 2(1 - g)$. For objects with boundaries or edges it can be expressed as $\chi = V - E + F$, where V, E and F represent the number of vertices, edges and faces of a polyhedron on the object. The integral over the surface of a material's Gaussian curvature K gives 2π times the Euler characteristic (the Gauss-Bonnet theorem), connecting geometry with topology.

1.2.2 Network Topology

In the context of supramacromolecular architectures metal-organic frameworks (MOF) and geometric hierarchies inherent to many soft- and biomaterials (e.g. intra-cellular structures such as endoplasmic reticulum), topology arises in a different guise, namely, network topology [2]. Here one focuses on local connectivity at a given node in the network, i.e. the number of links emanating from a particular node in the network. Two networks can have identical topology even though their physical interconnections or links, distances between the nodes and other attributes (e.g. transport of some quantity through the links) may differ. In biology network topology may refer to the network of biological interactions, for instance the metabolic network. Topologies commonly observed in biological networks include ring network, bus network and star network. As examples of the important role of network topology in materials, note that sound propagation in granular materials and mechanical properties of, e.g. siloxane, elastomers crucially depend on their network topology.

1.2.3 Geometry-Topology Interrelationship

As shown in Fig. 1.1 a cup can be continuously deformed into a sphere, thus they have $g = 0$ and $\chi = 2$. A cup with a handle is equivalent to a donut ($g = 1$, $\chi = 0$), but with two handles it is equivalent to a double-donut (or Swedish pretzel, $g = 2$, $\chi = -2$). In the same vein, a cup with three handles can be continuously deformed into a triple-donut (or German pretzel, $g = 3$, $\chi = -4$). A red blood corpuscle (RBC), a biological vesicle with one hole, two holes and three holes, respectively, represent $g = 0, 1, 2$ and 3 or $\chi = 2, 0, -2$ and -4 objects. Since topology is essentially elastic geometry, many different geometries may correspond to the same topology, i.e. same g and χ. Likewise, a variety of geometrically different networks may belong to the same topology, that is, a network with given geometry can be continuously deformed to obtain a network of different geometry.

1.3 Topological Taxonomy of Functional Materials

1.3.1 Nanocarbons

Carbon offers a rich variety of forms depending upon the covalent bonded hybridization, i.e. sp^3-, sp^2- and sp-bonded carbons. They exist in at least two natural allotropes (diamond, graphite) and various man-made or synthetic nanoscale forms (fullerenes, nanodiamond, carbon nanotubes and graphene). For the past few decades, there is an overwhelming interest in the family of nanocarbons due to their discernible structural characteristics at molecular scale and extraordinary physical (optical, mechanical, electronic) and chemical (electrochemical, biological etc.) properties attributed to their unique low-dimensional atomic scale lattice bonding structure. Therefore, they serve as functional building blocks for innovative nanotechnology such as in ultra-sensitive low-energy consumption electronics and mechanical devices, advanced cat-

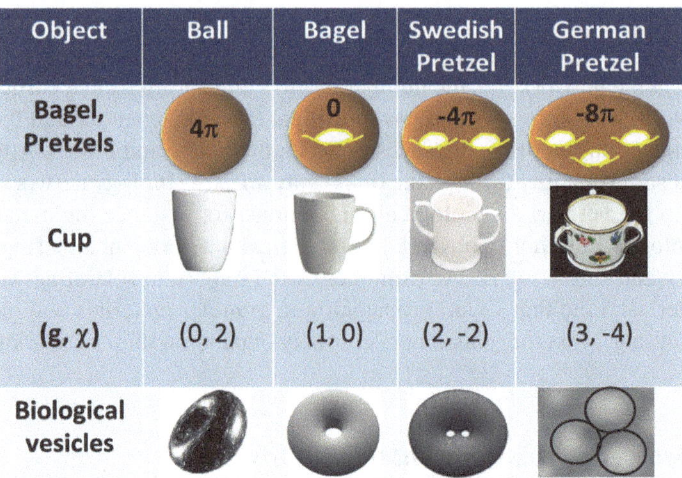

Fig. 1.1 Illustrations of topological objects with different genus, adapted in part from Haldane's 2016 Nobel lecture (top two rows) and [1], last row

alyst supports, sensitive biosensors, and microelectromechanical (micro-actuators) and electrochemical energy conversion and storage (e.g. fuel cell, batteries and supercapacitors) devices. Nanoscale carbons serve as a posterchild for the contexts where the interplay of geometry and topology is promising for basic and applied sciences. Nanocarbon allotropes exhibit numerous topologies with a variety of geometries ranging from planar (monolayer and multilayer graphene) to closed cage-like (e.g. fullerenes, hypo- and hyperfullerenes), open-ended (e.g. single-, double-, oligo- and multi-walled nanotubes), nano rings/nanotori, nanohorns, nanocones and peapods. All these distinct allotropes of nanocarbon serve as a fertile playing field for expounding the nontrivial notions of global topological attributes (see Table 1.1) [3–5]. Table 1.1 summarizes the global topology metrics of nanocarbons in terms of genus, g and Euler's characteristic, χ. Briefly, we indicate that nanocones, nanodisks and nanotubes closed at one end are topologically equivalent to a planar graphene sheet (g = 0). Fullerenes (C_{60}), hypofullerenes ($C_{36, 50,..}$), hyperfullerenes ($C_{70, 84,90,..}$) and capped nanotubes have the topology of a sphere (g = 0). Due to boundaries captured by χ, open ended nanotubes have a different topology (g = 1, χ = 0) than those of closed nanotubes (g = 0, χ = 2). Nanotori and nanorings forms of nanocarbon are topologically equivalent to a torus (g = 1, χ = 0). Furthermore, multi-walled carbon nanotubes, nano-onions and peapods (fullerenes nestled in single-walled carbon nanotubes like a beaded necklace) have complex topologies due to nestled spherical and cylindrical geometries. Among the negative Gaussian curvature ($K<0$) periodic carbons, Schwarzites have a complex topology with g = 3, $\chi = -8$ per unit cell. Similarly, a helicoid-shaped narrow graphene nanoribbon would have g = 0, χ = 2 [5].

Table 1.1 Topology of nanocarbon allotropes

Geometry	Topological characteristics	
	g (Genus)	χ (Euler)
Positive Gaussian curvature		
Mono-, few- and multi-layer graphene (HOPG and Kish Graphite)	0	2
Fullerenes and hypo-/hyperfullerenes	0	2
Single-walled carbon nanotube (SWCNT); open (closed)	1 (0)	0 (2)
Nanoring/nanohoop/nanotori	1	0
Nanohorn/nanocone	0	2
Double-walled (DWCNT), oligo-walled (OWCNT) and multi-walled carbon nanotube (MWCNTs)	Complex geometries	
Peapod		
Onion-like carbon (OLC)		
Negative Gaussian curvature	g (Genus) unit cell	χ (Euler)
Negatively curved carbons/Schwarzites (3D)	3	−8
Graphene nanoribbons (GNR)/helicoidal (2D); infinite (finite)	0 (0)	2 (1)

Our goal here is to relate measurable physical properties deduced from resonance Raman spectra of various nanoscale carbons to topological as well as geometric metrology characteristics. Spontaneous Raman spectroscopy (RS) has emerged inarguably as a powerful non-invasive analytical tool for structural characterization of carbon-based materials revealing both collective atomic/molecular motions and localized lattice vibrations (phonons) [3] besides defects (point or extended defects, stacking disorder, doping) and finite size of crystallites. The primary reason for this advantage is the strong Raman scattering response to the π states due to resonance enhancement, its simplicity for high-symmetry nanotubes and fullerenes, its easy access and noninvasive nature. Figure 1.2 shows micro-Raman spectra for various nanocarbons measured using excitation wavelength of 633 nm (or energy E_L = 1.92 eV) in backscattered configuration. Since all of these materials are sp^2 C derivatives, it is instructive to compare the Raman spectral features with planar highly ordered pyrolytic graphite (HOPG) or multilayer graphene (MLG) and monolayer graphene as they are two-dimensional building blocks for sp^2 C allotropes of every other dimensionality. Prominent bands of interest in first- and second-order Raman spectra are D, G and 2D bands occurring at ~1344 cm^{-1}, ~1585 cm^{-1}

and ~2670 cm^{-1}, respectively. The G band is associated with the tangential C-C stretch or the tangential displacement band having E_{2g} symmetry. For SWCNT, the G band decomposes into main peaks (G^+ at 1562 cm^{-1} and G at 1593 cm^{-1}) primarily due to the splitting of interlayer stretching mode attributed to curvature-induced re-hybridization of σ^*-π^* states which yields larger elastic constants and therefore better mechanical properties. The D band is a disorder-activated band with A_{1g} symmetry arising from various sources including in-plane substitutional heteroatoms, vacancies, grain boundaries, quantum confinement due to size effects, stacking disorder and other point and extended defects. Therefore, D band intensity in principle is proportional to the phonon density of states analogous to electronic density of states, applicable to all sp^2 C-based materials. It is worthwhile to note the near absence of D band in monolayer (and multilayer) graphene and HOPG indicative of presence of marginal defect number density. The stable sp^2 C spherical cage structures—fullerenes (C_{60})—are somewhat lower in yield and the effect of curvature and geometry is displayed in Raman spectral features for spheroids compared with HOPG and MLG. There are fewer lines for C_{60} (I_h symmetry) as compared to C_{84} (D_{2d} symmetry) possibly due to deviation from spherical geometry (oblate or prolate). While it is challenging to analyze complicated Raman spectra due to fullerenes, the downshift of $A_g(2)$ (pentagonal pinch mode; ~1470 cm^{-1}) and $H_g(8)$ band at ~1575 cm^{-1} is apparent. Figure 1.2 also shows Raman spectral features due to SWNR, SWNH and nanocone displaying similar phonon spectra to sp^2 C material systems, *albeit* they have some quantitative differences discussed below. It is imperative to mention that for low-dimensional carbons (nanocarbons), the 2D band (a second-order D band) is symmetry allowed by momentum conservation, therefore the overtone Raman feature is relatively sharp and comparable to G band intensity in contrast to disorder-activated D band and it becomes an intrinsic feature for sp^2 C materials (Fig. 1.2b).

The prominent Raman bands for representative nanocarbon materials are quantitatively analyzed in terms of position of D, G and 2D bands as possible topological metrics due to their sensitivity toward structural modification and mechanical deformation as well as charge transfer doping thus capturing "weaker" or "group" trends. Figure 1.3 shows the variation of G band position by itself (panel a) and with 2D band (panel b) providing subtle information on curvature induced shifts and the nature of intrinsic point defects (charged or residual). For instance, G band is marginally upshifted in SWCNT as compared with SWNR leading to microscopic compressive stress attributed to smaller nanotube curvature that is invoked. The G band for nanocone tip and SWNH shifts to higher value as compared to HOPG; this occurs due to curved cone surface and phonon confinement attributed to smaller crystallite sp^2 C domains. While the presence of G band is a direct indication of sp^2 C network, the shift (either decrease or increase) is a measure of (a) different sp^2-bonded C configurations; (b) curvature-induced re-hybridization and mixed hybridized character $sp^{2+\delta}$; (c) compressive or tensile microscopic stress/strain; and finally, (d) phonon confinement (localization of vibrational states) and (e) electronic character (n- or p-type). The tensile strain in graphene planes induces curvature by the introduction of pentagons in the hexagonal network governed by Euler's theorem.

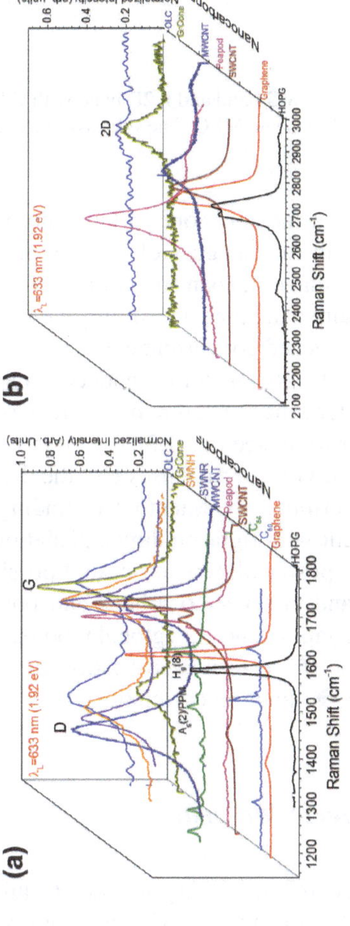

Fig. 1.2 Experimental, **a** first-order and **b** second-order Raman spectra for a range of nanocarbons

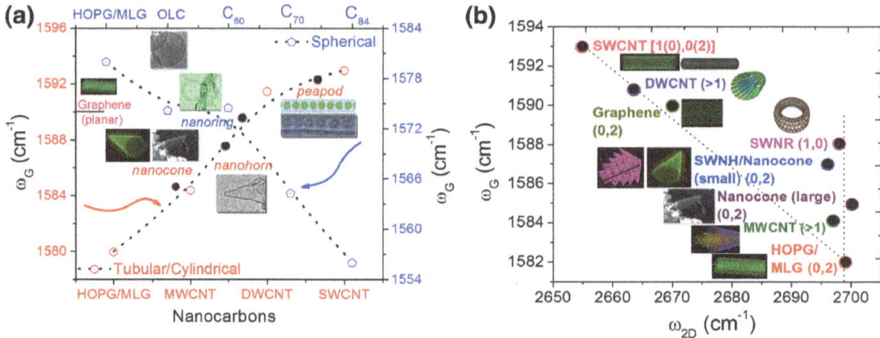

Fig. 1.3 Variation of the position of **a** G band and **b** 2D band with G band from Raman spectra for various nanocarbons along with HOPG and MLG. The values of (g, χ) are also shown in parentheses

While the Raman bands for the nanocone and nanohorn appear at almost similar positions, the nanoring lies in the category of a different geometry. A quantitative understanding of Raman lineshape and band position shifts occurring due to geometry and topology therefore require a detailed accounting for the changes in both phonon and electronic density of states and concurrent electron-phonon (lattice) interactions suggested by theoretical studies for nanotori, nanocones and nanohorns [6, 7]. We have made an attempt to determine the nature of the defects by plotting the 2D band position with the G band position (see Fig. 1.3b) [8]. It is safer to say that the defects are mainly *p*-/*n*-type (i.e. the G band increases and the 2D band decreases), which is quite encouraging. Furthermore, the quantitative findings obtained from Raman spectra are also in agreement with tight-binding calculations for the nanotubes [8]. This knowledge provides a powerful geometric (and possibly topological) metrology machinery to understand novel nanocarbons and points to an unprecedented emergent paradigm in materials science i.e. global topology (and curvature) ➔ process ➔ property ➔ function ➔ performance relationships in contrast to traditional microstructure ➔ property ➔ function correlations.

1.3.2 *Soft and Polymeric Materials*

Analogous to nanocarbons, hard-, soft- and polymeric (liquid crystal supermolecules and supramolecular chemistry) materials exist in a variety of complex topological phases and forms, summarized in Table 1.2. Some examples include semiconducting oxide and BN nanotubes, nanotori, helical gold nanotubes, mesoporous silica networks, Möbius conjugated organic materials, di-block and tri-block copolymers as well as smectic and nematic liquid crystals [2]. Also, foams have interesting network topology which can change its configuration at a local level, wherein the interface between two bubbles shrinks to zero length and subsequently expands to a finite

Table 1.2 Topology of soft-, polymeric, biological matter and supramolecular assemblies

Geometry	Topological characteristics		Suggested metrology
	g (Genus)	χ (Euler)	
Soap bubbles and foam	Network topology		Optical imaging
Liquid crystals (double and quadri-dislocations and disclinations; discotic, nematic with Schlieren texture)	Complex topology		Non-linear optical fluorescence microscopy, laser lithography
Di- and tri-block co-polymers (lamellar and tubular)	Complex topology		Small-angle x-ray and neutron scattering (SAXS/SANS)
Bio-membranes (lamellar and spherical)	0	2	Optical fluorescence microscopy and SAXS/SANS
Biological vesicles (w/ and w/o holes)	0/1, 1, 2	2/0, 0, −2	Optical fluorescence imaging
Zeolites (micro-/mesoporous, metallo-organic frameworks [MOFs])	3	−8	HRTEM, SAXS/SANS, x-ray and neutron tomography
Supramolecular assemblies	1	0	Optical fluorescence imaging and SAXS/SANS

length in another direction, thus resulting in a local topology change. For studying two- and three-dimensional microscopic structure evolution with regard to crystal grain growth and topological optimization of microstructure besides multicellular structures such as bubbles, foams, and biological tissues, network topology serves as an efficient tool [9].

As for polymeric liquid crystals which show complex topology—they are hard in that they have many interesting symmetries thus exhibit anisotropic elastic behavior, while their liquid properties enable a soft behavior such that these symmetries are disrupted by defect structures and their elasticity is dominated by fluctuations away from an ideal state. Analogous to all broken symmetry materials, liquid crystals admit topological defects, which are regions forced to be discontinuous by their topological behavior. Such defects are stable in the sense that they cannot be removed by local perturbation, rather they must either be moved out to the boundary of the sample or merged into other topological defects. Of utmost importance is the study of these topological defects, which are readily visualized. Since the behavior of materials is often dictated not by bulk properties, but by its defects, akin to the strength of a chain being determined by its weakest link. Historically, topological defects in ordered media were studied using the theory of homotopy classes in which homotopy groups of the order parameter space are calculated [10–13]. These ideas begin with identifying defects with small measuring loops or spheres around them. Thus, the "charge" of the defect is the topological metric (i.e. genus) of the configuration in the sample on such a measuring circuit [14]. Given the development of three-dimensional imaging techniques [14] and extensive simulations [15, 16], one has a way of seeing more globally the topological defects and other interesting topological features in these samples. These sets include the defects but usually include other points forming lines or sheets connecting the defects together as well. In general, the homotopy group approach is not justified when applied to smectic liquid crystals or other crystalline systems. However, in these smectics there are two classes of defects, dislocation-type and disclination-type, which are akin to critical points in the sense of local maxima, minima or saddle points.

1.3.3 Minimal Periodic Surfaces

1.3.3.1 Supramacromolecular Assemblies

Micelles, colloids, micro-emulsions, biological vesicles, microtubules and supramolecular photochemistry in restricted space belong to this class of materials [2]. Additionally, processes in many bio-macromolecules including DNA and RNA structure and protein folding, involve network, braid and knot topologies [17]. A network has connected nodes and lines in various ways. Lattices are a special kind of network in which the lengths are the same in each periodically repeated structure, called a unit cell. Two networks ('nets') have different distances between nodes and other characteristics yet may have identical topologies. As a special case, a square lattice is topologically equivalent to an oblique lattice or a rectangular lattice but not to a triangular Kagome-like lattice. Protein-protein interaction networks ('interactomes'), and mesoporous materials are examples in which correlations between the network structure and properties provide useful insights into design strategies.

Metal-Organic Framework, (MOFs, pronounced *moffs*), are compounds consisting of metal ions or clusters coordinated to organic ligands to form one-, two-, or three-dimensional structures (Fig. 1.4). They are a subclass of coordination polymers, with the special feature that they are mesoporous. The organic ligands included in them are sometimes referred to as "struts", one example being 1,4-benzenedicarboxylic acid (BDC). More formally, a metal–organic framework is a coordination network with organic ligands containing potential voids. A coordination network is a coordination compound extending through repeating coordination entities in one dimension, but with cross-links between two or more individual chains, loops, spiro-links, or a coordination compound extending through repeating coordination entities in two or three dimensions; and finally, a coordination polymer is a coordination compound with repeating coordination entities extending in one, two, or three dimensions [18]. The study of MOFs has been developed from the study of zeolites, except for the use of preformed ligands. MOFs and zeolites are

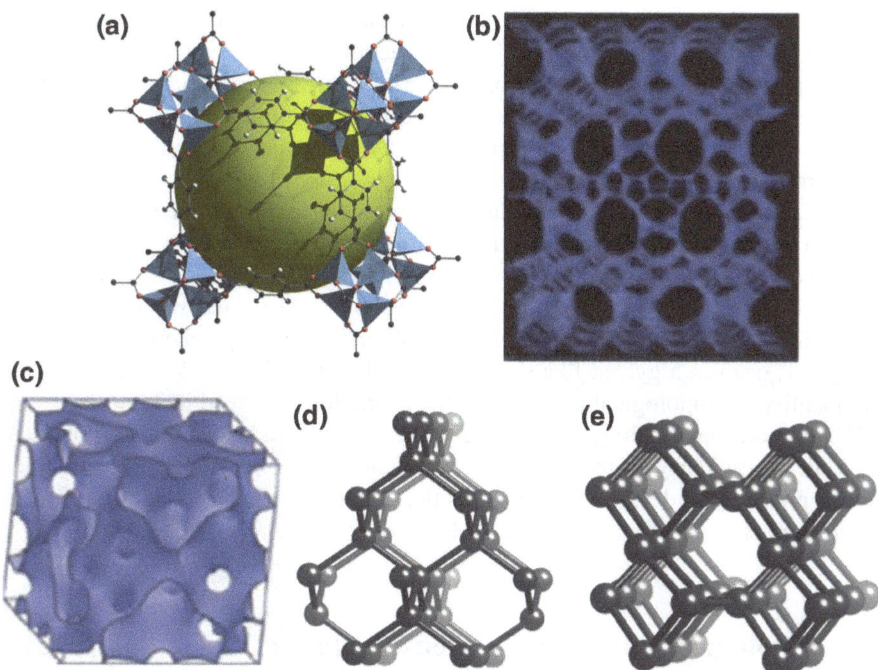

Fig. 1.4 Example of **a** MOF-5 and **b** zeolite catalyst. Corresponding **c** double gyroid and **d**, **e** network topologies with four-connected branching points (**dia**-net based on diamond structure) or vertices (chiral **qtz**-net based on quartz structure) [19]

produced almost exclusively by hydrothermal or solvothermal techniques, where crystals are slowly grown from a hot solution. In contrast, MOFs are constructed from bridging organic ligands that remain intact throughout the synthesis [19]. In some cases, the pores are stable during elimination of the guest molecules (often solvents) and could be used for the storage of gases such as hydrogen and carbon dioxide. Other possible applications of MOFs are in gas purification, gas separation, electro- and photocatalysis, as sensors, energy harvesters and supercapacitors [20].

1.3.3.2 Biophotonic Materials

The topology of mixed di- and triblock copolymers as well as single and double gyroids in butterfly wings, weevil chitin and bioinspired photonic bandgap crystalline materials, zeolites and other metallo-organic frameworks belongs to triply periodic minimal surfaces. Depending upon the relative concentration of the constituents, temperature and pressure, the topology of block copolymers and biomacromolecular systems can change from lamellar, globular, and tubular to gyroid and double gyroid structures [21–23]. The self- and directed-assembly of block copolymers result in complex topologies (third row, Table 1.2) as a function of relative concentration of different types of polymer blocks and the interaction energy between different types of monomers. Interestingly, the gyroid is a triply periodic minimal surface where mean curvature H = 0 that belongs to the family of P (primitive) and D (diamond) Schwarz surfaces and it separates space into two identical labyrinths of passages discovered by Alan Schöen in 1970.

Biomembranes can be flat (lamellar) or curved depending on the structures they enclose (fourth row) [24] and they can also morph into a spherical or toroidal topology, in some cases with g > 1 (fifth row). As indicated, MOFs [25] also form periodic minimal surfaces and gyroid-like structures (sixth row). Finally, supramolecular assemblies may emerge with a variety of topologies, including spheroidal, periodic with g = 1, or even a gyroid-like structure [26]. It is thus highly desirable to probe and quantify the topological attributes of the examples portrayed in Table 1.2.

We reiterate that topology finds its multifaceted presence in a variety of biological materials, particularly in the context of biophotonics. Some interesting examples include polarized iridescence in jewel beetle, butterfly chitin, bird keratin [27] and gyroid-type photonic crystals in diamond weevil and wing scales [28].

1.4 Topological Phases in Condensed Matter

Topological materials: Materials in which topological aspects, usually involving boundary effects such as surface or edge effects, alter the electronic, transport, magnetic and various other properties are christened as topological materials [29]. In particular, topological superconductors, topological insulators, topological crystalline insulators, and some Dirac materials (e.g. Weyl semimetals and graphene) are important members of this class. To understand topological invariants and the properties of these materials, topological field theory has been developed recently [30]. Finally, we note that some of these materials can possibly support the so called non-abelian particles (or elementary excitations) called anyons, Majorana fermions being an example, which can enable the current pursuit of topological quantum computing through the braiding statistics of Majorana fermions. The latter paradigm is a viable approach for fault-tolerant quantum computation currently under consideration [31].

Topological defects: In general, certain material defects that interpolate between two different orientation states, e.g. domain walls and those resulting from certain material discontinuity, for instance disclinations and dislocations, constitute the broad class of topological defects. Monopole-like excitations and out-of-plane vector configurations, e.g. skyrmions and vortices, also belong to this class [2]. Clearly, such defects alter the macroscopic properties of materials, e.g. strength, electronic transport and magnetic response. Since the presence of toplogical defects affects materials properties in unusual ways; it is imperative that we are able to control experimentally the density and generation of these defects.

1.4.1 Real-Space Topological Materials

While most of the topological materials discussed below such as topological insulators and Weyl semimetals deal with topology in the momentum or k-space, e.g. the topology of the Fermi surface, synthesizing topological crystals in real space is both intriguing and important. In fact, a Möbius strip has been synthesized using a crystalline ribbon of $NbSe_3$, which is a low-dimensional inorganic conductor exhibiting charge-density-waves (CDW). The width of such a strip is about one micron whereas the ring diameter is about 100 ms. Similarly, figure "eight" structures with a double twist, knot crystals and Hopf link materials have also been synthesized [32–34]. Interestingly, the various topological arrangements of $NbSe_3$ all show CDW phase transitions. Such topological variants can also be synthesized using $TaSe_3$ and TaS_3.

Changes in topology with a twist singularity have been observed in soap-film Möbius strips [35]. In this process the linking number of the film's Plateau border and the centerline is altered. Similarly, simulations based on a discrete, lattice based model have demonstrated the influence of a material's stretchability on the equilibrium shape of a Möbius strip [36].

1.4.2 Dirac Materials

The (nonrelativistic) Schrodinger equation describes conventional metals, semiconductors and insulators, for which the electronic energy dispersion of low-lying excitations is quadratic: $E_S = p^2/2\,m^*$. Here p denotes the electron momentum and m^* the effective mass. In contrast, there is a growing family of materials with electronic band structure that exhibits linear dispersion (Fig. 1.5), called Dirac materials [37, 38]. One salient feature of these materials is that their valence and conduction bands touch at a few isolated points known as the Dirac points (and the associated band attributes are called the Dirac cones). These points remain unaltered under perturbations, or equivalently are *topologically protected*, as a consequence of certain symmetries. For graphene, it is the sublattice symmetry, for topological insulators it is the time-reversal symmetry [39, 40] whereas for topological crystalline insulators [41, 42] it is the mirror (or a related crystalline) symmetry. In the case of two-dimensional Dirac materials (that are described by the relativistic Dirac equation) the electronic energy dispersion is linear in momentum, $E_D = c\,\sigma \cdot p + mc^2\,\sigma$. Here $\sigma = (\sigma_x, \sigma_y)$ are Pauli matrices and the Fermi velocity v_F replaces the speed of light in the material.

An important implication of the linear dispersion of Dirac materials is their enhanced sensitivity to applied magnetic field in two dimensions. Specifically, electronic energy level spacing is proportional to $\sqrt{B}$ in massless Dirac materials in contrast to B in usual materials. The electronic Dirac spectrum in topological insulators, d-wave superconductors and graphene has been measured using angle resolved photoemission spectroscopy (ARPES) as well as scanning tunneling spectroscopy (STS) [37]. Beyond graphene, silicon and germanium monolayer structures named

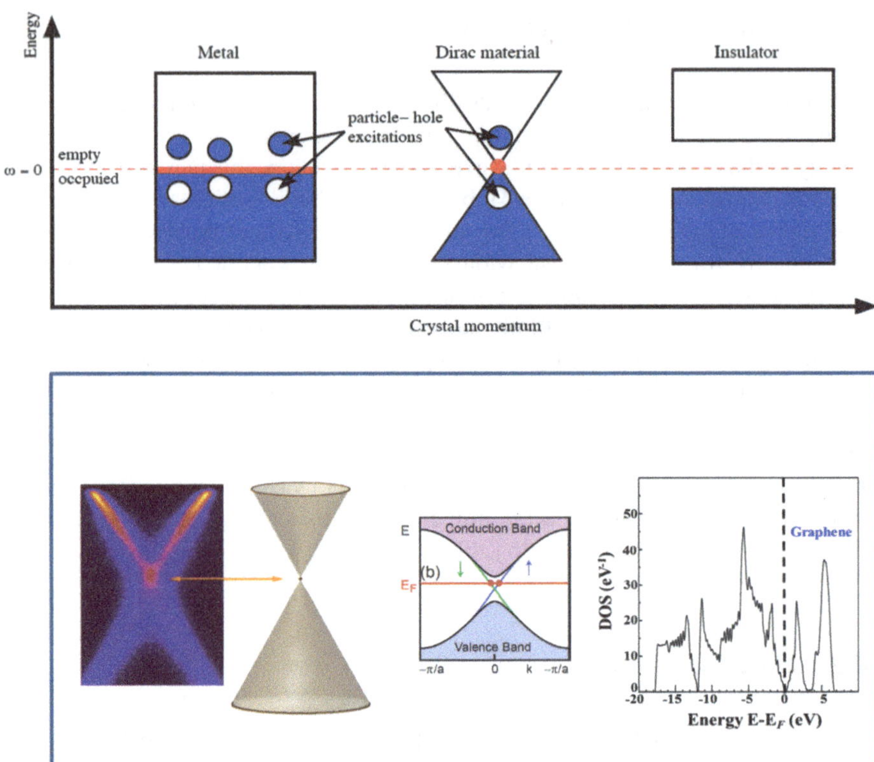

Fig. 1.5 (Upper panel) Schematic band diagrams for metals without a gap (left), Dirac materials exhibiting linear band dispersion as well as a Dirac cone (middle) and insulators with a band gap (right). (Lower panel) Angle-resolved photoemission spectroscopy (ARPES) results for graphene obtained at the Advanced Light Source at LBNL and the corresponding DFT calculations of density of states. Adapted in part from [37]

silicene [43], germanene [44], and sp-sp^2 carbon allotropes called graphynes [45] also exhibit Dirac cones and thus belong to the growing family of two-dimensional Dirac materials. Related honeycomb monolayers of phosphorus and tin called phosphorene [46] and stanene [47], respectively, represent novel two-dimensional topological materials. Even a lead based monolayer called plumbene has been theoretically proposed.

Nodal line insulators and semimetals: In some topological semimetals, such as the strongly spin-orbit coupled compound $PbTaSe_2$, valence and conduction band touch at one-dimensional Fermi lines known as nodal lines [48]. Unlike the (zero-dimensional) Weyl points, which are protected against perturbations that preserve translational symmetry, the protection of nodal lines requires additional crystal symmetries, e.g. mirror reflection. They have been studied using ARPES [48], and are also characterized by an integer topological invariant. Electronic structure calculations of these materials and relevant experiments are discussed in Chap. 6 in this book.

There are three kinds of spin half-integer particles (e.g. electrons) in nature called fermions that can occur in condensed matter and materials as well as in photonics: Majorana, Dirac and Weyl. For particles with mass and linear dispersion, graphene represents a prime example of Dirac particles. Materials comprising particles without mass but with linear dispersion are called Weyl materials or Weyl semi-metals, e.g. NbAs, NbP, TaAs and TaP [49, 50]. There has been an intense search for such materials in recent years. These materials can be viewed as three-dimensional analogs of graphene with broken time reversal and spatial inversion symmetry. Note that Weyl semimetals can be characterized by the experimental observation of Fermi arcs in ARPES [51]. Interestingly, they can also exhibit magnetic monopoles in the crystal momentum (or reciprocal) space. Particles that possess mass and are also their own antiparticles are termed Majorana fermions. There has been a great deal of experimental search for Majorana fermions [50]; it is anticipated that topological superconductors [52, 53] might possess them as quasi-particles. Beyond their fundamental significance, there is growing excitement about their role in topological quantum computing [31].

1.4.3 Topological Insulators and Topological Superconductors

Topological insulators (TI), which are metallic at the surface and insulating in the bulk, and related materials usually have a strong spin-orbit coupling [39, 40]. It is worth noting that until the discovery of quantum Hall effect (QHE) in 1980 it was believed that all fundamental laws of nature and phases could be understood in terms of symmetry (and geometry). However, QHE provided the first instance of a quantum state with no spontaneous broken symmetry. The behavior of QHE depends only on the system's topology and not its specific geometry thus opening up the frontier field of topological order.

The first two-dimensional TI was discovered in HgTe quantum wells [29]. This topological state is also known as the quantum spin Hall state. Band inversion is the main mechanism herein where the spin-orbit coupling inverts the usual ordering of conduction and valence band. Subsequently, three-dimensional topological insulators were discovered in materials such as Bi_2Se_3 Bi_2Te_3 and Sb_2Te_3. The band inversion in these materials occurs at the Brillouin zone center due to the spin-orbit coupling [29]. The topological surface state comprises a helical Dirac fermion in that the electron spin is perpendicular to its momentum; in other words it forms a left-handed helical texture in momentum space. No gap for the surface state can be introduced by a time reversal invariant perturbation.

Topological insulators can also exist *without* spin-orbit coupling in the presence of certain crystal point group symmetry. They are called topological crystalline insulators (TCI). SnTe is a prime example of a topological crystalline insulator [41, 42]. Lattice periodicity is not a requirement and even quasicrystals can display TI behavior under appropriate conditions. Recently, a two-dimensional quasicrystal in the presence of a uniform magnetic field exhibited chiral edge states (in analogy with Chern insulators in periodic lattices). Such materials have been christened as topological Hofstadter insulators [54].

The proximity of a superconductor to the surface of a topological insulator can result in a topological superconductor [52, 53]. The latter is characterized by the presence of Majorana zero modes. In essence, topological superconductivity involves edge-mode superconductivity in topological insulators. Note that two-dimensional topological superconductivity in InAs/GaSb [53] and three-dimensional superconductivity in $Cu_xBi_2Se_3$ [52] has been experimentally observed. Another example of such a material is Sr_2RuO_4 (possibly with chiral p-wave superconductivity). We emphasize that the Berry phase (in momentum space) plays a key role in TI and topological superconductors.

We can view TI in two ways: (i) as unusual band insulators with surface states (i.e. obtaining a 3D state from 2D), and (ii) as materials with a quantized magnetoelectric response (i.e. obtaining a 2D state from 3D). In essence, the notion of TI is a generalization of the idea of integer quantum hall effect (IQHE). In particular, a system of noninteracting lattice fermions with broken time-reversal symmetry can exhibit IQHE, which is characterized by a topological invariant called Chern number and is stable against disorder and interactions. Analogously, a system of noninteracting Bloch fermions with unbroken time-reversal symmetry corresponds to TI. Akin to fractional QHE there may well exist fractional TI [55].

1.4.4 Weyl Semimetals

Weyl semimetals refer to solid state crystals whose low energy excitations correspond to Weyl fermions [56–58]. The latter carry electrical charge even at room temperature. These materials are a topologically nontrivial phase of matter (Fig. 1.6). Historically, in 1929 Hermann Weyl showed the existence of a massless fermion as a solution of the Dirac equation, now known as the Weyl fermion. These materials are three-dimensional analogs of graphene in that Weyl semimetals show linear dispersion around certain nodes in the Brillouin zone called Weyl points, which always appear in pairs. They also exhibit Fermi arcs (Fig. 1.6), which are unclosed lines that start from one Weyl point and end at the other with opposite chirality (Figs. 1.7 and 1.8), in addition to chiral magnetotransport.

TaAs represents a typical example of a (type-I) Weyl semimetal (Fig. 1.7). Other related materials such as WTe_2 and $MoTe_2$ belong to what are known as type-II Weyl semimetals [59, 60]. For a material to be a Weyl semimetal it must break either the lattice inversion symmetry or the time-reversal symmetry. In the case when these two symmetries coexist, there may exist a pair of degenerate Weyl points resulting

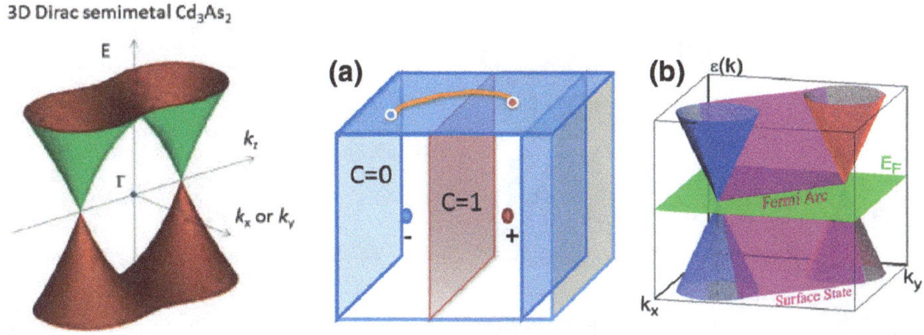

Fig. 1.6 A comparison of the double Dirac cone band structure for a Dirac semimetal (left panel) Cd_3As_2 and a Weyl semimetal TaAs (middle and right panels). For the Weyl semimetals (**a**) Fermi arcs connecting regions of different Chern numbers (C = 0 and C = 1) are depicted. (**b**) The corresponding Fermi level is also shown. Adapted in part from [56]

in what is known as a Dirac semimetal. Weyl fermions can be controlled by both the optical and electrical means (Fig. 1.9).

1.4.5 Other Topological Materials

In topological defects called magnetic monopoles (or magnetic charges) the effective magnetic field lines emanate from a point radially outward. Such defects have been likely observed in artificial spin ice (ASI) [61] as well as in current driven chiral magnets where skyrmion tubes merge or separate at an isolated number of points [62]. ASI is an assembly of nanomagnets in a particular lattice whereas skyrmion is a spin texture in which the spin orientation goes from 0 to π. Equivalently, the spin texture covers the unit sphere once. One could think of ASI as a *magnetolyte* in analogy with charges in an electrolyte. In these materials one expects a monopole-antimonopole pair to exist, which is connected by the so called Dirac string. There is a flux in the interior of a Dirac string which renders the presence of monopoles consistent with the requirements of Maxwell's equations. Both skyrmions and ASI are discussed in detail in Chaps. 4 and 5 in this book.

Penta-Graphene: Based on total energy calculations a new two-dimensional metastable carbon allotrope, composed entirely of pentagons (that resemble Cairo pentagonal tiling), has been proposed [63]. It was motivated by the recent proposal of T12-carbon phase, which can be chemically exfoliated to produce a single layer penta-graphene. This allotrope exhibits dynamical, thermal and mechanical stability in addition to a large band gap, ultrahigh mechanical strength and negative Poisson's ratio. It can withstand temperatures as large as 1,000 K. However, it still remains to be experimentally synthesized.

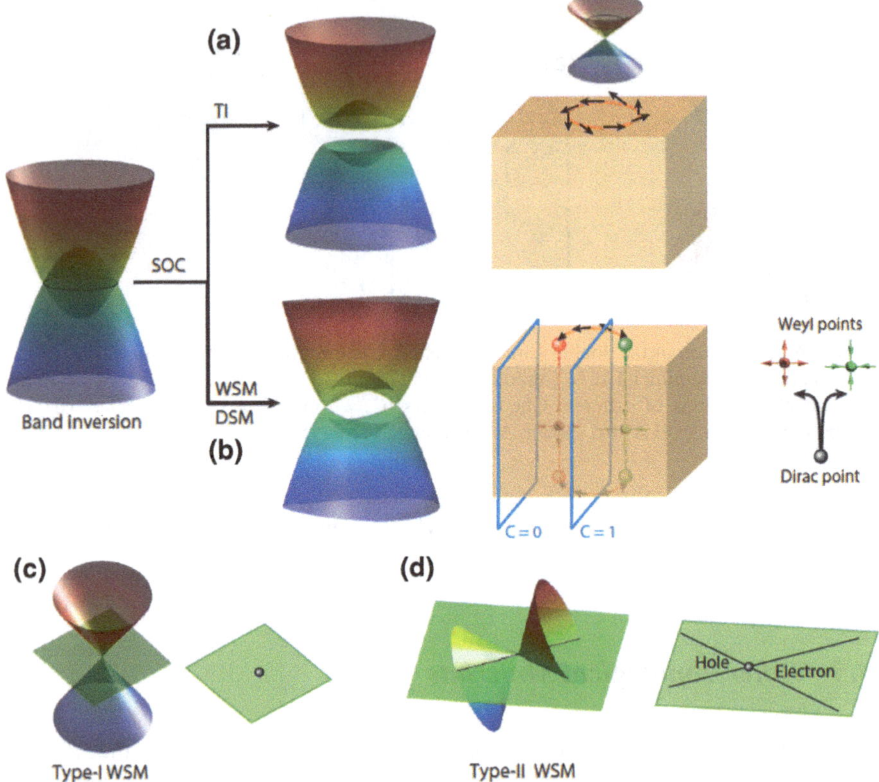

Fig. 1.7 Spin-orbit coupling (SOC) induced band inversion leading to the formation of topological insulators (TI), Dirac semimetals (DSM) and Weyl semimetals (WSM). **a** A full gap is opened in a TI resulting in metallic surface states. **b** In DSM and WSM bulk bands are gapped except at some isolated points with linear dispersion called Dirac points and Weyl points, respectively—they constitute a three-dimensional analog of graphene. **c** Type-I WSM in which the Fermi surface shrinks to zero at the Weyl points when these points are sufficiently close to the Fermi energy. **d** Type-II WSM: the Weyl points represent the touching points between the electron and hole pockets in the Fermi surface as a result of the strong tilting of the Weyl cone. Adapted from [59]

Rolled-up penta-graphene leads to penta-tubes: Carbon nanotubes solely comprising pentagons, which demonstrates the structural versatility of penta-graphene. Phonon calculations and ab initio molecular dynamics (AIMD) simulations demonstrate the dynamic and thermal stability of penta-tubes, respectively. Unlike carbon nanotubes, penta-tubes are semiconducting independent of their chirality [63]. Stacking of penta-graphene layers leads to a three-dimensional stable structure called AA-T12 carbon, which is also semiconducting and has properties quite different from T12-carbon. Both electronic structure and phonon dispersion have been calculated for this layered carbon allotrope. Analogous calculations also suggest a tetragonal phase of metallic three-dimensional boron-nitride [64].

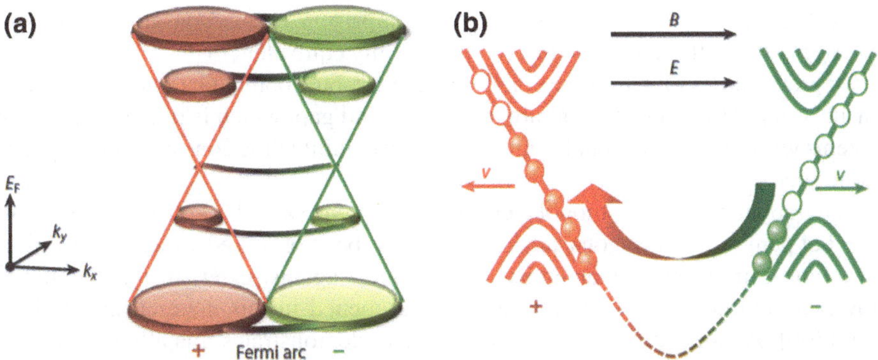

Fig. 1.8 Fermi arcs in the Fermi surface of the surface band structure of a Weyl semimetal. **a** A pair of Weyl cones (in two different colors) representing two different chiralities exist at nonzero Fermi energy or at zero Fermi energy. The Fermi arcs connect these two cones. **b** The chiral anomaly in these materials can be understood in terms of the zeroth Landau level in the quantum limit. E and B represent applied electric and magnetic fields. Adapted from [59]

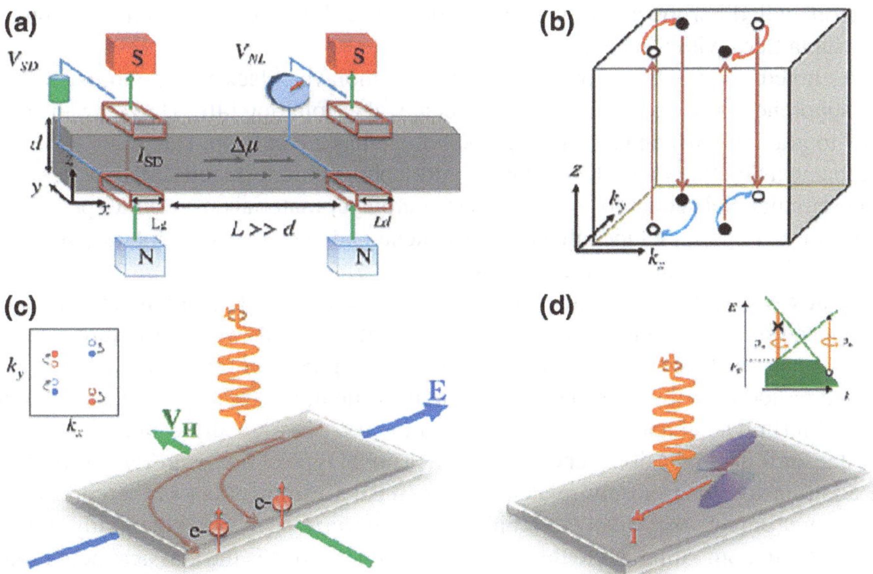

Fig. 1.9 Control of Weyl fermions by electrical and optical means. **a** A nonlocal electrical transport device utilizing axial current from the chiral anomaly. **b** Electrons exhibit unusual paths on the surface of a Weyl semimetal. **c** Time-reversal symmetry can be broken by shining an intense circularly polarized light. **d** A large photogalvanic current breaks inversion symmetry as well as any mirror symmetry in the presence of circularly polarized light. Adapted from [60]

Hepta-graphene: Based on density-functional-theory (DFT) calculations a dynamically stable, seven-membered carbon allotrope called hepta-graphene has been predicted [65]. It has a rectangular unit cell comprising ten carbon atoms and four hydrogen atoms. Its band structure is topologically equivalent to that of strongly distorted graphene, i.e. it has Dirac cones which are robust both under compressive and tensile strain. However, shear strain leads to a band gap, which is tunable. Note that systems without the hexagonal symmetry rarely exhibit Dirac cones; hepta-graphene is one such example.

Phagraphene: Another monolayer allotrope of carbon called phagraphene composed of pentagons, hexagons and heptagons has been proposed which also exhibits Dirac cones in a rectangular lattice and is robust against external uniaxial stress. However, similar to hepta-graphene a gap is opened in this material under shear stress [66]. A calculation of the phonon spectra demonstrates this allotrope is also dynamically stable (akin to hepta-graphene).

Phosphorene and its nanoribbons: Remarkably, the phosphorus analog of graphene called phosphorene has been synthesized, which is a promising candidate for thermoelectric applications [67]. Corresponding electronic, optical and transport properties have been studied for phosphorene as well as phosphorene nanoribbons including exciton effects. Interestingly, a related single-element based, monolayer material called borophene [68] has been found to exhibit Dirac cones in addition to two sublattices with a substantial ionic character.

As noted above, the graphene analog of monolayer black phospohorus, called phosphorene was isolated in 2014 by mechanical exfoliation [46]. However, in contract to graphene which is gapless, phosphorene has a band gap. Due to its superior mechanical flexibility and electrostatic control, phosphorene is well suited for flexible nano-circuits. Subsequently, other similar monolayer materials were either proposed or synthesized that include silicene, germanene and stanene [69]. Notably, stanene is a two-dimensional topological insulator.

Time crystals: Recently, the concept of time crystals (both quantum and floquet) has gained significant attention [70]. In simple terms, crystals whose structure repeats in time can be called time crystals. Such crystals repeat in time because they are kicked periodically by laser pulsing or magnetic field, i.e. they are intrinsically out of equilibrium and break time translation symmetry. They are also called space-time crystals or four-dimensional crystals and are a novel type of *non-equilibrium matter*. In addition to being closely related to dynamical Casimir effect in the context of zero-point energy, they likely exhibit *topological order* which is potentially useful for quantum computing. Note that topological order violates the classical belief that ordering requires symmetry breaking. Time-dependent electromagnetic fields driving a crystalline material can tune its topological properties and may cause it to become a Floquet topological insulator. Thus the notion of a time crystal can be extended to Floquet time crystals [70].

1.5 Metrology and Techniques

1.5.1 High-Resolution Electron Microscopy

Measurements that can provide information about the underlying topological characteristics (e.g. genus, local connectivity) of a material through different experimental probes such as fluorescence, optical means, etc. constitute what we mean by topological metrology. In addition to Raman measurements, small angle X-ray (SAXS) and neutron scattering (SANS), scanning electron microscopy (SEM), transmission electron microscopy (TEM), stimulated Brillouin spectroscopy (SBS), various non-linear optical imaging techniques such as three-photon excitation fluorescence polarizing microscopy [71], X-ray tomography [72], electron holography and tomography [73] and Lorentz TEM [74] provide metrological means to study various aspects of materials topology.

1.5.2 Nonlinear Optical Imaging

In the context of liquid crystals and colloids many optical imaging techniques have been invoked to study topological defects such as Schlieren texture and even a more elaborate defect called a Hopf fibration [2]. In a uniaxial nematic crystal, the Schlieren defect is essentially the "director field"; it can be observed using a polarizing microscope. On the other hand, the Hopf fibration is an exotic texture which resembles a series of rings that are wrapped around a torus. It has been observed in chiral nematic liquid crystals with the aid of holographic optical tweezers in conjunction with fluorescence polarizing microscopy. Similarly, anisotropic optical absorption techniques have been invoked to study the electronic structure of many of the correlated topological materials including a giant, nonlinear optical response in Weyl semimetals [75].

1.5.3 X-Ray Tomography and Electron Holography

X-ray tomography is useful for characterizing porous media and porous networks including gyroid structures [76]. Similarly, for studying magnetic topological defects (e.g. vortices and skyrmions) and magnetic microstructures electron holography [73] and Lorentz TEM [74] are very useful techniques. Depending on the length scale of the magnetic nanostructure under consideration, magnetic force microscopy can be used as a complementary imaging technique.

1.5.4 X-Ray and Neutron Scattering

In order to link materials topology with metrology, small angle X-ray scattering (SAXS) and small angle neutron scattering (SANS) techniques can be used to reveal information about the structural and topological phases about nanoscale ordering in materials. Note that SAXS and SANS have been effectively used to study biomembranes, vesicles, certain supramolecular assemblies and di- and triblock copolymer morphologies in addition to understanding the structure of a variety of mesoporous materials such as zeolites and MOFs [2].

1.5.5 Elasticity and Deformation Energy Characterization

Biological vesicles and block copolymers having complex topologies with a genus up to g = 3 have been observed. In contrast, synthetic vesicles with very large values of g (~50) can occur [83]. To model these systems, we start with the Helfrich-Canham curvature (Fig. 1.10) free energy [84]: $E_s = \int dS\,[\frac{\kappa_b}{2}(H - H_o)^2 + \kappa_o K]$, where κ_b = bending rigidity, κ_0 = Gaussian rigidity, dS = surface element. In addition, H_0 denotes spontaneous mean curvature, K and H are Gaussian and mean curvature, respectively. By using only topological means (e.g. Bogomol'nyi decomposition, which is usually invoked to study topological invariance and indicates that the Helfrich-Canham energy is greater than or equal to $4\pi\kappa_o$ times a genus dependent term), one can then calculate the elastic energy of deformation as a function of genus for vesicles [2], as shown in Fig. 1.11b. The energy increases proportionally with genus and eventually attains the value of 8π, consistent with a mathematical extrapolation called Wilmore conjecture. From topological analysis, one concludes that the spontaneous bending energy contribution from any deformation of the vesicles from their metastable shapes comprises two different topological sets: shapes of spherical topology (g = 0) and shapes of non-spherical topology ($g > 0$). One can readily apply these ideas to other topologies and materials. In a similar way, the deformation associated with negative curvature periodic minimal surfaces, e.g. double gyroids and Schwarzites, can be calculated, in particular under hydrostatic stress, if one assumes that only the lattice parameter changes under deformation. Analogously, graphene and carbon nanoribbons can exist in helicoidal shape. Their axial deformation can also be calculated by varying the pitch of the helicoid. In these cases, the elastic energy is proportional to the material's bulk (or axial) modulus and the Gaussian curvature.

Figure 1.11 shows the variation of Gaussian curvature K as a function of the surface parameter τ for three different values of the other surface parameter σ, and for the special direction when $\sigma = \tau$ [Through mathematical formalism for IPMS in the complex plane, the Gaussian curvature K can be explicitly expressed as a function of real variables τ and σ, [4, 5]. The expression for K(σ,τ) is invariant under the exchange of σ and τ and therefore, the two such figures must be identical [5]]. Note that for σ = 0, 0.5 and τ there is a minimum in K indicating elastically "soft" directions on the

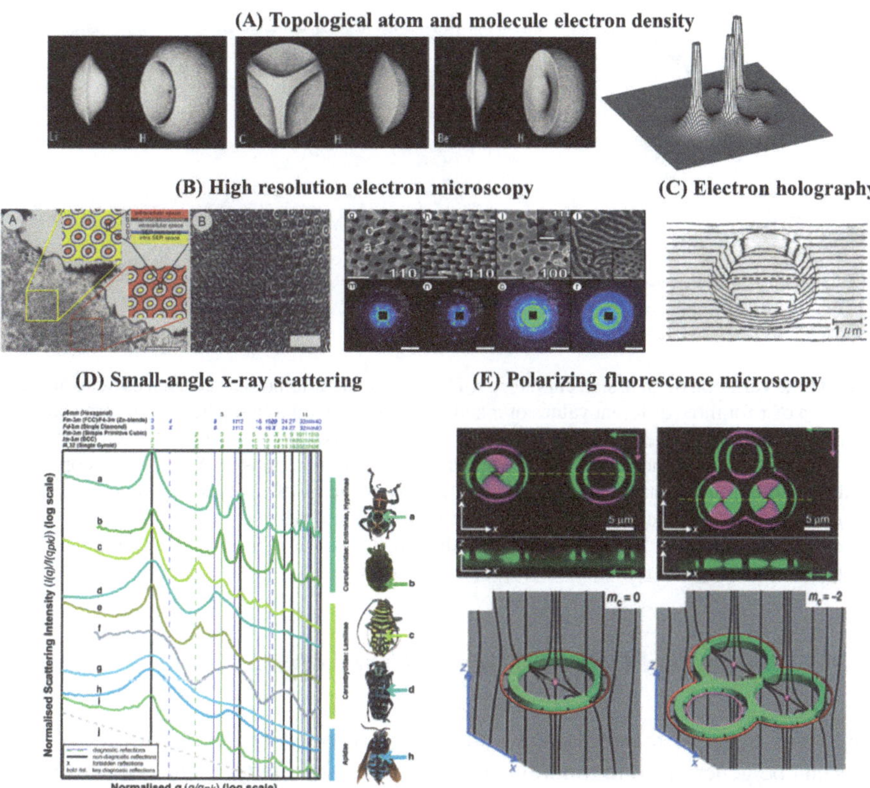

Fig. 1.10 Illustrations of various topological metrologies of **a** electron density distribution property of topological atoms and molecules [77, 78]. **b** Butterfly wing scale photonic nanostructure development using cross-section TEM depicting complex in-folding of the plasma membrane and SER membrane. The developing nanostructure shows the diagnostic motif of two concentric rings roughly in a triangular lattice. Yellow and red boxes highlight areas revealing different sections through the (110) plane of a polarized pentacontinuous core-shell double gyroid (color insets) [79]. Representative structural morphology of arthropod cuticular nanostructures and SAXS two-dimensional patterns from the photonic scales or setae [80], high-resolution electron microscopy revealing topological structures. **c** Interferogram of a toroidal ferromagnet measured using electron holography [81]. **d** Structural diagnoses of representative SAXS profiles of arthropod cuticular photonic nanostructures. **e** single- and triple handlebar (g = 1, 3, respectively) textures of colloidal particles that are non-spherical and dispersed in liquid crystals as obtained by three-photon excitation fluorescence polarizing microscopy (3PEF-PM) in addition to optical tweezers [82]

periodic minimal surface. However, for $\sigma = 1$ we observe a monotonically increasing "kink like" variation of K with τ. Also for the $\sigma = 0$ case, there are two values ($3\pi/50$ and $3\pi/5$) for which a local maximum is observed indicating elastically "hard" or stiff directions. Also note that the elastic energy density is directly proportional to K. Thus these curves provide a guide to the deformation energy behavior of Schwarzite surfaces.

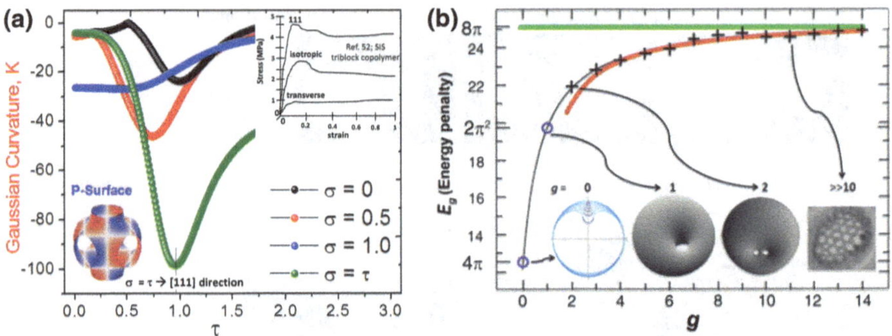

Fig. 1.11 **a** Variation of Gaussian curvature K for a minimal periodic surface such as Schwarzite as a function of τ for three different values of σ and for the special direction when $\sigma = \tau$ corresponding to various directions on the unit patch of the P-surface. The inset shows the stress-strain behavior in three different curves namely, [111], isotropic and transverse directions for SIS triblock copolymer forming a double gyroid phase [85]. **b** Elastic deformation energy of vesicles (E_g) versus genus [86] where blue circles are exact values; crosses are numerical estimates computed with a (Brakke's) surface evolver. Note that for large values of g the red curve is an estimated fit to the deformation energy. Finally, the green line is the asymptotic value of 8π

1.5.6 Topological Correlators and Other Metrics

Analogous to usual materials, the notions of two-point and higher order correlations can be generalized to topological correlations. The latter could conceivably be inferred from a combination of scattering techniques. Similarly, for nets and ramified structures one could envision obtaining network topology correlations (about local connectivity) from appropriately chosen momentum range in scattering experiments. Geometric measures such as curvature-curvature correlations (both in mean and Gaussian curvature) can complement the insights gained via topological correlations.

1.6 Computational Topology of Materials

Combining notions from topology and numerical algorithms, the field of computational topology has emerged. One of the key ideas in this context is the discovery of topology through algorithms. Although this topic belongs to the realm of computer science and mathematics, extending it to study the properties of topological materials opens a new avenue of investigation. This is of particular interest for materials involving network like structures, double gyroids, etc.

1.6.1 Topological Databases and Visualizing Topology

Over the past two decades topological databases and computer algorithms such as EPINET and TOPOS [87] have been developed. They are vey useful in designing extended crystalline frameworks and architectures. In the context of Materials by Design, Directed Materials for Energy and Environment, and Materials Genome Initiative (MGI), it is now imperative to create databases for the properties of topological materials which can in turn be used to predict new topological materials with desired physical properties targeting specific applications (i.e. Topological Materials Informatics). On the other hand, the analysis of various databases using techniques from topology is called topological data analysis (TDA), and it is a growing area of research which can fruitfully be applied to study materials.

Many tools have been developed to visualize the topology of networks and that of vector fields (e.g. spin configuration, flow fields, etc.). Some examples include TOPO, Otter, TorusVis, Kiwi, RadialNet, among others. They also enable us to display various topological aspects of a structure such as genus and network node connectivity. Also, one can explore how local topology evolves or changes under parametric variation in a material. Adopting these tools to understand the structure and properties of topological materials can lead to an entirely new way of understanding materials we have discussed in this chapter.

1.6.2 Miscellaneous Topics

In this chapter, we have tried to cover a broad variety of topological materials and their properties. However, our exposure is not comprehensive as we have not addressed the fields of topological photonics, topological plasmonics (and metamaterials), mechanical metamaterials (or auxetic materials) as well as the use of topology in characterizing materials microstructure. For the sake of completeness we briefly mention these topics here.

Topological photonics: Inspired by the observation of topological phases in condensed matter (e.g. TI and TCI, see Fig. 1.12) and materials science, analogs of such phases have been realized in the photonic context ushering in the field of topological photonics [88]. An ingenious design of wave vector space topologies is enabling the creation of interfaces supporting entirely new states of light with many useful properties. For instance, one can create unidirectional waveguides in which light flows around large imperfections without back-reflection (akin to interfacial electron transport without dissipation in topological insulators in condensed matter [39, 40]). There has been significant progress in the realization of a whole slew of topological effects in photonic crystals, photonic quasicrystals, coupled resonators and metamaterials. In the near future one expects to discover topological mirrors and new applications of interacting photons by invoking nonlinearity and entanglement. Some of the technological advantages here involve decreased power consumption,

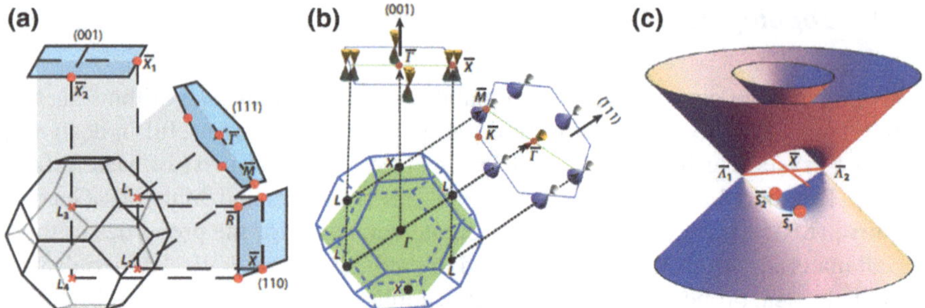

Fig. 1.12 Topological crystalline insulator. **a** High-symmetry points in the Brillouin zone and three projected surfaces for the rock-salt crystal structure. **b** Dirac cones in the (001) and (111) surface Brillouin zones. **c** Calculated dispersion for the (001) double Dirac cone surface state. Adapted from [42]

improved coherence in quantum links, avoiding use of isolators in photonic circuits, etc.

Topological plasmonics: Plasmonic excitations in Dirac materials including single and bilayer graphene and topological insulators are important both for a fundamental understanding of these materials as well as for their application in optoelectronic devices [89]. Recently an analogy between the usual two-dimensional magnetoplasmon [90] and p-wave topological superconductors has emerged. Analogs of photonic topology can be envisioned not only for surface plasmons but also for other bosonic systems such as magnons, phonons, excitons and exciton-polaritons. The key idea is that topological effects can be exploited to substantially improve the robustness of plasmonic, photonic and other devices in the presence of imperfections and various types of disorder.

Microstructure and topology: Statistical topology of cellular networks using Poisson-Voronoi cells has been recently developed along with a topological framework for local structure analysis and grain-growth microstructure characterization [91, 92]. Specifically, within a unified mathematical framework, local structure in both ordered and disordered materials can be classified by using the topology of the Voronoi cell associated with a particle (Fig. 1.13). For a given set of particles the Voronoi cell of a particle refers to a region in real space that is closer to the particle than to any other. This topological description of local structure offers many advantages for structural analysis compared to continuous descriptions. It also enables to identify which particles are associated with defects as opposed to belonging to specific (crystalline) phases. This versatile approach is also applicable to highly defected solids and glass-forming liquids. Moreover, through the distribution of different topological types the Voronoi topology aids the characterization of disordered systems in a statistical manner.

Another important concept that helps our understanding of microstructure is that of hyperuniformity [93]. One could think of unusual amorphous states of materials that lie between crystal and liquid as disordered many-particle hyperuniform sys-

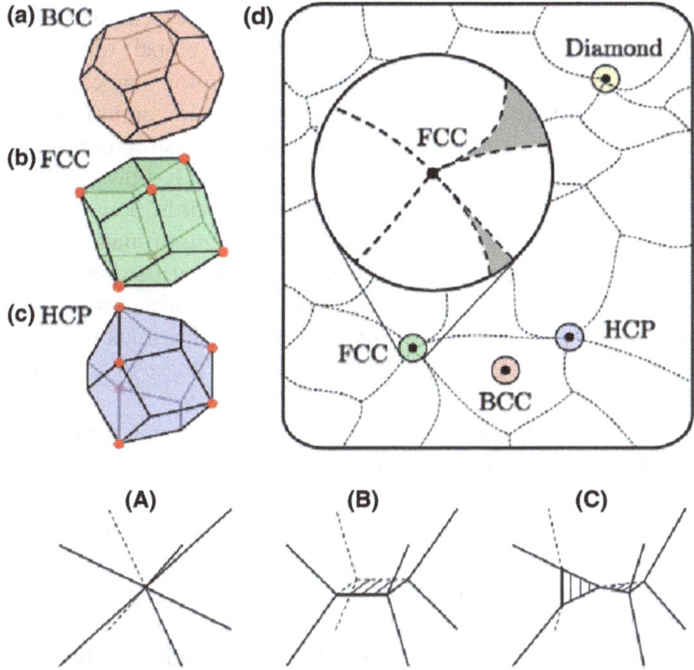

Fig. 1.13 Upper panel: **a**, **b** and **c** Voronoi cells of particles for the BCC, FCC and HCP crystals. Vertices where more than four Voronoi cells meet are indicated by red circles. Near these vertices small perturbations in particle positions lead to topological changes. **d** Space of all possible configurations of n neighbors; it can be divided into regions of constant Voronoi cell topology. Inset depicts the neighborhood around an FCC point. Lower panel: Sharing of an unstable vertex by six Voronoi cells in HCP or FCC crystals. A small perturbation can morph the vertex (in A) to either a four-sided face (in B) or a pair of contiguous triangular faces (in C). Adapted from [92]

tems. Put another way, in a hyperuniform system density fluctuations are completely suppressed at very large length scales, which means that the structure factor $S(\mathbf{k})$ tends to zero as the wave vector $\mathbf{k}$ vanishes. It provides a unified framework to classify and categorize certain disordered configurations, crystals and quasicrystals. It is also important for understanding local fluctuations in the interfacial area of two-phase media.

1.7 Conclusion

Based on a series of eclectic examples, we elucidated that the topological concepts are beginning to take a firm foothold in materials science [2]. Our hope is that the emerging shift from a structure ➔ property ➔ functionality traditional paradigm to a new topology/geometry ➔ property ➔ functionality emerging paradigm will assist materials scientists to study various usual and topological materials using the powerful interrelated concepts of (local) geometry and (global) topology. We illustrated this paradigm through a variety of materials including nanocarbon allotropes, soft matter, supramacromolecular assemblies, vesicles, biomembranes and MOFs.

An additional question is to understand the processing-structure-property (PSP) aspects of materials science for topological materials, i.e. how do synthesis and processing affect the topology of a material. In other words, how do we develop a processing-topology-property (PTP) based understanding of materials, which may lead to insights into interfaces between disparate topological materials and eventually to advanced manufacturing. This consideration naturally leads to another research frontier that is emerging: *topological dynamics*, i.e. keeping track of the time evolution of either the topology of or the relevant topological phase in a given material of interest.

The power and role of topology is turning out to be crucial in materials science in terms of unraveling certain types of defects (e.g. skyrmions, monopoles, hopfions, vortex lines) and in understanding an emerging class of topological materials with linear electronic dispersion such as Dirac materials and Weyl semimetals. Graphene, its two-dimensional siblings, e.g. silicone, germanene and phosphorene [46], in addition to topological insulators and topological superconductors are paving the way for unusual and technologically important properties. A variety of external perturbations including magnetic doping, electric and magnetic fields, light, disorder, temperature gradient, field gradients, strain and variation in film thickness can be used to alter and probe these fascinating materials. Inclusion of electronic correlations may lead to yet more exotic states of matter, e.g. fractional topological insulators [55].

We also discussed a variety of characterization techniques for topological materials ranging from optical imaging to electron holography, from fluorescence polarizing microscopy to Lorentz TEM, as well as small angle X-ray and neutron scattering. One aspect of this field that requires substantial progress and is now ripe for new ideas is that of topological metrology including use of topological correlations (both in real- and k-space) to characterize different material properties. The field of topological databases and informatics is still in its infancy. The importance of topology is thus ineluctable in materials science given the enormity of expected breakthroughs spanning fundamental and invaluable insights into modern and novel technologies.

Acknowledgements The authors acknowledge stimulating and helpful discussions with several colleagues around the world over the past decade. This work was supported in parts by the U.S. Department of Energy and Western Kentucky University Research Foundation Inc.

References

1. S. Gupta, A. Saxena, MRS Bull. News Feature, Oct. 18 (2016); F. Ghahari, D. Walkup, C. Gutiérrez, J.F.R.- Nieva, Y. Zhao, J. Wyrick, F.D. Natterer, W.G. Cullen, K. Watanabe, T. Taniguchi, L.S. Leitov, N.B. Zhitenev, J.A. Stroscio, Science **356**, 845 (2017)
2. S. Gupta, A. Saxena, MRS Bull. **39**, 265 (2014)

3. S. Gupta, A. Saxena, J. Raman Spectrosc. **40**, 1127 (2009)
4. S. Gupta, A. Saxena, J. Appl. Phys. **109**, 074316 (2011)
5. S. Gupta, A. Saxena, J. Appl. Phys. **112**, 114316 (2012)
6. V. Meunier, Ph Lambin, A.A. Lucas, Phys. Rev. B **57**, 14886 (1998)
7. J.C. Charlier, G.M. Rignanese, Phys. Rev. Lett. **86**, 5970 (2001)
8. M.W. Iqbal, A.K. Singh, M.Z. Iqbal, J. Eom, J. Phys.: Condens. Matter **24**, 335301 (2012)
9. V. Mennella, G. Monaco, L. Colangeli, E. Bussoletti, Carbon **33**, 115 (1995)
10. G. Toulouse, M. Kléman, J. Phys. Lett. **37**, L149 (1976)
11. G. E. Volovik, V. P. Mineyev, Zh. Eksp. Teor. Fiz. Pis'ma Red. **24**, 605 (1976)
12. J. Wright, N.D. Mermin, Rev. Mod. Phys. **61**, 385 (1989)
13. X. Xing, J. Stat. Phys. **134**, 487 (2009)
14. B. Senyuk, Q. Liu, S. He, R.D. Kamien, R.B. Kusner, T.C. Lubensky, I.I. Smylukhov, Nature **493**, 200 (2013)
15. B.G. Chen, P.J. Ackerman, G.P. Alexander, R.D. Kamien, I.I. Smylukhov, Phys. Rev. Lett. **110**, 237801 (2013)
16. G.S. Settles, *Schlieren and Shadowgraph Techniques: Visualizing Phenomena in Transparent Media* (Springer, Berlin, 2001)
17. A.T. Skjeltorp, *Knots and Applications to Biology, Chemistry and Physics* (Springer, NY, 1996)
18. S.R. Batten, N.R. Champness, X.-M. Chen, J.G.-Martinez, S. Kitagawa, S.L. Öhrström, M. O'Keeffe, M.P. Suh, J. Reedijk, Pure Appl. Chem. **85**, 1715 (2013)
19. A.K. Cheetham, G. Férey, T. Loiseau, Angew. Chem. Inter. Ed. **38**, 3268 (1999)
20. J. Cejka (ed.), *Metal-Organic Frameworks Applications from Catalysis to Gas Storage* (Wiley-VCH, 2011)
21. S.T. Hyde, G.E. Schröder-Turk, Interface Focus **2**, 529 (2012)
22. T. Ishøy, K. Mortensen, Langmuir **21**, 1766 (2005)
23. N. Hadjichristidis, S. Pispas, G. Floudas, *Block Copolymers: Synthetic Strategies, Physical Properties and Applications* (Wiley, NY, 2003)
24. E. Sezgin, H.-J. Kaiser, T. Baumgart, P. Schwille, K. Simons, I. Levental, Nat. Protoc. **7**, 1042 (2012)
25. D. Zhao, D.J. Timmons, D. Yuan, H.-C. Zhou, Acc. Chem. Res. **44**, 123 (2011)
26. J. Katsaras, T. Gutberlet (eds.), *Lipid Bilayers—Structure and Interactions* (Springer, Berlin-Heidelberg, 2001)
27. H.L. Leertouwer, B.D. Wilts, D.G. Stavenga, Opt. Exp. **19**, 24061 (2011)
28. B.D. Wilts, K. Michielsen, J. Kuipers, H. De Raedt, D.G. Stavenga, Proc. R Soc. B **279**, 2524 (2012)
29. B. Yan, S.-C. Zhang, Rep. Prog. Phys. **75**, 096501 (2012)
30. X.-L. Qi, T.L. Hughes, S.-C. Zhang, Phys. Rev. B **78**, 195424 (2008)
31. A. Roy, D.P. DiVincenzo, *Topological Quantum Computing*, arXiv:1701.05052
32. S. Tanda, T. Tsuneta, Y. Okajima, K. Inagaki, K. Yamaya, N. Hatakenaka, Nature **417**, 397 (2002)
33. T. Tsuneta, S. Tanda, J. Cryst. Growth **267**, 223 (2004)
34. T. Matsuura, M. Yamanaka, N. Hatakenaka, T. Matsuyama, S. Tanda, J. Cryst. Growth **297**, 157 (2006)
35. R.E. Goldstein, H.K. Moffatt, A.I. Pesci, R.L. Ricca, Proc. Natl. Acad. Sci. (USA) **107**, 21979 (2010)
36. D.M. Kleiman, D.F. Hinz, Y. Takato, E. Fried, Soft Matter **12**, 3750 (2016)
37. T.O. Wehling, A.M. Black-Schaffer, A.V. Balatsky, Adv. Phys. **63**, 1 (2014)
38. J. Cayssol, Comp. Rend. Physique **14**, 760 (2013)
39. M.Z. Hassan, C.L. Kane, Rev. Mod. Phys. **82**, 3045 (2010)
40. X.L. Qi, S.-C. Zhang, Rev. Mod. Phys. **83**, 1057 (2011)
41. L. Fu, Phys. Rev. Lett. **106**, 106802 (2011)
42. Y. Ando, L. Fu, Annu. Rev. Condens. Matter Phys. **6**, 361 (2015)
43. B. Aufray, A. Kara, S.B. Vizzini, H. Oughaddou, C. LeAndri, G. Le Lay, Appl. Phys. Lett. **96**, 183102 (2010)

44. M.E. Davila, L. Xian, S. Cahangirov, A. Rubio, G. Le Lay, New J. Phys. **16**, 095002 (2014)
45. B.G. Kim, H.J. Choi, Phys. Rev. B **86**, 115435 (2012)
46. H. Liu, A.T. Neal, Z. Zhu, Z. Luo, X. Xu, D. Tomanek, P.D. Ye, ACS Nano **8**, 4033 (2014)
47. Y. Xu, B. Yan, H.-J Zhang, J. Wang, G. Xu, P. Tang, W. Duan, S.-C. Zhang, Phys. Rev. Lett. **111**, 136804 (2013)
48. G. Bian et al., Nat. Commun. **7**, 10556 (2016)
49. S.-Y. Xu et al., Science **349**, 613 (2015)
50. B.Q. Lv et al., Phys. Rev. X **5**, 031013 (2015)
51. V. Mourik, K. Zuo, S.M. Frolov, S.R. Plissard, E.P.A.M. Bakkers, L.P. Kouwenhoven, Science **336**, 1003 (2012)
52. S. Sasaki, M. Kreiner, K. Segawa, K. Yada, Y. Tanaka, M. Sato, Y. Ando, Phys. Rev. Lett. **107**, 217001 (2011); M. Sato, Y. Ando, Rep. Prog. Phys. **80**, 076501 (2017)
53. V.S. Pribiag, A.J.A. Beukman, F. Qu, M.C. Cassidy, C. Charpentier, W. Wegscheider, L.P. Kouwenhoven, Nat. Nanotechnol. **10**, 593 (2015)
54. D.-T. Tran, A. Dauphin, N. Goldman, P. Gaspard, Phys. Rev. B **91**, 085125 (2015)
55. M. Levin, A. Stern, Phys. Rev. Lett. **103**, 196803 (2009)
56. O. Vafek, A. Vishwanath, Ann. Rev. Cond. Mat. Phys. **5**, 83 (2014)
57. X. Wan, A.M. Turner, A. Vishwanath, S.Y. Savrasov, Phys. Rev. B **83**, 205101 (2011)
58. A.A. Burkov, L. Balents, Phys. Rev. Lett. **107**, 127205 (2011)
59. B. Yan, C. Felser, Ann. Rev. Cond. Mat. Phys. **8**, 337 (2017)
60. S. Jia, S.-Y. Xu, M.Z. Hasan, Nat. Mater. **15**, 1140 (2016)
61. S. Zhang, I. Gilbert, C. Nisoli, G.-W. Chern, M.J. Erickson, L. O'Brien, C. Leighton, P.E. Lammert, V.H. Crespi, P. Schiffer, Nature **500**, 553 (2013)
62. P. Milde, D. Kohler, J. Seidel, L.M. Eng, A. Bauer, A. Chacon, J. Kindervater, S. Muhlbauer, C. Pfleiderer, S. Buhrandt, C. Schutte, A. Rosch, Science **340**, 1076 (2013)
63. S.H. Zhang, J. Zhou, Q. Wang, X.S. Chen, Y. Kawazoe, P. Jena, Proc. Natl. Acad. Sci. (USA) **112**, 2372 (2015)
64. S. Zhang, Q. Wang, Y. Kawazoe, P. Jena, J. Am. Chem. Soc. **135**, 18216 (2013)
65. A. Lopez-Bezanilla, I. Martin, P.B. Littlewood, Sci. Rep. **6**, 33220 (2016)
66. A. Lopez-Bezanilla, J. Phys. Chem. C **120**, 17101 (2016)
67. J. Zhang, H.J. Liu, L. Cheng, J. Wei, J.H. Liang, D.D. Fan, J. Shi, X.F. Tang, Q.J. Zhang, Sci. Rep. **4**, 6542 (2014)
68. Z. Wang, X.-F. Zhou, X. Zhang, Q. Zhu, H. Dong, M. Zhao, A.R. Oganov, Nano Lett. **15**, 6182 (2015)
69. C.-C. Liu, H. Jiang, Y. Yao, Phys. Rev. B **84**, 195430 (2011)
70. D.V. Else, B. Bauer, C. Nayak, Phys. Rev. Lett. **117**, 090402 (2016)
71. B.G. Chen, P.J. Ackerman, G.P. Alexander, R.D. Kamien, I.I. Smalyukh, Phys. Rev. Lett. **110**, 237801 (2013)
72. L. Salvo, M. Suery, A. Marmottant, N. Limodin, D. Bernard, Comput. Rend. Phys. **11**, 641 (2010)
73. P.A. Midgley, R.E. Dunin-Borkowski, Nat. Mater. **8**, 271 (2009)
74. A.K. Petford-Long, M. De Graef, *Lorentz Microscopy, Characterization of Materials,* 1–15 (Wiley Online Library, 2012)
75. L. Wu, S. Patankar, T. Morimoto, N.L. Nair, E. Thewalt, A. Little, J.G. Analytis, J.E. Moore, J. Orenstein, Nat. Phys. **13**, 350 (2017)
76. H.K. Chae, D.Y. Siberio-Perez, J. Kim, Y.-B. Go, M. Eddaoudi, A.J. Matzger, M. O'Keefe, O.M. Yaghi, Nature **427**, 523 (2004)
77. R.F.W. Bader, C. Matta, J. Phys. Chem. A **108**, 8385 (2004)
78. R.F.W. Bader, D.E. Fang, J. Chem. Theory Comput. **1**, 403 (2005)
79. V. Saranathan, C.O. Osujib, S.G.J. Mochrieb, H. Nohb, S. Narayanan, A. Sandy, E.R. Dufresneb, R.O. Pruma, Proc. Natl. Acad. Sci. (USA) **107**, 11676 (2010)
80. V. Saranathan, A.E. Seago, A. Sandy, S. Narayanan, S.G.J. Mochrie, E.R. Dufresne, H. Cao, C.O. Osuji, R.O. Prumand, Nano Lett. **15**, 3735 (2015)

81. A. Tonomura, H. Umezaki, T. Matsuda, N. Osakabe, J. Endo, Y. Sugita, Phys. Rev. Lett. **51**, 331 (1983)
82. B. Senyuk et al., Nature **493**, 200 (2013)
83. X. Michalet, D. Bensimon, Science **269**, 666 (1995)
84. R. Lipowsky, Encyclopedia of Applied Physics **23**, 199 (1998)
85. B.J. Dair, A. Avgeropoulos, N. Hadjichristidis, E.L. Thomas, J. Mater. Sci. **35**, 5207 (2000)
86. J. Benoit, A. Saxena, T. Lookman, J. Phys. A **34**, 9417 (2001)
87. http://www.topos.samsu.ru and http://www.epinet.anu.edu.au/reference
88. L. Liu, J.D. Joannopoulos, M. Soljacic, Nat. Photonics **8**, 821 (2014)
89. T. Stauber, J. Phys.: Condens. Mater **26**, 123201 (2014)
90. D. Jin et al., Nat. Commun. **7**, 13486 (2016)
91. E.A. Lazar, J.K. Mason, R.D. MacPherson, D.J. Srolovitz, Phys. Rev. Lett. **109**, 095505 (2012)
92. E.A. Lazar, J. Han, D.J. Srolovitz, Proc. Natl. Acad. Sci. (USA) **112**, E5769 (2015)
93. S. Torquato, Phys. Rev. E **94**, 022122 (2016)

Chapter 2
Topology and Geometry in Condensed Matter

Rossen Dandoloff

2.1 Topology

2.1.1 Introduction

There is an old joke among mathematicians. It goes that way [1]: A mathematician was asked: What is a topologist? Someone for whom there is no difference between a doughnut and a coffee cup with a handle.

In general, topology studies continuum properties of spaces that are not affected by continuous deformations. Such deformations may be e.g. stretching and bending. Obviously e.g. cutting and gluing do not belong to the allowed deformations. Usually these properties are studied on what is called topological spaces i.e. a collection of subspaces that are open sets. These open sets satisfying certain conditions represent a topological space. Some of the most important topological properties are (a) connectedness, which simply counts the number of holes in the space and (b) compactness which means a subset of the Euclidean space that is closed and bounded; closed means that it contains all its boundary points and bounded means that all its points are at some distance to a given point that is less than some fixed maximal distance. Some examples a given by a closed interval, a rectangle, an ellipse circle, a sphere or a finite set of points. An ellipse e.g. is topologically equivalent to a circle (into which it can be deformed by continuous deformation e.g. stretching) and a sphere is equivalent to an ellipsoid. Similarly, the set of the numbers 0, 4, 6 and 9 are topologically equivalent - they have one hole each. The numbers 1, 2, 3, 5 and 7 are also topologically equivalent - they have 0 holes each and the number 8 (has two holes) is not topologically equivalent to either set of numbers.

R. Dandoloff (✉)
Laboratoire de Physique Théorique et Modélisation, Université de Cergy-Pontoise, 95302 Cergy-Pontoise, France
e-mail: rdandoloff@yahoo.com

S. Gupta and A. Saxena (eds.), *The Role of Topology in Materials*, Springer Series in Solid-State Sciences 189,
https://doi.org/10.1007/978-3-319-76596-9_2

In fact topology is the newest branch of geometry. It studies different sorts of spaces and especially the question what distinguishes different geometries. Felix Klein has suggested that the allowed transformations that keep certain kind of geometry unchanged is in fact its main mark. For example in the ordinary Euclidean geometry one is allowed to translate and rotate different objects, but bending and stretching are not allowed. Projective geometry on the other hand recognises different views of the same object as an allowed "transformation" within the projective geometry. The circle and the ellipse are projectively equivalent: all depends on the point of observation of a circle that may look like a circle or as an ellipse. Topology allows any continuous transformation that is reversible in a continuous way. Let us take the ellipse - it is equivalent to a circle or a square because one can continuously transform it into a circle or a square in a reversible way. If during the transformation one needs to cross two lines this is not any more a reversible transformation: an example is the Fig. 2.8. Topology as the almost most fundamental form of geometry is used in almost all branches of mathematics. It turns out that there is an even more fundamental form of geometry - the homotopy theory. It was formulated around 1900 by Poincare. Two geometric objects are called homotopic if they can continuously be transformed from one into the other without cutting and gluing. The number of allowed transformations is very big and one can deal with them as it is done in algebra. One can use homotopy to classify different geometrical objects and there are many applications to physics too: spin systems, liquid crystals etc.

Here we will start with the application of homotopy to physics. In order to do this we will use the notion of *order parameter* which is widely used in physics. The order parameter is usually a geometric object (unit vector, tensor etc.) which best characterizes the state of the physical material that we are studying. For example if we are studying a three dimensional ferromagnet, the most important property is its magnetization, which is represented by an unit vector field. The unit vectors of this vector field may point in any direction (at sufficiently high temperature where the magnets are not oriented in any particular direction). These vectors may be mapped to a unit sphere as shown on the Fig. 2.1. The unit sphere is called a *target space*. If the ferromagnet is two dimensional, and the magnetization vectors lye in the plane, then the target space is a circle with unit radius.

In order to fully take advantage of the topology and the mapping from the physical to the order parameter space, we will introduce the so called *compactification* of the physical space. Let us explain this on a simple example. Consider a two dimensional plane with a ferromagnet field on it which at the infinity points to the same direction, say perpendicular to the plane and upward. From the point of view of the ferromagnetic field the infinity of the physical plane is characterized by only one single vector that points up. So, we may bring al the infinity in one point, but then our physical plane will look like a sphere where the north pole of the sphere is the point which represents the infinity of the plane.

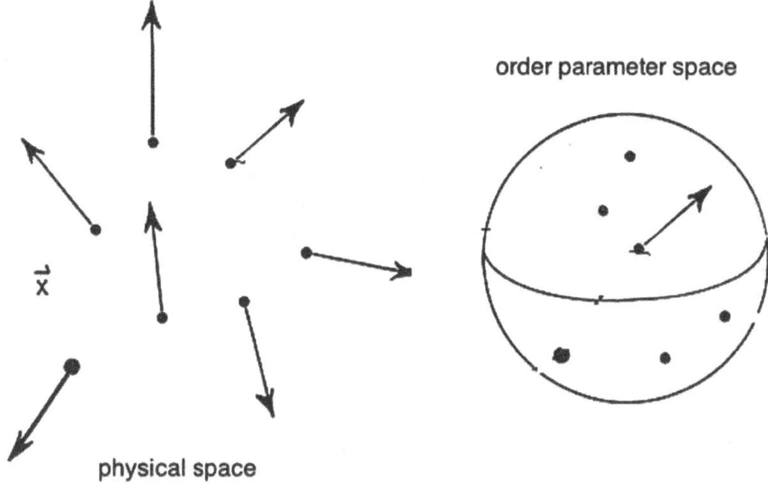

Fig. 2.1 Mapping from physical space to order parameter space

2.1.2 Classification of Vector Fields with Homogeneous Boundary Conditions

The mapping from the compactified physical space (e.g. the sphere S^2) onto the target space (S^2 in the case of the ferromagnetic field), allows to classify the different vector field configurations into separate *homotopy* classes. The notation is the following:

$$\pi_2(S^2) = \pi_2(S^2) = \mathbf{Z}(the\, group\, of\, the\, relative\, integers) \qquad (2.1)$$

where $n = 0, 1, 2, \ldots$ is an integer which labels the corresponding vector field configuration. Equation (2.1) means how many times the sphere may wrap another sphere. As an example let us consider $n = 1$ - this configuration is such that the mapping of the vector field on the target space covers the sphere S^2 just once, for $n = 2$ the target space S^2 has been covered twice. It is clear now that it is not possible by continuous transformation to deform the configuration with $n = 1$ (here the tips of the vectors mapped onto the target space cover the sphere once) into configuration with $n = 2$ (here the tips of the vectors mapped onto the sphere cover S^2 twice).

Let us consider now an one dimensional physical space (an infinite line) with a ferromagnetic field (a unit vector field) on it. We will consider first a magnetization which lies in the plane perpendicular to the line, i.e. the magnetisation vector may point in any direction perpendicular to the physical line. It is obvious now that the target space is an unit circle S^1. If the boundary conditions are homogeneous i.e. the magnetisation vectors at + infinity and at − infinity of the physical line, are parallel, then we may compactify the line into a circle. Now the different homotopy classes of configurations are given by the following equation:

$$\pi_1(S^1) = \pi_2(S^2) = \mathbf{Z} \tag{2.2}$$

This equation tells us how many times the circle may wrap around another circle. The class $n \in \mathbf{Z}$ means that the magnetization vector points at the same direction at $\pm\infty$ and turns once around the line going from $-\infty$ to $+\infty$. Let us note however that not all spin configurations with homogeneous boundary conditions lead to different homotopy classes. As an example let us take the line with spins that may point in any direction i.e. the target space now is s^2. Now the topological classification is given by the following homotopy:

$$\pi_1(S^2) = 0 \tag{2.3}$$

Here any closed curve (the mapping from the circle to the surface of the sphere) on S^2 may be shrunk to a point.

2.1.3 Classification of Defects in Vector Fields (Mainly Spin Fields)

In order to better illustrate the role of topology in the classification of defects [2] we will concentrate only on two dimensional spin fields where spins lye in the plane i.e. the target space is the circle S^1. The case of spin fields (vector fields) may be generalised to describe liquid crystals as well as there the order parameter is a headless vector. Now, here the idea is to surround the defect by a closed contour and to map the vectors on that contour onto the target space, see Fig. 2.2. On Fig. 2.2 we see that the closed curve on target space may be shrunk to a point. This means that there is no defect inside the loop. The situation would have been different if the vector field on the circle would have been radial in any point - then the closed curve on the target space would have wrapped once the circle and the result would have been the following homotopy equation:

$$\pi_1(S^1) = \pi_2(S^2) = \mathbf{Z} \tag{2.4}$$

The defect here represents a source of the vector field. On the other hand if the vector field is allowed to come out of the plane, the target space becomes S^2 and as we have seen, any closed curve on the sphere can continuously be shrunk to a point, meaning that there is no defect enclosed by the loop on the plane.

2.1.4 Defects and Homogeneous Boundary Conditions

Let us for a moment come back to our discussion of homogeneous boundary conditions and see what does it mean in view of the preceding discussion of topological

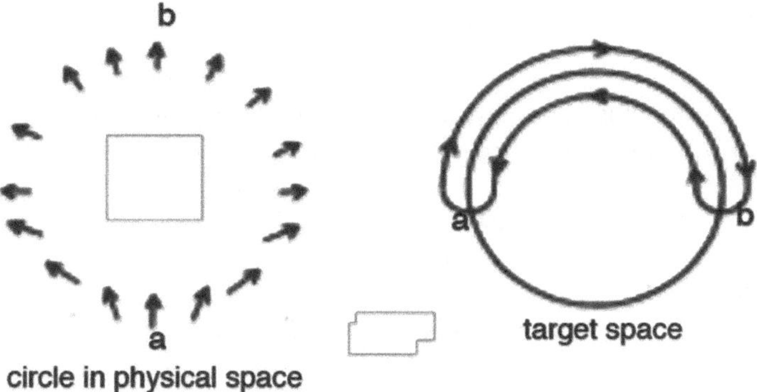

Fig. 2.2 Mapping from a closed contour around the defect to the order parameter space

classification of defects. If we consider the plane as our physical space, the homogeneous boundary condition for a vector field on it means that a loop at infinity (where the vectors point all in the same direction) will map to a single point on the target space. Now, let us suppose that there is a point defect somewhere on the plane. If we draw a closed loop around this defect and map the vectors on the target space we will get a closed line which we will not be able to shrink to a point. We may deform our loop on the physical plane to a loop at infinity and the map on the target space would not be shrinkable to a point. This contradiction tells us that imposing homogeneous boundary conditions also means that we exclude all point defects from the vector field.

2.2 Geometry

2.2.1 Energy

So far we have seen that very often spin fields with homogeneous boundary conditions fall into different homotopy classes. Now we will see what is the consequence for the energy of these different classes. For simplicity let us consider a ferromagnetic field – the energy is lowest when all vectors are parallel.

We will consider first the plane R^2 which represents the simplest 2D manifold. We will impose on the vector field on R^2 homogeneous boundary conditions: $lim_{r\to\infty}\vec{n} \to \vec{n}_0$. With these boundary conditions we may compactify the plane into the sphere S^2. Now the vector field configurations that may appear on the plane may be classified in homotopy classes $\pi_2(S^2) = \mathbf{Z}$ [3]. This topological classification in general is not related to the energy of the system. We may write the Hamiltonian for the vector field on R^2 as follows:

$$H = \int (\nabla \vec{n})^2 d^2x, \quad \vec{n}^2 = 1. \tag{2.5}$$

Nevertheless topology does give some indications about the energy of a field configuration in each homotopy class of equivalence using the Bogomolny inequalities [4]:

$$(\partial_i \vec{n} - \epsilon_{ij} \partial_j \vec{n})^2 \geq 0, \tag{2.6}$$

and therefore

$$H \geq \int \vec{n} \cdot (\partial_x \vec{n} \wedge \partial_y \vec{n}) dx dy. \tag{2.7}$$

It is obvious from (2.6) that when

$$\partial_i \vec{n} = \pm \epsilon_{ij} \partial_j \vec{n}. \tag{2.8}$$

minimum energy is reached in each class. Equation (2.8) are called "self-dual equations". Homotopy is useful for establishing different classes of vector field configurations and may help establish some inequalities regarding the energy in each configuration but geometry in general is not much more helpful in establishing the energy of a field configuration. Especially a geometry without an internal length (this is the case e.g. for the plane R^2). Nontrivial filed configurations on the plane R^2 may be scaled (shrunk) to a point without affecting the energy of the configuration. This happens because the Hamiltonian is symmetric under homothety (stretching of the space). Let us consider what happens with the Hamiltonian under stretching of the space by a factor λ: $x \to \lambda x$ and $y \to \lambda y$

$$E_\lambda = \int \int \left(\left(\frac{\partial n}{\partial \lambda x} \right)^2 + \left(\frac{\partial n}{\partial \lambda y} \right)^2 \right) d\lambda x d\lambda y = \int \int \left(\left(\frac{\partial n}{\partial x} \right)^2 + \left(\frac{\partial n}{\partial y} \right)^2 \right) dx dy = E. \tag{2.9}$$

The energy of the field configuration is invariant under stretching. As we mentioned above this means that the whole configuration may be shrunk into a point. These topological configurations are energetically metastable. The deeper reason for this to happen is that there is no internal length (or characteristic length) in the problem. Naturally there is no length in topology.

2.2.2 *Geometry with Intrinsic Length: The Cylinder*

As a geometry with intrinsic length we will consider the cylinder. The intrinsic length of the cylinder is its radius ρ_0. We will consider a cylindrically symmetric vector field (spin field) on the cylinder with radius ρ_0. As an immediate consequence of the presence of the intrinsic length ρ_0 we note that the homothety does not apply here and the energy does depend on ρ_0. The order parameter for the classical Heisenberg model is the unit vector which covers the sphere S^2. It is easier to work with the Euler angles θ and ϕ on the unit sphere because they incorporate the constraint ($\mathbf{n}^2 = 1$). Our independent vector fields will be (θ, Φ) where $\vec{n} = (\cos\theta,\ \sin\theta\ \cos\Phi,\ \sin\theta\ \sin\Phi)$. Here θ is the co-latitude and Φ is the azimuthal angle. In cylindrical coordinates (ρ, x, φ) we can write the Hamiltonian in the following way [5]

$$H_{isotropic} = J \iint_{cylinder} \left[(\partial_x\theta)^2 + \sin^2\theta\,(\partial_x\Phi)^2 + \frac{(\partial_\varphi\theta)^2}{\rho_0^2} + \frac{\sin^2\theta}{\rho_0^2}(\partial_\varphi\Phi)^2 \right] \rho\, dx d\varphi, \tag{2.10}$$

where J is the spin-spin coupling constant.

We will consider our vector field with homogeneous boundary conditions at both ends of the cylinder, because then we can compactify the cylinder using topological considerations. Homogeneous boundary conditions in this case mean $\lim_{x\to\infty} \theta \equiv 0[\pi]$. We ask also that $\lim_{x\to\infty} \frac{d\theta}{dx} = 0$. This second condition is required in order to have finite energy on the infinite cylinder. The fact that $\frac{d\theta}{dx}$ goes to zero should insure the convergence of the integral in (2.10). With the homogeneous boundary conditions at both ends of the cylinder we can make coinside all points at infinity and compactify the infinite cylinder into a sphere (the ends of the cylinder become the two poles of the sphere). Then we can map the sphere (the compactified cylinder) onto S^2 (the order parameter manifold) and so we get $\pi_2(S^2)=\mathbf{Z}$. The result is that the spin configurations on the infinite cylinder can be classified in different classes of topologically non-trivial spin distributions [3, 5]. Inside each class, the spin configurations are topologically equivalent because they belong to the same homotopy class.

In this example we will consider only solutions with cylindrical symmetry. They will be sufficient for our purposes. For the angles θ and Φ the following conditions must apply:

$$\Phi = \varphi \quad , \quad \frac{\partial\theta}{\partial\varphi} = 0. \tag{2.11}$$

The Hamiltonian (2.10) then becomes

$$H_{isotropic} = 2\pi\rho_0 J \int_{-\infty}^{+\infty} \left[\left(\frac{d\theta}{dx}\right)^2 + \frac{\sin^2\theta}{\rho_0^2} \right] dx. \tag{2.12}$$

After variation of the Hamiltonian $\delta H = 0$, the Euler–Lagrange equation leads to

$$\frac{d^2\theta(x)}{dx^2} = \frac{1}{2\rho_0^2}\sin 2\theta. \tag{2.13}$$

This equation represents the sine-Gordon equation whose solutions are solitons. This second order differential equation appears in a big variety of physical prob-

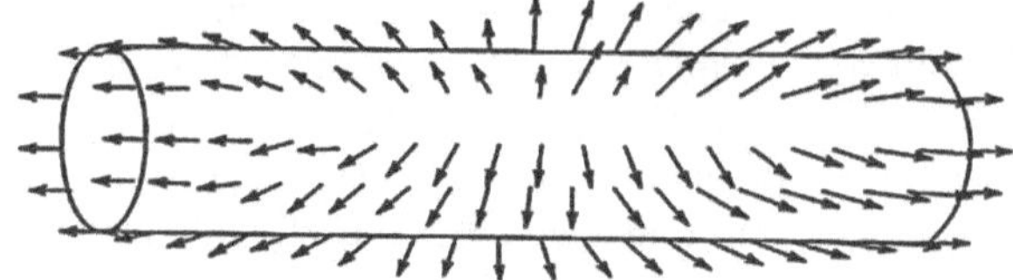

Fig. 2.3 Cylindrically symmetric $0 \rightarrow \pi$ twist soliton on an infinite cylinder

lems e.g. charge-density-wave in different materials, splay waves on membranes, Bloch wall motion in magnetic crystals, magnetic flux in Josephson lines, propagation of dislocations in crystals, torsion coupled pendula, two-dimensional models of elementary particles, etc.

One solution for a configuration which belongs to the first homotopy class and representing a single spin twist, is given by:

$$\theta = 2 \arctan \exp \frac{x}{\rho_0}. \tag{2.14}$$

A schematic representation is given in Fig. 2.3 The characteristic length ρ of the cylinder appears explicitly in the solution. In this solution ρ_0 represents the width of the twist soliton.

2.2.3 Geometry with Intrinsic Length: Plane with a Disc Missing

Now we will consider yet another example of a geometry with intrinsic length namely a non simply connected plane R^2. Here the intrinsic length will be the radius ρ_0 of the disk $D^2_{\rho_0}$ cut off from the plane. First we will consider spins on $R^2 \backslash D^2_{\rho_0}$ and then the same classical spin field but in a perpendicular to the plane magnetic field **B**. In a cylindrical coordinate system (ρ, ϕ) he Hamiltonian (the nonlinear sigma model) reads:

$$H = 2\pi \int_{\rho_0}^{\infty} d\rho \left[\rho\theta_\rho^2 + \frac{\sin^2\theta}{\rho} \right]. \tag{2.15}$$

Here we are using the Euler angles representation for the unit vector $\mathbf{n} = (\sin\theta\cos\Phi, \sin\theta\sin\Phi, \cos\theta)$, (the spins lie on a unit sphere S^2). Without loss of generality we will assume cylindrical symmetry for the spin configurations: $\theta = \theta(\rho)$ and $\Phi = \phi$. The solutions of the Euler–Lagrange (EL) equation will give the configurations with lowest energy:

$$\theta_\rho + \rho\theta_{\rho\rho} = \frac{\sin\theta\cos\theta}{\rho}. \tag{2.16}$$

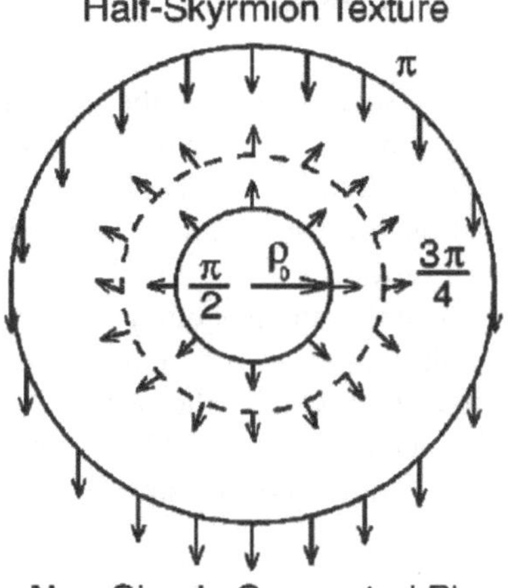

Fig. 2.4 Field on a plane with a disc missing

Here we take $\theta(\rho_0) =$ constant. and define a new radius coordinate $\overline{\rho} = \ln(\rho/\rho_0)$ which will allow us to reduce the EL equation to a simple sine-Gordon equation:

$$\theta_{\overline{\rho}\overline{\rho}} = \frac{\sin 2\theta}{2}. \tag{2.17}$$

A novel exact half-skyrmion appears to be the solution of this sine-Gordon equation on the non-simply connected plane shown on Fig. 2.4.

$$\theta(\rho, \rho_0) = 2\tan^{-1}\frac{\rho}{\rho_0}. \tag{2.18}$$

This solution depends on ρ_0, the intrinsic length in the problem and can not be shrunk to a point like the usual Belavin-Polyakov skyrmion [3]. It is located at $\rho_c = \rho_0 \cot(\pi/8) = \rho_0(1 + \sqrt{2})$ with energy 4π (instead of 8π) and topological charge density

$$q(\rho) = \frac{\theta_\rho \sin\theta}{4\pi\rho} = \frac{1}{\pi}\frac{\rho_0^2}{(\rho^2 + \rho_0^2)^2}. \tag{2.19}$$

Equivalently, the components of the unit vector field are:

$$n^x = \frac{2x\rho_0}{\rho^2 + \rho_0^2}, \quad n^y = \frac{2y\rho_0}{\rho^2 + \rho_0^2}, \quad n^z = \frac{\rho^2 - \rho_0^2}{\rho^2 + \rho_0^2}. \tag{2.20}$$

2.2.4 Interaction Between Geometry and Physical Field

We have seen so far that the intrinsic length of the underlying manifold appears in the solutions for the vector field distributions i.e. the underlying manifold influences the vector field. The opposite is true too. In order to illustrate this we will turn our attention to yet another exact solution of the sine-Gordon equation. We will consider now a periodic solution of this equation which represents a the soliton lattice [6]. We note here that the Bogomol'nyi argument can be applied for any period (for which the unit vector covers S^2) of this periodic solution. We note here that for one period on the rigid cylinder the self-dual equations (2.6) are still valid and therefore any function that satisfies (2.6) will satisfy the sine-Gordon equation as well (2.13). The periodic solution of the sine-Gordon equation is given by:

$$\theta = \arccos\left[\mathrm{sn}\left(\frac{x}{k\rho_0}, k\right)\right]. \tag{2.21}$$

Let us discuss this periodic solution. The constant k is the modulus of the Jacobi elliptic function sn (sine-amplitude); the period of the solution is given by $4d = 4\rho_0 k K(k)$ where here $K(k)$ is the complete elliptic integral of the first kind. The periodic solution transforms into the single twist soliton (2.14) solution of the sine-Gordon equation In the limit $k \to 1$, as $\lim_{k\to 1} K(k) \to \infty$, the half period $2d$ tends to infinity and at the the boundaries, we get the homogeneous conditions we have discussed in Sect. 2.2. We can now calculate the energy per soliton (over half period $2d$, as $\theta(\pm d) \equiv 0[\pi]$):

$$H_{isotropic} = \frac{8\pi J}{k}\left[E(k) - \frac{k'^2 K(k)}{2}\right], \tag{2.22}$$

In this equation k' is the complementary modulus ($k'^2 = 1 - k^2$) and $E(k)$ is the complete elliptic integral of the second kind. In the low soliton density limit, i.e. $k \to 1$ [then $E(k) \to 1$], we can expand the exact solution (2.22) and obtain the energy per soliton which reads:

$$H_{isotropic} = 8\pi J + 32\pi J exp\left(-\frac{2d}{\rho_0}\right) + \ldots = 8\pi J + 2\pi J k'^2 + \ldots. \tag{2.23}$$

The first term in this expansion represents the self energy of a soliton over one period (which corresponds to a soliton that stretches from $-\infty$ to $+\infty$. The second term represents an additional energy that corresponds to the repulsive interaction between solitons. The periodic solution (2.21) is an exact solution of the sine-Gordon equation but nevertheless does not satisfy the self duality equations and that is the reason why the energy per soliton in the periodic solution does not reach the minimum energy per soliton $H^1_{isotropic} = 8\pi J$. For a single soliton on a cylinder we have some sort of "equipartition" relation $\rho_0{}^2\ (\partial_x\theta)^2 = \sin^2\theta$ between "kinetic" energy on the

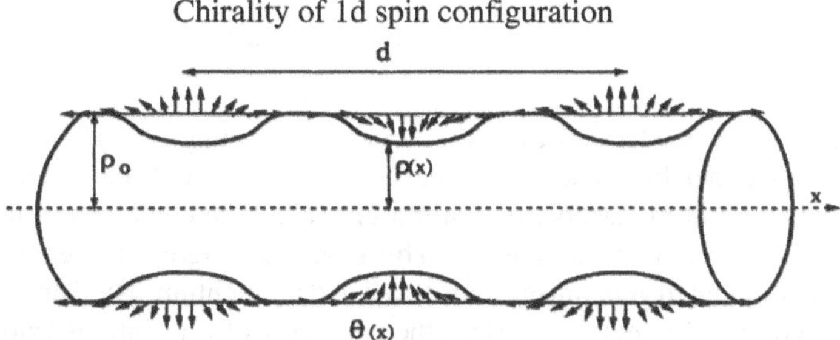

Fig. 2.5 Periodic spin soliton and periodic deformation of the cylinder

left and "potential" energy on the right. For the periodic solution the corresponding relation is:

$$\rho_0{}^2\ (\partial_x\theta)^2 = \sin^2\theta + \frac{k'^2}{k^2}. \tag{2.24}$$

Now we see that on the right hand side of this equation there is an additional "potential" energy $(\frac{k'}{k})^2$ that corresponds to an exponential repulsive interaction between the solitons. The soliton likes to stretch from $-\infty$ to $+\infty$ but the presence of additional soliton does not allow this to happen - this is the geometric frustration that appears in this case. In the limit of a single twist soliton ($d \to \infty$ and $k' = 0$) $(\frac{k'}{k})^2 = 0$ and we get the energy $H^1_{isotropic}$ of the single twist soliton: the interaction term vanishes. There is another possibility that may allow us to diminish the magnetic energy per soliton on the cylinder. If we allow the cylinder to deform as shown on Fig. 2.5 this will diminish the magnetic energy per soliton but will require some elastic energy for the deformation of the cylinder. Allowing for elastic deformations of the cylinder means that ρ will become x-dependent which will modify the Hamiltonian and the associated equations as discussed in [6].

2.2.5 Chirality of 1d Spin Configurations

We have seen that the usual homotopy classification of Heisenberg spins (target space is S^2) in one dimension is trivial: $\pi_1(S^2) = 0$. However, there is another non-trivial topological classification of Heisenberg spins based on chirality [7]. In order to find out if there are other topological structures in the one dimensional case one has to better analyse the Heisenberg hamiltonian. As the vector field is normalised to unity we once again will use the Euler angles representation for $\mathbf{n} = (\sin\theta\cos\phi, \sin\theta\sin\phi, \cos\theta)$. Using θ and ϕ variables the hamiltonian can be written in the following form:

$$H = J \int_{-L}^{+L} (\theta_s^2 + \sin\theta^2 \phi_s^2) ds \tag{2.25}$$

here s stands for $\frac{d}{ds}$ and s represents the coordinate along R^1. This hamiltonian is not invariant under homothety transformation $s \to \lambda s$ and that's why the spin configurations are not metastable like in the $2D$ case. The equations of motion for this spin hamiltonian have been established by Tjon and Wright [8] where ϕ and $\cos\theta$ represent the conjugated generalised coordinate and momentum. The Poisson bracket gives $[\phi(x), \cos\theta(y)] = \delta(x - y)$. Here the generator of translations (momentum) is given by the following expression:

$$P = \int_{-L}^{+L} (1 - \cos\theta)\phi_s ds \tag{2.26}$$

The momentum operator verifies the Poisson brackets: $[\phi(s), P] = -\phi_s$ and $[\cos\theta(s), P] = -\frac{d}{ds}\cos\theta(s)$ [8]. It turns out that P is a constant of the motion for this Hamiltonian [8].

Now, for our analysis of the possible spin configurations we will map the unit vector $\mathbf{n}$ to the unit tangent of a space. Now, it turns out that different space curves represent different spin configurations. Here the boundary conditions we will use are that at $\pm L$ the spins will be parallel. In this case different curves representing different spin configurations will tend to a straight line as $s \to \pm L$. Of special interest for us is the writhe of a curve (which characterises its chirality of the). For a closed curve it is defined as follows [9]:

$$Wr = \frac{1}{4\pi} \oint ds \oint ds' \frac{(\mathbf{r}(s) - \mathbf{r}(s').(\mathbf{n}(s) - \mathbf{n}(s'))}{|\mathbf{r}(s) - \mathbf{r}(s')|^3} \tag{2.27}$$

In this case we distinguish two classes of homotopy equivalent curves: one where two ends of the curve are rotated to each other by 2π and the other where the two ends are rotated by 4π. The reason for the appearance of these two classes is the fact that the group $SO(3)$ is non simply connected manifold and closed loops in $SO(3)$ fall into two classes: those who can be contracted to a point and those for which this is impossible. For example a triad evolving on such a space curve from $s = -L$ to $s = L$ traces out a closed curve on $SO(3)$ [10]

We will apply a theorem by Fuller which allows to express Wr as an integral of a local quantity. We will express Wr with respect to a reference curve C_0 (which for simplicity is taken to be a straight line):

$$Wr = Wr_0 + \frac{1}{2\pi} \int_{-L}^{+L} \frac{\mathbf{n}_0 \times \mathbf{n}.\frac{d}{ds}(\mathbf{n}_0 + \mathbf{n})}{(1 + \mathbf{n}_0.\mathbf{n})} ds \tag{2.28}$$

here Wr_0 is the writhe of the reference curve. A simple calculation gives the following expression for the writhe:

$$Wr = \frac{1}{2\pi}\int_{-L}^{+L}(1-\cos\theta)\phi_s ds \tag{2.29}$$

Let us note here that the writhe Wr for the spin configurations (quantity that characterises the chirality of the spin configuration) coincides with the total momentum P. We have seen that the total momentum P is a conserved quantity - it follows that Wr is a conserved quantity too. This will lead us to a new class of possible excitations for the continuous classical spin Heisenberg model. The ground state configuration is represented by $\theta = 0$ and let us note that the curves whose ends are rotated by 4π also belong to the same class of configurations [10]. In order to calculate the lower bound for the energy of a configuration that doesn't belong to the ground state configuration (curves whose ends are rotated by 2π) we need closed curves. Let us take first a space curve representing a spin configuration that goes from $-L$ to $+L$. This curve is completed by a straight line between $-L$ and $-\infty$ and between $+L$ and $+\infty$ and is closed by a semi-circle at infinity in order to form a closed curve. Note that on the straight segments and on the semi-circle at infinity the curvature k is zero. The writhe is zero for the straight segments when $s \in \pm(L, \infty)$ as well as for the infinite semi-circle. This geometrical construction does not change the writhe of the actual curve. Such a curve belongs to a whole class of configurations which deform smoothly from one to another and who are separated from the ground state class by a jump in the writhe Wr. Let us first note that for closed curves [11]:

$$\oint k ds \geq 2\pi \tag{2.30}$$

We note that the curvature $k \neq 0$ only for $s \in (-L, +L)$, and then the above inequality is equivalent to:

$$\int_{-L}^{+L} k ds \geq 2\pi \tag{2.31}$$

We will use the following Cauchy–Schwarz inequality:

$$\left(\int_{-L}^{+L} k ds\right)^2 \leq 2L\int_{-L}^{+L} k^2 ds \tag{2.32}$$

and the following expression for the curvature in Euler angles: $k^2 = \theta_s^2 + \sin^2\theta\phi_s^2$. Then we can present the following obvious inequality for the energy of the spin chain:

$$H = J\int_{-L}^{+L}(\theta_s^2 + \sin^2\theta\phi_s^2)ds = J\int_{-L}^{+L} k^2 ds \geq \frac{J\left(\int_{-L}^{+L} k ds\right)^2}{2L} \geq J\frac{4\pi^2}{2L} = J\frac{2\pi^2}{L} \tag{2.33}$$

The energy is limited from below for this class of configurations for which both ends of the representing space curve are rotated at 2π. It is clear that for an infinite chain

$L \to \infty$ the lower bound goes to 0. On the other hand the barrier which separates P from the zeroth class remains. The above result is consistent with the inequality for the elastic energy of thin rod whose ends are rotated by 2π relative to each other [10] where the result is based on the same property of the rotation group $SO(3)$. In this case the thin rod has only bending rigidity J and no torsional rigidity.

2.3 Quantum Potential, Thin Tubes, Knots

Yet another interesting application of geometry concerns quantum theory. Especially the appearance of induced quantum potential on curved surfaces and on curves (thin tubes) (the effective potential that appears on curved surfaces or membranes is $V_{eff} = -\frac{\hbar^2}{2m}(M^2 - K)$ where $K = k_1 k_2$ is the Gaussian curvature and k_1 and k_2 are the sectional curvatures and $M = 1/2(k_1 + k_2)$ is the mean curvature and the effective potential on curves or in thin tubes is $V_{eff} = -\frac{\hbar^2}{2m}\frac{k^2}{4}$, where k is the curvature of the axis of the tube [12]. We will briefly discuss the appearance of quantum potential in a thin tube and give a "hand-waving" argument in favour of it. The argument is based on Heisenberg's uncertainty principle, see Fig. 2.6.

It is the obvious that as $\Delta p_s \leq \Delta p_x$, the corresponding energies are related as follows: $E_s = \frac{\Delta p_s^2}{2m} \leq E_x = \frac{\Delta p_x^2}{2m}$.This shows that the free particles prefer to be localised in the bent region of the thin tube.

Let us now, as an example, consider a trefoil knot (knots appear often in polymers). It has turned out that one may create a qubit using the geometry of a tight trefoil knot [13].

In mathematics knots are represented as closed, self-avoiding *curves* embedded in a three-dimensional space. Any knot can be tied on a thread (or a curve) of any length and when we pull on both ends of the thread the knot transforms into a point which means that all conformations are essentially equivalent. Yet another problem is the lack of characteristic length with does not allow for the introduction of energy scale. Physics deals with real material knots. The thread has a finite diameter and pulling the thread on both ends does not transform the knot into a point. The diameter of the physical thread plays the role of a characteristic length.

In our example we will use a trefoil knot where the thread will have a circular cross-section with a finite radius. Then we will pull the knot tight. At some point we

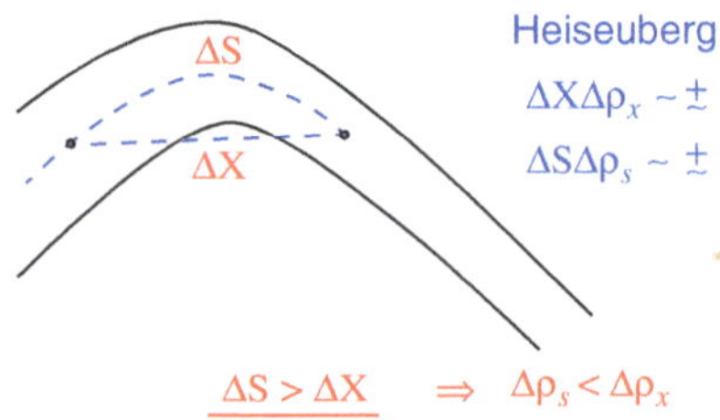

Fig. 2.6 Bent thin tube

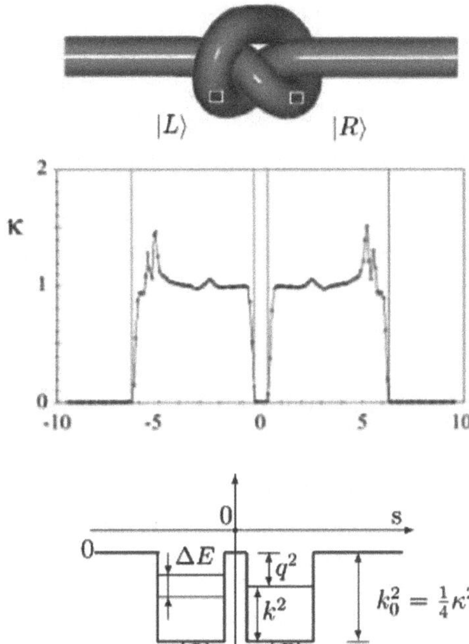

Fig. 2.7 Tight open knot and the curvature of its center ligne

Fig. 2.8 A double well potential for the tight trefoil knot

will note that we can not pull any further without changing the cross-section of the knot. The final confirmation we have reached is called *tight open knot*. We may get a *tight closed knot* simply by "gluing" together the loose ends of the *tight open knot*. As seen on Fig. 2.7 there is a plane of symmetry which separates the left from the right part of the trefoil knot. This most symmetrical conformation of the trefoil knot appears to be also the most energetically (elastic energy) favourable.

There is no analytical expression for the curvature of the center line of a tight knot, but there are experimental measurement, which are presented on Fig. 2.7 [14]. There is compete left-right symmetry of the curvature and there is a flat region ($k = 0$) in the middle.

As we have seen, the curvature of a space curve or the centerline of a thin tube is related to the induced quantum potential. The curvature presented in Fig. 2.8 can be modelled to represent the following double well potential for the tight knot. See Fig. 2.8. Finally the tuneling possibility between the two wells splits the localised level into two and so the tight trefoil knot represents quantum mechanically a two level system which may be used as a microscopic qubit.

The combination of curved and straight nanobars may produce the same double well potential as shown in Fig. 2.9. One such combination is shown in Fig. 2.9. The possibility to create almost any given quantum potential using a combination of suitable curved nanobars is very big.

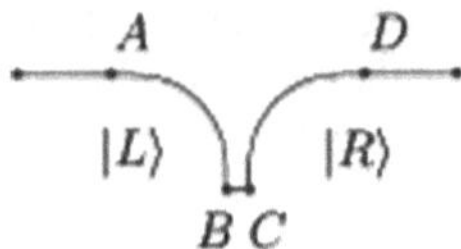

Fig. 2.9 Circular and straight nanobars create a double-well potential

The sectors $A - B$ and $C - D$ represent a quarter of a circle ($V_{eff.} = -\frac{\hbar^2}{2m}\frac{k_0^2}{4}$, with $k_0 = \frac{1}{r_0}$, where r_0 is the radius of the circle) and the sector $B - C$ is a straight line ($V_{eff.} = 0$).

2.4 Conclusions

By using topology and geometry we established a link between the order parameter of the concrete material and the underlying geometry and make predictions regarding the bounds of energy imposed by topological and geometrical constraints without solving the very complicated equations of motion. Classification of defects in different materials is also possible. This is true for the classical as well as for the quantum level of consideration. Further investigations on the quantum level, using topology, may include the quantum effective potential which is geometric in its nature and may play an important role in understanding of some properties of nanostructures. Finally we mention that topology and geometry can not replace solving of the microscopic equations of motion but may bring additional insights for the understanding of the fundamental properties of matter.

References

1. P. Renteln, A. Dundes, Foolproof: a sampling of mathematical folk humor. Not. Am. Math. Soc. **52**, 24–34 (2005)
2. G. Toulouse, M. Kleman, J. Phys. Lett. (Paris) **37**, L-149 (1976); M. Kleman, L. Michel, G. Toulouse, J. Phys. Lett. (Paris) **38**, L-195 (1977)
3. A.A. Belavin, A.M. Polyakov, JETP Lett. **22**, 245 (1975)
4. E.B. Bogomol'nyi, Sov. J. Nucl. Phys. **24**, 449 (1976)
5. S. Villain-Guillot, R. Dandoloff, A. Saxena, Phys. Lett. A **188**, 343 (1994)
6. R. Dandoloff, S. Villain-Guillot, A. Saxena, A. Bishop, Phys. Rev. Lett. **74**, 813 (1995)
7. R. Dandoloff, Adv. Condens. Matter Phys. **2015** (2015). (Article ID 954524)
8. J. Tjon, J. Wright, Phys. Rev. B **15**, 3470 (1977)
9. M.D. Frank-Kamenetskii, A.V. Vologodskii, Sov. Phys. Usp. **24**(8), 679 (1981)
10. J. Baez, R. Dandoloff, Phys. Lett. A **155**, 145 (1991)
11. W. Fenchel, Math. Ann. **10**, 238 (1929)
12. R.C.T. da Costa, Phys. Rev. A **23**, 1982 (1981)
13. V. Atanasov, R. Dandoloff, Phys. Lett. A **373**, 716 (2009)
14. P. Pieranski, S. Przybyl, A. Stasiak, Eur. J. Phys. E **6**, 123 (2001)

Part II
Condensed Matter Materials Physics

Chapter 3
Topology-Induced Geometry and Properties of Carbon Nanomaterials

Hiroyuki Shima and Jun Onoe

Abstract Nanoscale graphene sheets (i.e., sp^2-bonded monoatomic carbon layers) often exhibit drastic changes in their geometry and properties when non-hexagonal carbon rings are embedded into the original hexagonal lattice. This chapter gives a short review on the physics of sp^2 nanocarbon materials with curved geometry, together with a concise explanation of the mechanism as to how the presence of topological defects causes the anomalously curved geometry. A special emphasis will be placed on the geometry-property correlation inherent in quasi-one-dimensional fullerene polymers and single-walled carbon nanocoils, each of which shows geometric curvature or torsion induced by local change in the carbon network topology.

3.1 Introduction

3.1.1 Carbon as a Building Block

Carbon is one of the most wealth-generating elements on the Earth. In view of biology, carbon is the root of all living matters, allowing many other chemical elements (hydrogen, oxygen, nitrogen, etc.) to react with it to produce substantial organic compounds. From the physico-chemical perspective, carbon is highly versatile in the sense that it accepts different atomic orbital hybridizations. For instance, hybridization of the $2s$ orbital with all three $2p$ orbitals (p_x, p_y, and p_z) in a carbon atom results in four sp^3 hybrid orbitals with tetrahedral geometry. The sp^3 orbitals spanning the three-dimensional (3D) space allow the realization of diamond, one of the most

H. Shima
Department of Environmental Sciences, University of Yamanashi, 4-4-37, Takeda, Kofu, Yamanashi 400-8510, Japan
e-mail: hshima@yamanashi.ac.jp

J. Onoe (✉)
Department of Energy Science and Engineering, Nagoya University, Furocho, Chikusa-ku, Nagoya 464-8603, Japan
e-mail: j-onoe@energy.nagoya-u.ac.jp

S. Gupta and A. Saxena (eds.), *The Role of Topology in Materials*, Springer Series in Solid-State Sciences 189,
https://doi.org/10.1007/978-3-319-76596-9_3

attractive materials in the world. Similarly, the *s* orbital and two *p* orbitals hybridize to form three sp^2 orbitals with the trigonal planar structure, allowing the formation of graphene, a two-dimensional (2D) mono-atomic sheet made purely from carbon. Finally, the *s* orbital hybridizes with one of the *p* orbitals to form two sp hybridized orbitals, whose frontal lobes face away from each other forming a one-dimensional (1D) straight line.

A direct consequence of the versatility in the hybridized orbitals, each of which spans a different class of spatial dimension (3D, 2D, and 1D), is the diversity of carbon-based nanomaterials with various morphologies [1, 2]. Of the intriguing nanocarbon family, those made by monoatomic carbon layers (i.e., sp^2 graphene sheet) have long been the hot topic in the community of materials science [3]. It is definitely sure that carbon nanotube is the greatest example among the epoch-making nanocarbons made by wrapping a graphene sheet [4, 5]. As well, graphene nanoribbon, which is a stripe of graphene sheet with a few tens Å in width, is also the one spotlighted from the discovery in the last decade [6].

Basically, the profound nature of those sp^2 nanocarbons originates from the perfect hexagonal structure of the monoatomic sheet. And further interesting aspect is that, it may be richer and variegated if the hexagonal symmetry of the crystalline lattice is broken locally by introduction of structural defects [7]. So what kind of symmetry breaking is relevant to the physics of sp^2 carbon? See below an overview of the subject.

3.1.2 Defect in sp^2 Nanocarbon

Ideally, an sp^2 graphene sheet is a perfect 2D crystal in which the carbon atoms perfectly align into hexagons. But in reality, a graphene sheet does not show either perfect horizontality nor perfect hexagonal symmetry. In fact, it inevitably contains a considerable number of defects.

From structural viewpoints, defects in a graphene sheet (or more generally, in a monoatomic layer of sp^2 carbon) are classified into four groups:

(1) Topological defect;
(2) Presence of sp^3 chemical bonds due to hybridization;
(3) Vacancy and/or dislocation;
(4) Non-carbon impurity.

The class of topological defects, written at the top of the list above, is the main topic of this chapter. A topological defect is produced by introducing non-hexagonal rings (e.g., pentagons or heptagons) into the perfect graphene sheet with hexagonal symmetry. We will see later the presence of topological defects may provide a significant alteration in the geometry and properties of sp^2 nanocarbons. Here, the term "topology" means the local connectivity of carbon atoms via covalent bonding, though it has a more abstract meaning as a mathematical jargon [8]. When being inserted in a graphene sheet, a topological defect breaks locally the structural order of the hexagonal lattice, altering the connectivity of carbon atoms in a limited region. We will see in this chapter that the presence of topological defects can cause a significant alteration in the geometry and properties of the originally planar sp^2 nanocarbon sheets.

Emphasis should be placed on the fact that the defect-induced alteration in geometry can allow a new class of anomalously-shaped nanocarbon materials endowed with surface curvature and/or torsion. Examples include, but are not restricted to, peanut-shaped fullerene polymers [9], helix-shaped carbon nanocoils [10], and gyroid-shaped carbon Schwarzite [11]. Many intriguing properties of the three anomalously-shaped nanocarbons have been unveiled in the last decade. Against the backdrop, the present chapter aims at a bird's eye view of the latest achievements on the new family of sp^2 nanocarbon materials, together with a concise explanation as to how the topology of local atomic structure correlates with global geometry of the sp^2 nanocarbon materials.

3.2 Topology-Induced Geometry in sp^2 Nanocarbon

3.2.1 Surface Curvature Generation in Graphene Sheets

Topological defects play a crucial role in tailoring equilibrium structures of graphene sheets [12, 13]. This is because insertion of topological defects into a graphene sheet causes a change in local connectivity of carbon atoms, resulting in a global change in the geometry of the sheet from the planar to curved structure as explained below.

Figure 3.1 provides a schematic diagram of surface curvature generation by topological defect insertion [14, 15]. Suppose that we are given a monoatomic graphene sheet with perfect hexagonal symmetry, as displayed in the middle panel of Fig. 3.1. Next, we remove a $\pi/3$ wedge from the original sheet and pull the two new edges to connect them with each other. The resulting structure is displayed in the left panel of Fig. 3.1. A particular attention should be paid to that the local atomic connectivity is altered only at the central carbon rings, from hexagon to pentagon, while other hexagon rings surrounding the central non-hexagon ring remain unchanged. This diagram thus shows that artificial insertion of a pentagon ring into the perfect hexagonal sheet yields a transformation from the flat to a positively curved graphitic layer.

A contrasting situation is illustrated in the right panel of Fig. 3.1. In that situation, one $\pi/3$ wedge hatched in the middle panel is replaced forcibly by an already-joined two $\pi/3$ wedges. As a consequence, the originally flat sheet (middle panel) is transformed to a saddle-shaped graphitic sheet that has a heptagonal ring at the center (right panel). Similarly to the previous case, local atomic configuration remains unchanged except for the central non-hexagonal ring.

We have known that artificial change in the local atomic connectivity at the center of a planar graphitic sheet results in a positively or negatively curved graphitic

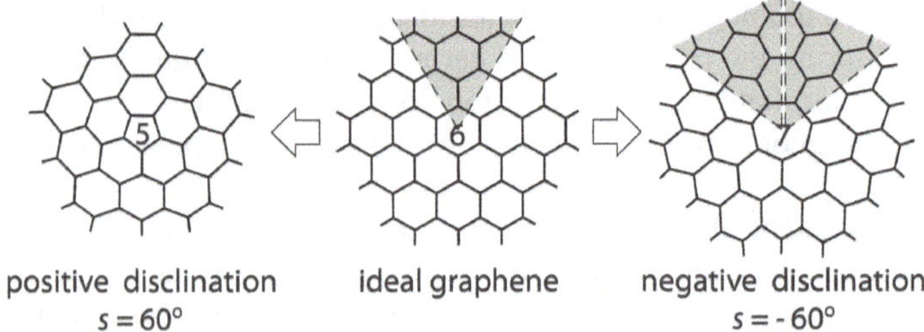

Fig. 3.1 Diagram of surface curvature generation by topological defect insertion. A portion of an initially flat graphene sheet (middle) becomes endowed with positive (left) or negative (right) surface curvature when a $\pi/3$ wedge (shaded area) of the original sheet is removed or added, respectively. Reprinted with permission from [15]

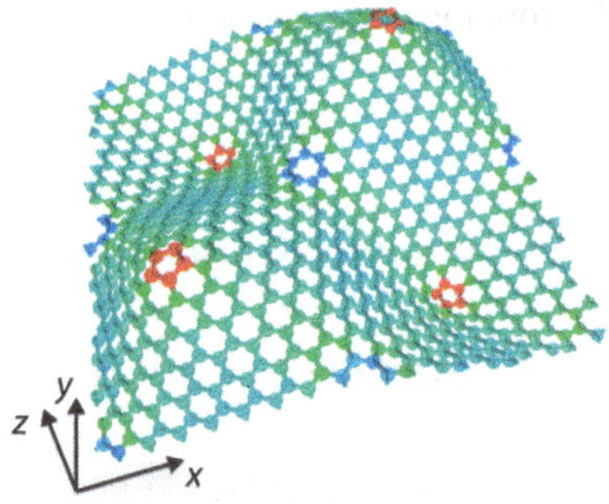

Fig. 3.2 A wrinkled graphene structure. It was theoretically designed through a controlled distribution of topological defects. Reprinted with permission from [17]

sheet. So what happens if the central hexagon is replaced by larger rings (e.g., octagon, nonagon, or more) or smaller rings (square and triangle)? In principle, a severely curved sheet with positive or negative surface curvature can be built through such local connectivity alteration; however, most of them turned out to be unstable mechanically and thus it will be difficult to synthesize them in reality [16].

The topology-induced change in the surface geometry implies the ability of morphological control of sp^2 nanocarbons by artificial insertion of topological defects. In fact, numerical simulations have suggested that collective behaviors of topological defects can be utilized to design wrinkled graphene structures in controllable manner. Figure 3.2 shows a thermodynamically stable structure of a defective graphene sheet obtained by molecular dynamics simulations [17]. The numerically obtained structure shows concavo-convex shape, involving pentagons at crests and valleys and heptagons at saddle points. The simulation revealed the mechanical stability of the periodically curved structure with 4 nm in wavelength and 0.75 nm in amplitude.

It was theoretically predicted that the topology-induced curvature in a graphene sheet leads to a significant modification of the physical properties [18] such as electronic band structure [19–21], the charge carrier transport [22, 23], and spin-orbit couplings [24]. In particular, negatively curved graphene sheets embedding heptagonal and octagonal defects have been shown to significantly enhance the capacity

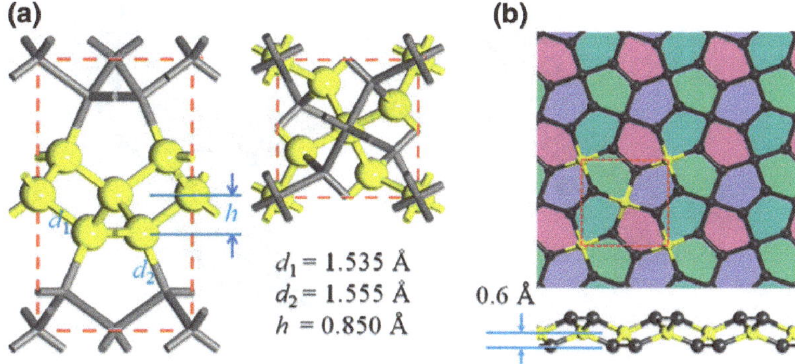

Fig. 3.3 **a** Three-dimensional crystal structure of T12-carbon, from which the two-dimensional pentagon-based sheet (highlighted in yellow) will be exfoliated. Views from the [100] and [001] directions are displayed. **b** Two-dimensional all-pentagon-made monoatomic carbon sheet, called "penta-graphene". Top and side views are shown. Reprinted with permission from [27]

of graphene based electrodes [25, 26], because of the high surface area compared with the planar counterpart. More recently, another graphene sheet consisting of only pentagons (so called "penta-graphene") has been proposed theoretically as shown in Fig. 3.3 [27]. It is interesting to note that penta-graphene exhibits a negative Poisson ratio.

3.2.2 Plastic Deformation of Carbon Nanotubes

Aside from the "planar" graphitic materials, artificial introduction of topological defects into "tubular" sp^2 nanocarbons is also an interesting subject; defect insertion can cause drastic changes in the structure and property of the tubular systems. For instance, X-branched and T-branched junctions made of carbon nanotubes were synthesized by inserting topological defects into the joint area of two or more different nanotubes, as demonstrated in Fig. 3.4 [28–30]. These branched nanocarbons will find useful applications in nanoelectronics [31] and fiber-reinforced composites [32, 33]. Other interesting examples of anomalously shaped nanotubes include carbon nanocoils [34] and bamboo-shaped carbon nanotubes [35–37]. The structural anomaly observed in these systems relies on the periodic insertion of defects into the originally straight nanotubes, which causes periodic variation in torsion and/or curvature that allow coiled and/or twisted structures as metastable states.

Figure 3.5 demonstrates spatial variation in the tube radius caused by topological defect insertion into a carbon nanotube [38, 39]. The symbol "P" depicted in the electron microscope image indicates an apex at which one pentagon is embedded into a graphitic cylinder. The presence of the pentagon causes mechanical strain in the hexagonal carbon network around the pentagon. To relieve the strain, the cross-

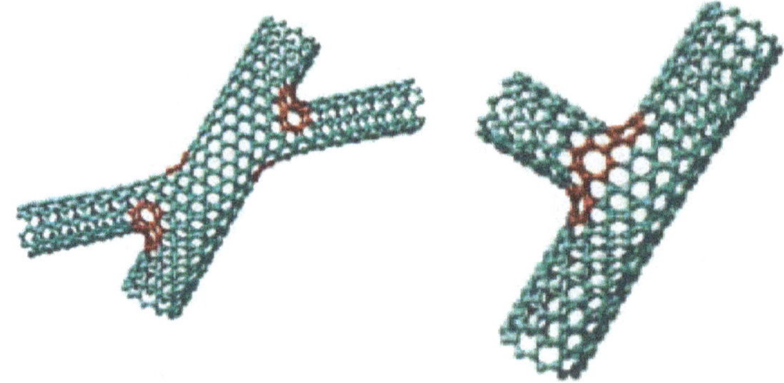

Fig. 3.4 Left: Molecular models of the X-junction between a thin and wide nanotube. Heptagonal rings are indicated in red. Right: Atomic models of the T-junction. Reprinted with permission from [28]

section of the tube slightly deviates from the initial circular shape at the position indicated by "P", resulting in an oval shape. Another strain-induced deformation occurs at a local dent, indicated by "H" in the microscope image of Fig. 3.5. At the dent, one heptagon is embedded into the graphitic sheet, generating negative surface curvature near the point. As a result of the paired defects, a pentagon and a heptagon, the tubular diameter of the carbon nanotube decreases gradually from the trunk to the tip of the nanotube.

Interestingly, topological defects in carbon nanotubes (and other nanocarbons having a closed shape) shed light on a beautiful interplay between mathematics and atomic structures. A typical example is a capped structure of a nanotube obtained by defect insertion [40]. Figure 3.6 shows an enlarged view of the tip of a capped carbon nanotube with chirality of (30,0). Two different capped structures, shown in the panels (a) and (b), are obtained by inserting six pentagon defects at the tip. An intriguing fact is that the axial symmetry of the resulting capped structure is dependent on the relative configuration of the topological defects, even if the same number of defects is inserted. Indeed in Fig. 3.6, the exactly six pentagonal defects yield five-fold symmetry in the case of (a) and six-fold symmetry in (b) with respect to the tubular axis, in which more surprisingly, it is prohibited to choose the other number of pentagons to be embedded; only the six pentagon insertion is allowed for capping a carbon nanotube regardless of its tube radius and chirality. This restriction with respect to the number of pentagons is a consequence of a mathematical theorem, called the Gauss-Bonnet theorem [41]. The theorem states that the number of non-hexagonal rings (i.e., topology) is related to the curvature of the embedding graphene sheet (i.e., geometry). From the technological perspective, the closed-cap formation driven by topological defect insertion has been expected to offer the possibility for the control of the chirality of carbon nanotubes during growth [42].

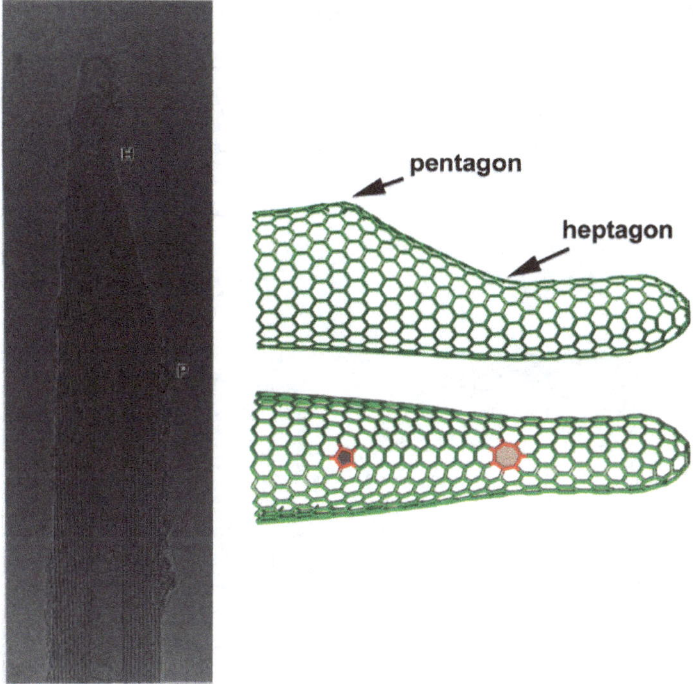

Fig. 3.5 Left: Electron microscope image of a closed cap at the tip of a multi-walled carbon nanotube. The successive presence of topological defects changes the diameter of the concentric tubes gradually toward the tip [38]. Right: Atomistic structure of a capped carbon nanotube, in which one pentagon and one heptagon are incorporated. The overall picture looks similar to the image shown in the left panel. Reprinted with permission from [39]

3.3 Stone-Wales Defect

3.3.1 Symmetry Breaking by C–C Bond Rotation

A Stone-Wales (SW) defect, or also called a 5-7-7-5 defect, is the lowest energy topological defect in sp^2 nanocarbons [43]. It is formed by a 90-degree rotation of a C–C bond in the hexagonal carbon network. The rotation causes atomic reconfiguration at the neighbor from the set of four hexagons to the set of two pentagons and two heptagons; see the top panels in Fig. 3.7. This 90-degree bond rotation is often called a Stone-Wales (SW) transformation. It should be noted that the bond rotation breaks the hexagonal symmetry of the perfect graphene sheet. Thus the resulting SW defects scatter carrier transport, degrading the high mobility of carriers, which is the hallmark of Dirac fermion behavior in graphene [44]. Yet it is not always a bad thing. In fact, the symmetry breaking induced by SW defect insertion is also manifested by

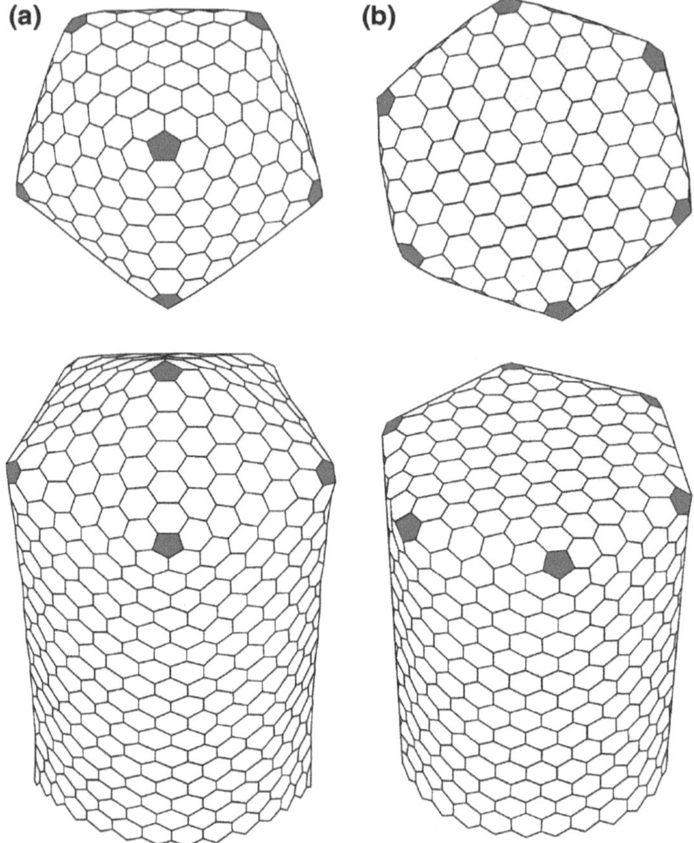

Fig. 3.6 Capping a (30,0) zigzag carbon nanotube by inserting six pentagonal carbon rings (marked by shadow) at the tip. Depending on the defect configuration, two different kinds of axial symmetry arise in the capped structure: **a** five-fold and **b** six-fold symmetry. Top views and overall views are displayed in the upper and bottom panels, respectively. Reprinted with permission from [40]

the band gap opening in the electronic spectrum of graphene [45]. This finding will expand the range of possible applications of graphene in nanoelectronics.

3.3.2 Formation Energy

Let us remind that SW transformation is accompanied by the consecutive breaking of two single C–C bonds. Since the breaking of a single C–C bond in graphene requires expending energy of nearly 5 eV, the height of the potential barrier preventing the SW transformation is on the order of a few eV or more [46]. If we choose thermal activation to overcome the high-energy barrier, it would be necessary to use extremely

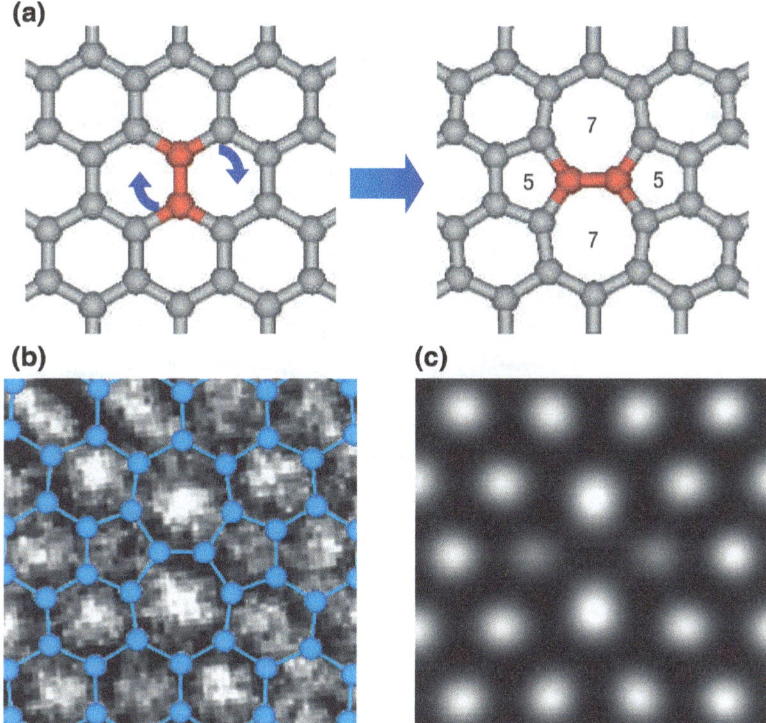

Fig. 3.7 Top: Atomic configuration of a graphene sheet before (left) and after (right) rotating the C–C bond marked in red. Bottom left: Electron microscopic images of a SW defect. Bottom right: Simulated results of the atomic configuration obtained by density functional theory. Reprinted with permission from [53]

high temperatures. Therefore, earlier experimental attempts at artificial production of SW defects have been based on the high-energy beam irradiation or the action of mechanical stresses at the stage of the synthesis of the sp^2 nanocarbons, as will be argued in Sect. 3.4.

3.3.3 Out-of-Plane Displacement

One may think that a graphene sheet is flat and purely two-dimensional, even when a SW defect is present in the sheet. But this is not true; in fact, the lowest-energy atomic configuration of the SW defect has a sine-wave-like form [47]. In this wavy structure, the two atoms involved in the rotated C–C bond move out of plane in the opposite direction. This buckling of the C–C bond at the core of the SW defect gives rise to the vertical displacement of many atoms around the defect, resulting in a sine-like shape of the cross section as illustrated in Fig. 3.8 [47].

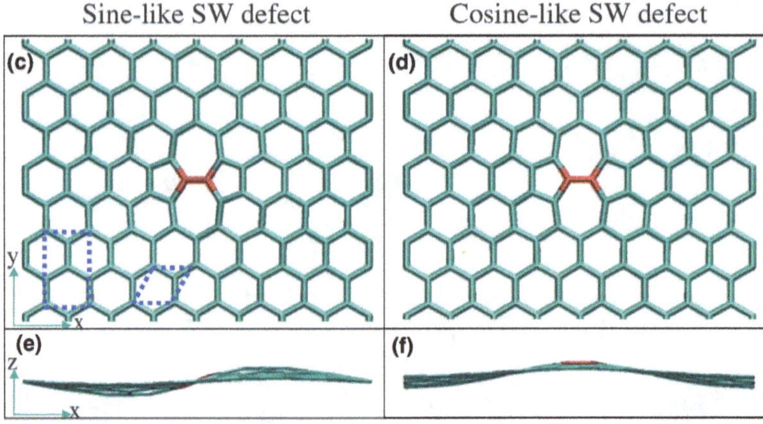

Fig. 3.8 The structure of two buckled SW defects in graphene. Left: Top and side views of the sine-like buckled SW defect, being identified as the most stable defect structure. Right: Top and side views of the cosine like buckled SW defect. The rotated bonds at the defect core are highlighted. Dashed lines indicate 1×1 rectangular and hexagonal unit cells. Reprinted with permission from [47]

It may be non-trivial for readers why the SW defect in graphene does not remain flat in equilibrium but buckles out of the plane. One plausible mechanism is based on the release of the excess strain energy induced by the C–C bond rotation [47]. Prior to the bond rotation, the equilibrium C–C bond length in perfect graphene is 1.42 Å. If we rotated a C–C bond with keeping the original flat shape of the embedding hexagonal sheet, the separation between the rotated atoms would be compressed to 1.32 Å. Here, the flat structure cannot release the compression efficiently, because in-plane motion of C atoms in graphene is too much expensive as compared to out-of-plane motion. Instead of the in-plane deformation, the SW defect exhibits out-of-plane deformation, as a result of which the compressed C–C bonds can be expanded enough to release the in-plain strain energy. In fact, displacement of the SW defect in the direction normal to the graphene sheet results in the pulling of neighboring atoms out of the plane, because it is energetically favorable for the C–C bonds that surround the SW defect to remain as close to a planar sp^2-bonded network as possible.

SW defects also affect the side-wall curvature of carbon nanotubes. Figure 3.9 illustrates local deformation of a nanotube caused by SW defect insertion into the graphitic sidewall [48]. It follows from Fig. 3.9 that the magnitude of out-of-plane displacement is maximized at the interface between two pentagons. Because of the significant out-of-plane displacement, addition reactions are most favored at the C–C double bonds of these positions [49].

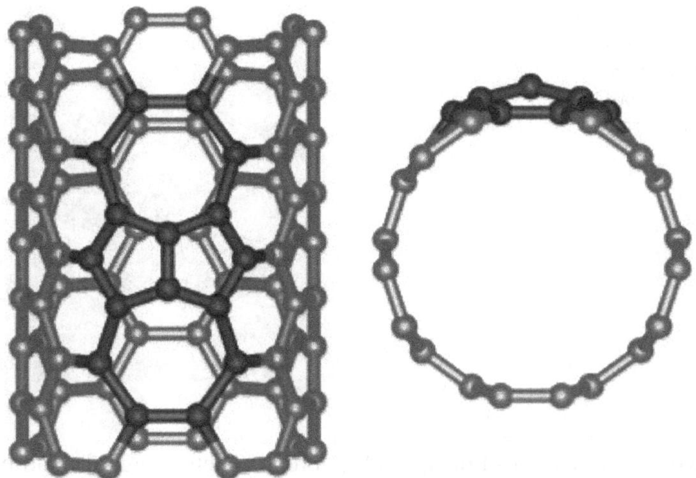

Fig. 3.9 Stone-Wales defect on the sidewall of a single-walled carbon nanotube. Reprinted with permission from [48]

3.3.4 Microscopic Observation

From an experimental viewpoint, it is difficult to observe SW defects by ordinary microscope techniques, despite a few indirect evidences have been provided to date [50–52]. The difficulty stems mainly from the high activation barrier for the SW transformation, whose energy scale is estimated as a few eV [30]. Due to the high activation barrier, detection of native topological defects in a pristine sp^2 nanocarbon is statistically unrealistic. Hence most of earlier experiments for identifying topological defects with an atomistic-scale precision were based on artificial generation of the defects, with the aid of high-energy beam irradiation and/or high-temperature heat treatment.

The first direct observation of SW defects was successful on single-walled carbon nanotubes in 2007 [53]. In the seminal work, high-resolution transmission electron microscopy with atomic accuracy was used to obtain the image that is displayed in the left-bottom panel of Fig. 3.6. It is interesting to remind that the defects shown in the image were those artificially introduced in a pristine non-defective carbon nanotube. Being exposed to high temperature ($\approx$2300 K) in vacuum, a portion of hexagonal lattices at the sidewall of nanotubes breaks out, which leads to the fusion of adjacent nanotubes into a unified single-walled nanotube with large tube diameter [54]. The large-diameter nanotube thus obtained suffers from substantial amount of structural imperfection. When rapidly cooled, therefore, a large number of SW defects as well as other kinds of topological defects are involved in the resulting nanotube; typically a few SW defects can be detected per 10 nm length (and/or per 10 nm long).

Aside from those found in carbon nanotubes, SW defects were also found in a flat monolayer graphene [55]. Figure 3.10 shows the microscope image of SW

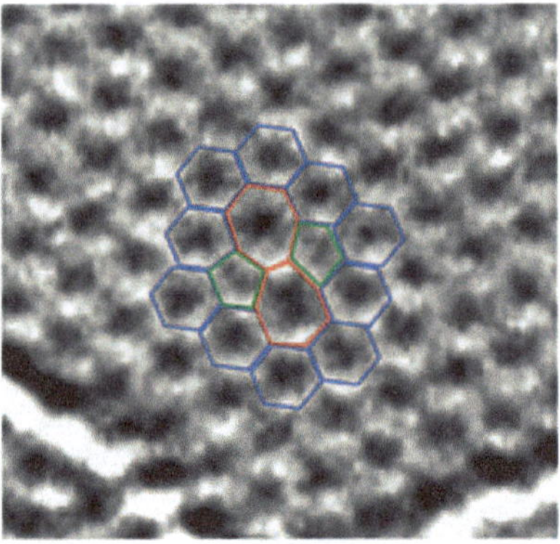

Fig. 3.10 Microscope image of a Stone-Wales defect with atomic configuration superimposed. Reprinted with permission from [55]

defects involved in the two-dimensional graphitic sheet; pentagons and heptagons are colored in green and red, respectively. The image provides clear evidence for SW defects existing in the examined graphene layer.

Quite recently, statistical atomic kinetics during SW transformation in graphene was formulated, enabling the decoupling of the two contributions from high-energy beam irradiation and thermal excitation to the activation of the C–C bond rotation [56]. The result indicated the exceptionally high rate of healing for SW defects generated by irradiation, compared with the healing rate of thermally-induced SW defects. This implies the complexity in the route of reaction processes toward SW defect realization in a graphene sheet.

3.4 Defect of 5–7 Paired Type

3.4.1 Dissociation of a SW Defect

It is interesting to note that SW defects on the sidewall of carbon nanotubes can show a curious mechanical response to the axial load. Upon application of the axial load to the defective carbon nanotube, a SW defect can dissociate into two separated 5–7 pairs. This tensile-driven dissociation of a SW defect is caused by additional transformations of C–C bonds that locate in the vicinity of the 5–7 pairs [57]. Now we suppose that the axial load remains exerting continuously on the nanotube. Then we will observe further successive transformations of C–C bonds around the 5–7 pairs. Because of the successive transformations, the 5–7 pairs start to migrate along a helical path that twines around the tube [58].

The migration of the 5–7 pairs along the helical path involves a number of atomic reconfigurations around it, thus requiring a sufficient amount of energy supply to proceed. This requisite energy is often supplied in the form of thermal excitation and the work done by axial strain; it can be also supplied by bend deformation [59, 60].

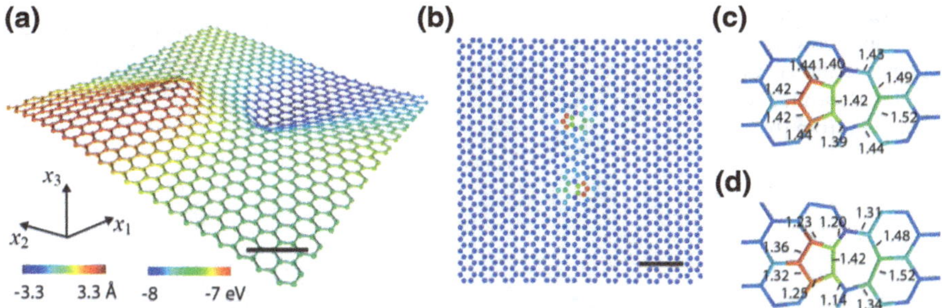

Fig. 3.11 **a**, **b** A perspective view (**a**) and a top view (**b**) of the atomic configuration involving two 5–7 paired defects in a graphene sheet. **c**, **d** Bond structures around the defect core in 3D (**c**) and 2D projection (**d**). The color represents the scale of the out-of-plane displacement in **a** and potential energy in **b**–**d**, respectively (scalebar:1 nm). Reprinted with permission from [61]

The chirality alteration is an important consequence of the 5–7 pair migration. When the 5–7 pairs complete the full distance from one end to the other end of the given nanotube, the charity of the tube domain covered by the helical path will be altered. This chirality change is accompanied by the change in the tube diameter. Therefore, the defect migration may serve as a driving force for a ductile elongation behavior of carbon nanotubes under large axial strain, as was observed in experiments [57].

3.4.2 As a Seed of Surface Curvature

We have seen in Sect. 3.3 that the presence of SW defects causes a surface curvature in an originally flat graphene sheet. Similar curvature generation occurs in a graphene sheet containing topological defects of an isolated 5–7 pair, i.e., an adjacent pair of pentagonal and heptagonal rings. Figure 3.11 illustrates the wrinkle of an initially flat graphene produced by inserting two 5–7 defects [61]. As shown in the middle panel, the two 5–7 defects are placed apart from each other with a separation equivalent to several hexagonal rings. Then the two defects generate the surface curvature, giving rise to a large wrinkle near the defect core (see the left panel). The vertical amplitude of the out-of-plane displacement is estimated to be up to 3.3 Å. The middle panel of Fig. 3.11 shows the potential energy of carbon atoms around a 5–7 defect, where atoms involved in the defect exhibit a higher energy than those far from the defect. In particular, three atoms on the heptagon side have the highest energy. The right two panels display the 3D distributions of bond lengths around a defect (top) and the corresponding 2D projection (bottom). It can be clearly observed that the 2D projection significantly underestimates the length of covalent bonds around the 5–7 defect.

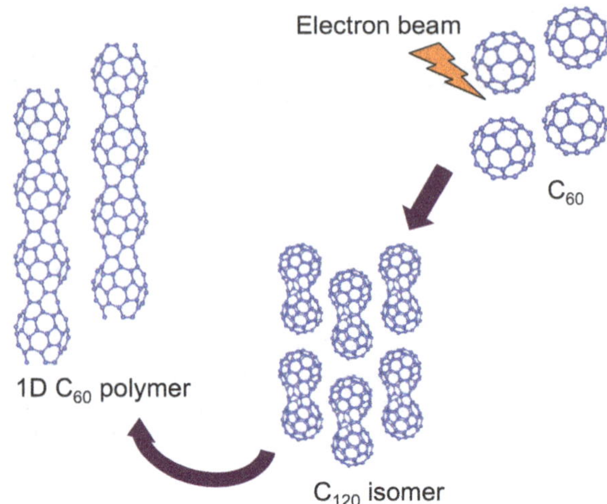

Fig. 3.12 Schematic illustration showing the synthesis of the 1D C_{60} polymer. Reprinted with permission from [65]

3.5 Peanut-Shaped C_{60} Polymers

3.5.1 Fusion of C_{60} Molecules

In the previous sections, we have seen that alteration in local atomic configuration of sp^2 nanocarbons can lead to a drastic change in the global geometry. Artificial control of topology in the carbon sheet, therefore, may enable to realize anomalously shaped sp^2 nanomaterials endowed with surface curvature. This section focuses on a particular class of such anomalously shaped sp^2 nanocarbons, called peanut-shaped fullerene (C_{60}) polymer [9]. It is a quasi-one dimensional (1D) nanostructure obtained by a series of fusions between C_{60} molecules [62, 63]. Figure 3.12 shows a schematic diagram of the synthesis. When irradiated by the high-energy electron beam, two C_{60} are coalesced with each other via the generalized SW transformation [64]. They are then transformed into C_{120} isomers, followed by the synthesis of the 1D C_{60} polymer [65].

It has been unveiled that the 1D C_{60} polymer exhibits unique properties differing from those of the other nanocarbon allotropes (fullerenes, nanotubes, and graphenes) [66]. Experimental findings on the Peierls transition [67] and the infrared (IR) absorption intensity [68, 69] have indeed suggested the intriguing correlation between the shape and physical properties of the 1D C_{60} polymers. The most interesting properties in view of topology-geometry correlation would be the surface curvature effects on the electronic properties. Indeed, the 1D C_{60} polymers commonly show the periodic modulation in the surface curvature of the sidewall along the tubular axis. This curvature modulation stems from periodic variation in the tube diameter, which results from periodic insertion of topological defects into a rolled-up graphene sheet. We will see in the subsequent sections that the periodically modulating curvature of the

1D C_{60} polymers drastically changes the nature of collective excitations in the mobile electrons compared with the case of the flat sp^2 nanocarbons.

3.5.2 TLL State in C_{60} Polymers

The 1D C_{60} polymer is a long, thin, and hollow cylinder whose radius varies periodically in the axial direction. Generally, in low-dimensional systems, Coulombic interaction between mobile electrons plays a crucial role in their quantum nature and often creates new collective states. Particularly in one-dimensional systems, such as the 1D C_{60} polymer, interacting electrons cannot be regarded as a Fermi-liquid but as a Tomonaga-Luttinger liquid (TLL) [70]. In the TLL state, the single-particle excitation spectrum has an energy gap near the Fermi level, and this gap often causes a power-law anomaly in the measurement data of the system. The hallmark of TLL states is the power law behavior of the single-particle density of states (DOS), designated by $D(\omega, T)$, near the Fermi energy E_F. Here, ω indicates the energy measured from E_F, and T denotes the absolute temperature. Such power-law anomalies were experimentally confirmed in carbon nanotubes [71], which are also a quasi-one dimensional nanocarbon. These facts naturally raise a question as to whether or not the periodic curvature modulation inherent to the 1D C_{60} polymers affects the TLL behavior of the systems if it occurs [72]. Given a significant effect is detected, it evidences the topology-induced alteration both in the geometry and property of the 1D C_{60} polymers.

The mentioned above problem was resolved experimentally in 2012 by photoemission spectral (PES) measurements [73]. The measurements revealed the PES spectra in the vicinity of E_F in the temperature range of 30–350 K. As the temperature decreased, the DOS $D(\omega, T)$ converged to a power-law whose dependence on the binding energy ω near E_F is described by

$$D(\omega, T = 0) \propto |\hbar\omega - E_F|^{\alpha}$$

in an energy range of 18–70 meV with the exponent α of ca. 0.66. In a similar manner, the DOS just on E_F (i.e., $\omega = 0$) was found to obey the power law with respect to T as

$$D(\omega = 0, T) \propto T^{\alpha}$$

in the range of 30–350 K with α being ca. 0.59. It should be emphasized that the above mentioned values of α, nearly equal to 0.6, are quite larger than that of single-walled carbon nanotubes ($\alpha \sim 0.5$) [71]. It is thus concluded that the feasible increment in α obtained in the experiment can be attributed to the effect of surface curvature modulation that are produced by the periodically inserted topological defects along the tube axis.

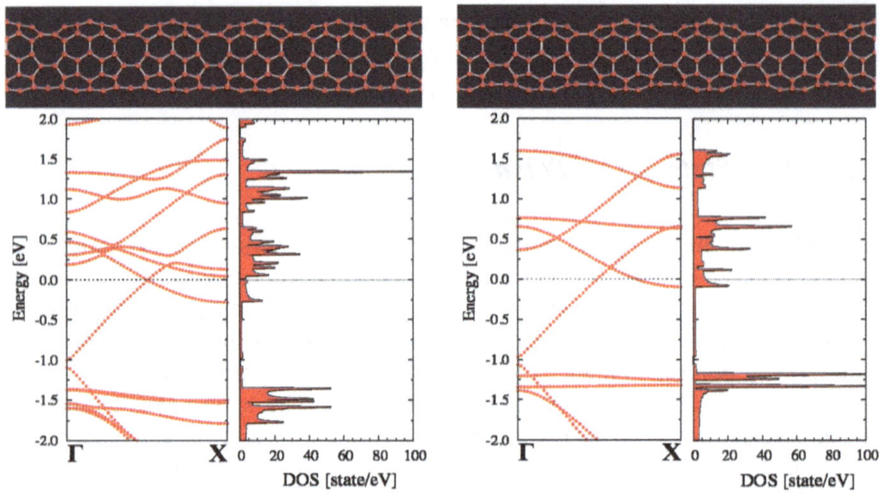

Fig. 3.13 Top: Two different atomic models of the 1D C_{60} polymers deduced from first-principles calculations. Constricted portions are occupied by a set of heptagons and octagons in the left-side model, and by a set of five octagons in the right-side model. Bottom: Dispersion curves and the densities of states of electrons in the two atomistic systems. Reprinted with permission from [72]

3.5.3 Topology-Based Understanding

Theoretical interpretation regarding the TLL behavior of the 1D C_{60} polymer has been proposed in 2016 [72], shortly after the experimental finding reported in 2012 [73]. The aim of the theoretical work was to describe the low energy behavior of local DOS in the 1D C_{60} polymer using the bosonization procedure. This procedure requires the knowledge above the Fermi velocity of carriers, which can be evaluated from the electronic band structures. To secure the quantitative accuracy, the band structures as well as the atomic configuration of the 1D C_{60} polymers were deduced from the first-principles calculations [74], from which the correlation between topological defects and the TLL nature can be unveiled.

Figure 3.13 shows the energetically stable atomic models of the 1D C_{60} polymers deduced from the first-principles calculations [74]. To find the two most stable configurations, more than fifty kinds of 1D C_{60} polymers having different atomic structures were analyzed. Through the analysis, it turned out that the total energy and the stability of the models depend on the number of non-hexagonal rings (i.e., pentagon, heptagon, and octagon) embedded and their relative positions. One of the two most stable models, depicted in the left-upper panel of Fig. 3.13, is characterized by that the constricted portions are occupied by a set of heptagons and octagons. On the other hand, the constricted parts of the other stable model depicted in the right-upper panel are composed of five octagons. The difference in the formation energy between the two models was found to be less than 0.7 eV; in actual fabrication of the 1D C_{60} polymers, therefore, both atomic configurations are possibly realized.

The bottom panels of Fig. 3.13 show the electronic energy bands of the two models. It follows from the graphs that more than one dispersion curves cross the Fermi level. A difference in the number of Fermi-level crossing points between the two models is a consequence of the slight difference in the lattice symmetry at the waist part as mentioned above. Another remarkable feature is that for both models, the X-shaped band crossing is a little slanted, asymmetric at the crossing point. This slanted level crossing is in contrast with the fully symmetric level-crossing observed in the metallic single-walled carbon nanotubes, indicating the effect of periodic insertion of topological defects in the rolled-up sp^2 monoatomic sheet.

The simulation data for the stable lattice structure and the associated electronic band structure made it possible to obtain an analytic formula for DOS of the 1D C_{60} polymers. The formula successfully reproduces the experimental data of PES, thus describing the effect of topological defects on the TLL nature. Furthermore, the formula tells us that different Fermi-level-crossing bands result in different power-law dependence in the PES. This conclusion implies an uncovered crossover in the spectra at energy on the order of 100 meV, beyond which the value of the exponent shifts significantly.

3.5.4 *Curvature-Based Understanding*

We have argued in Sect. 5.3 that the atomistic model involving topological defects provides the theoretical understanding of the TLL states realized in the 1D C_{60} polymers. Periodic insertion of defects causes variation in the surface curvature of the sp^2 monoatomic sheet, resulting in the upward shift in the power-law exponent compared with that of single-walled carbon nanotubes.

The story does not end. If the wavefunctions of the mobile electrons are expanded over the whole system, as those in 1D C_{60} polymers [75], there is another theoretical approach that is useful to describe the low-energy excitations of the electrons moving on curved surfaces. In this approach, the discrete atomic structure is approximated as a continuum thin surface, and the electrons' motions are assumed to be restricted to the surface by a confining force perpendicular to the surface [76]. Due to the strong confinement, quantum excitation energies in the direction normal to the surface are elevated much higher than those in the tangential direction. As a result, the particle motion normal to the surface can be disregarded; this approximation allows us to define an effective Hamiltonian for propagation along the curved surface.

It is well known that, given a curved surface, the effective Hamiltonian describing the quantum motion on the surface involves an effective scalar potential. The sign and magnitude of the effective potential depend on the local surface curvature. An important consequence derived from the effective potential term in the Hamiltonian is that quantum particles confined to a thin curved layer should behave differently from those on a flat plane. Namely, the presence of nonzero surface curvature impacts the quantum motion of electrons in the curved system. In this context, it is expected

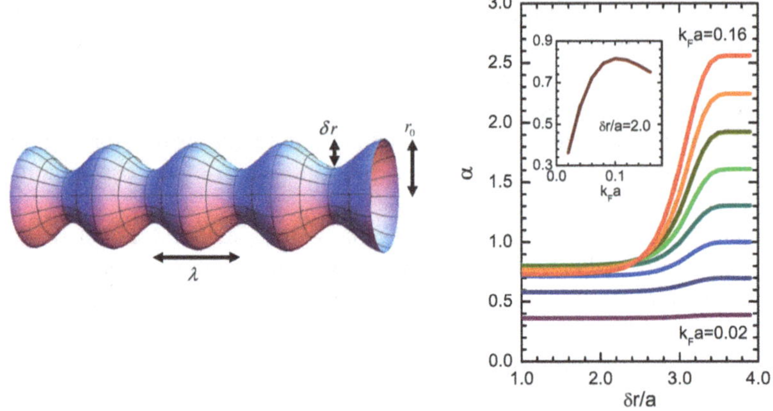

Fig. 3.14 Left: Schematic illustration of a quantum hollow cylinder with periodic radius modulation. Right: Surface curvature effect on the TLL power-law exponent. Reprinted with permission from [77]

that the periodic surface curvature inherent in the 1D C_{60} polymers engenders sizable effects on their TLL properties.

The problem was addressed theoretically in 2009 using the continuum approximation [77]. There, the 1D C_{60} polymers were mapped onto thin, long, and hollow cylinders; the cylindrical radius is assumed to be varied in a periodic manner, as analogous to the envelope surface of the real C_{60} polymers. The tube radius $r(z)$ was assumed to be periodically modulated in the axial z direction as

$$r(z) = r_0 - \frac{\delta r}{2} + \frac{\delta r}{2}\cos\left(\frac{2\pi}{\Lambda}z\right)$$

where the parameter r_0 and δr were introduced to express the maximum and minimum of $r(z)$ as r_0 and $r_0 - \delta r$, respectively; see Fig. 3.14 for the schematic illustration of the deformed tube. The values of the parameters, $r_0 = 4.0$ Å and $\Lambda = 8.0$ Å, suffice to reproduce the actual shape of the 1D C_{60} polymers, while the value of δr may vary according to the time duration of electron beam irradiation applied to pristine C_{60} molecules at the stage of synthesis. It is important to note that the value of δr determines the degree of curvature-induced scalar potential. Therefore, the central aim is to be placed on where (or not) the variation in δr causes a certain shift in the TLL power-law exponent α.

Let us remind that given a TLL state, the single-particle DOS at zero temperature, $D(\omega)$, near E_F exhibits a power-law singularity of the following form:

$$D(\omega) \propto |\hbar\omega - E_F|^{\alpha}, \quad \alpha = \frac{u + u^{-1}}{2} - 1$$

The explicit form of u is derived using the bosonization procedure as follows:

$$u = \sqrt{\frac{2\pi\hbar v_F + g_4(q) - g_2(q)}{2\pi\hbar v_F + g_4(q) + g_2(q)}}$$

Here, $g_4(q)$ and $g_2(q)$ are the wavenumber(q)-dependent coupling constants, and v_F is the Fermi velocity, determined by the slope of the electronic dispersion curve at the Fermi level. Note that the variation in δr causes a change in the dispersion curve; thus v_F is a function of δr. As a consequence, the exponent α should be dependent on δr, as clearly observed in Fig. 3.14. The plot shows the numerical result of the δr-dependent α for different Fermi wavenumber k_F. The salient feature of Fig. 3.14 is the significant increase in α with increasing δr. Such surface curvature-driven shift in α is consistent with the experimental finding that α for the 1D C_{60} polymers is larger than that for straight-shaped carbon nanotubes. Eventually, the upward shift in α can be attributed to the effects of geometric curvature as well as topological defect insertion to the constituent sp^2 monoatomic sheet

3.5.5 Electron-Phonon Coupling in C_{60} Polymers

It is plausible that topological defect insertion, or equivalently, periodic modulation in surface curvature, may affect the degree of electron-phonon (e-ph) interaction in the 1D C_{60} polymers. This is because the spatial profiles both of lattice vibration amplitude and the probability density of electron's wavefunctions will be dependent on the atomistic-level topology and global geometry of the system. In general nanocarbon allotropes, the e-ph interaction plays the key role in the collective motion of carriers, as proved in superconductivity [78] and charge density waves (CDW) [79]. Thus the theoretical prediction as well as experimental detection of the strength of e-ph interaction in the 1D C_{60} polymers will facilitate the understanding of their physico-chemical properties and of the contribution from the topological effects on them.

To proceed with consideration, we remind that in sp^2 nanocarbons, the strength of e-ph coupling is highly dependent on the global geometry of the structures. Denoting the e-ph coupling strength by λ, it is known that $\lambda \sim 0.6$ for K_3C_{60} having spherical structure [80], $\lambda \sim 0.006$ for single-walled carbon nanotubes having monolayered tubular structure [81] and $\lambda \sim 5.4 \times 10^{-4}$ for multi-walled nanotubes having concentric-layered cylindrical structure [82]. The wide variation in the values of λ, despite the common sp^2 nature, is plausibly attributed to the difference in the morphology of extended π-conjugation in the system. So what value of λ should be obtained for the 1D C_{60} polymers, having an intermediate geometry between the "spherical" C_{60} molecules and the "tubular" carbon nanotubes?

In order to estimate λ, carrier relaxation dynamics of the 1D C_{60} polymers was examined using femtosecond time-resolved pump-probe spectroscopy [83]. The measurement data of the femtosecond-transient refractivity was theoretically analyzed, leading to the conclusion that the magnitude of λ of the 1D C_{60} polymers to be 0.02. This result indicated that topology-induced curvature in the sp^2 nanocarbons significantly affects the magnitude of λ for nanocarbon allotropes.

3.6 Carbon Nanocoil

3.6.1 Benefit from Coiled Structure

Topological defect insertion into sp^2 nanocarbon allows to realize not only "curved" structures but also geometrically "twisted" ones. Carbon nanocoil is a typical example of such twisted sp^2 nanocarbon, resembling a "telephone cord" attached to a traditional phone receiver [84]. Carbon nanocoil promises to show both mechanical flexibility reflecting the coil morphology and the mechanical toughness originating from the sp^2 bonding. Furthermore, unique spiral structures of carbon nanocoils imply their versatile applications [85] including ultra-sensitive contact with resolution as high as femtograms [86].

From a historical perspective, the existence of carbon nanocoils was theoretically predicted in advance of the experimental realization. The prediction was first reported in 1993 [10]; it is interesting to note that this was just two years after the seminal finding on carbon nanotubes. In the theoretically proposed structures, the coiling arose as a consequence of periodic insertion of topological defects into a perfect graphitic cylinder. Figure 3.15 presents three different kinds of proposed structures [10], in which the pentagons and heptagons inserted are marked by black and shadow, respectively. It should be noted that in the proposed structures, topological defects are not "defects" but essential building blocks for realizing the atomic coiling network. Furthermore, the presence of non-hexagonal rings gives a physico-chemical impact on carbon nanocoils; it was numerically found that local change in the atomic bonding around the non-hexagonal rings leads to an enhancement of molecular hydrogen absorption on the outer surface [87]. The enhanced hydrogen absorption on the body surface indicates the possibility of developing a new class of hydrogen storage devices based on carbon nanocoils and/or other defective sp^2 layered materials.

3.6.2 Atomistic Modeling

A simple atomistic modelling of carbon nanocoils was suggested in [88]. The modelling is based on insertion of heptagons and pentagons in a periodic manner into a hexagonal carbon network. First let us suppose a piece of straight carbon nanotube as depicted in Fig. 3.16a. Next, a pair of pentagons is incorporated onto one side of the piece (at the position colored in blue), and a pair of heptagons onto the other side (red). After relaxing the defective tube segment, we will see that it is bent around the non-hexagonal rings, through which the defect-induced strain energy is relieved [10,

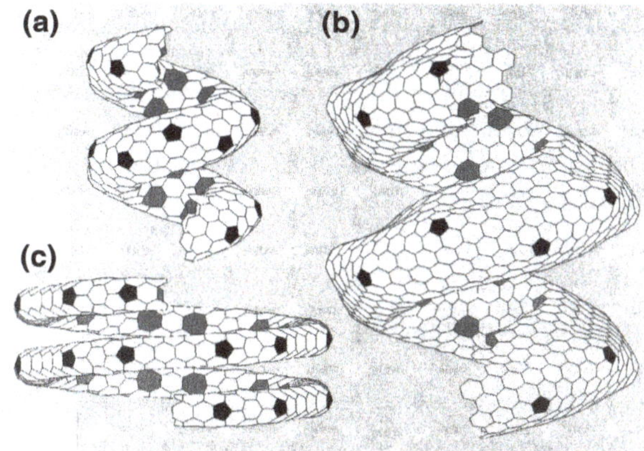

Fig. 3.15 Various atomic models of carbon coils with the lowest cohesive energy per atom. Pentagons and heptagons (shaded) appear in the outer and inner ridge lines, respectively, amid a background of the hexagonal lattice. Reprinted with permission from [10]

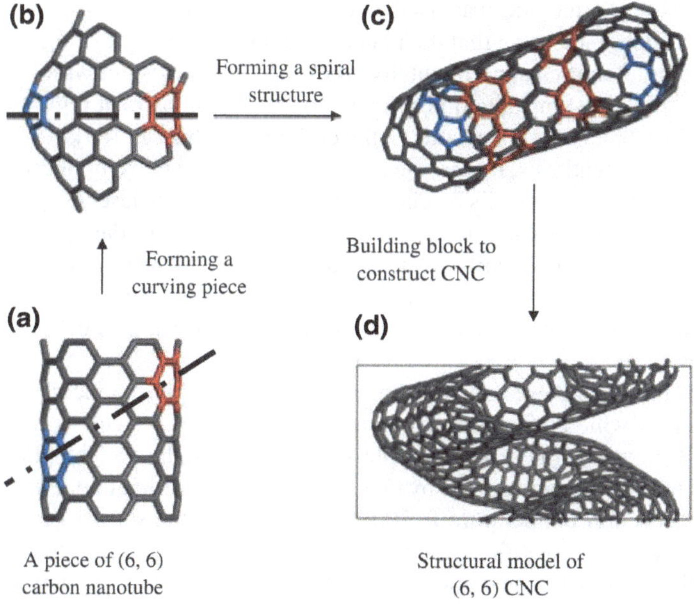

Fig. 3.16 Procedures of constructing a carbon nanocoil of a (6,6) type from pieces of a (6,6) carbon nanotube. Reprinted from [88]

89]. Similarly to previous arguments, pentagons generate positive curvature leading to a local cone shape, while heptagons generate negative curvature leading to a local saddle shape. The close relation between the defect position and the local change in the surface curvature is illustrated in Fig. 3.16b.

To obtain a spiral structure, we have only to connect the many building blocks already prepared one by one with a certain twisting angle. Figure 3.16c demonstrates a joint of two segments; repeating the connecting procedure, we finally obtain a seamless nanocoil as demonstrated in Fig. 3.16d. By changing the tube length at the two ends of the building block segment, or by varying the nanotube diameter, we can control coil diameter, coil pitch, and tubular diameter of a carbon nanocoil.

3.6.3 *Experimental Realization*

It may be surprising that experimental fabrication of carbon nanocoils has a relatively short story. The first electron microscopy observation of carbon nanocoils was reported in 1994 [90], serving as the pathfinder for the subsequent efforts both of high-quality production and of physical property investigation in carbon nanocoils. In the seminal work [90], catalytic decomposition of C_2H_2 worked successfully to synthesize carbon nanocoils with 30 nm in pitch and 18 nm in tubular diameter. Electron diffraction method was used to reveal the multiwalled, hollow, and polygonized structure, indicating that the nanocoils consisted of short straight segments. These structural features experimentally confirmed are consistent with the theoretical prediction that the introduction of pentagon-heptagon pairs at regular distances in a straight carbon nanotube results in the coiling morphology [89].

Since the first synthesis, many researchers have tried to make up these materials. Production of carbon nanocoils by chemical vapor deposition, laser evaporation, and opposed flow flame combustion method has been reported to date [5]. In addition to the "multi"-walled nanocoils with tubular diameters of several tens nanometers or more [91], indirect evidence of ultrathin "single"-walled carbon nanocoils (with both tubular diameter and pitch length down to 1 nm) was achieved by electron microscopy [92].

Among many syntheses so far, the triple-walled carbon nanocoil may be the thinnest one of which the direct microscopic image was obtained as a proof [93]. Figure 3.17 shows the image of the triple-walled nanocoil, having the fiber diameter less than 5 nm. It was synthesized by chemical vapor deposition under the following conditions: reaction temperature is 700 °C, the ratio of the source gas (acetylene, C_2H_2) to the dilution gas (nitrogen, N_2) is 0.01, and the gas pressure to be 0.67 kPa. A low C_2H_2 gas flow rate and a low partial gas pressure were important in reducing the fiber diameter.

3.6.4 *Theoretical Prediction*

For general twisted nanomaterials, geometric torsion is thought to result in significant impact on the coherent transport [94–97] and spin–orbit coupling [98, 99] of electrons migrating over the system. Namely, quantum transport through a nanoscale material

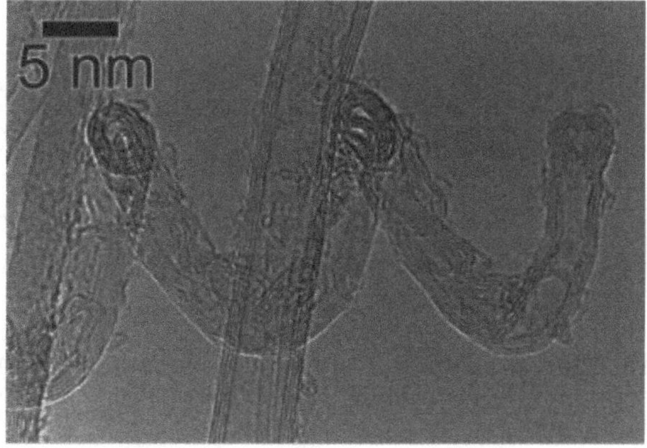

Fig. 3.17 Electron microscope image of a triple-walled carbon nanocoil. Reprinted with permission from [93]

with torsion will be essentially different from that in a curved nanomaterial with no torsion. This is because geometric torsion of the host material generates an additional quantum potential that affects the quantum motion of carriers in the system. It is thus interesting whether such torsion effect is feasible in the quantum nature of carbon nanocoils; this remains an open question, especially from an experimental viewpoint, mainly because of the difficulty in synthesizing few-walled carbon nanocoils with fine crystalline structure.

Numerical efforts have also been underway to unveil the curvature (and/or topology) effect on the quantum transport in carbon nanocoils. For instance, tight-binding simulations have shown that the quantum conductance is quantized due to the translational symmetry in the coiled direction [100]. Yet the conductance behaviors differ greatly from those of pristine metallic carbon nanotubes; instead, it looks similar to those of carbon nanotube superlattices, as a manifestation of the periodic insertion of topological defects.

The thermal properties of carbon nanocoils are also an interesting subject. Non-equilibrium molecular dynamics simulations have been used to reveal that the thermal conductivity in carbon nanocoils is reduced up to 70% compared with that of the corresponding straight single-walled carbon nanotubes [101]. The extreme reduction is due to the phonon scattering by coupled defects and folding. Shortly afterward, phonon thermal transport in carbon nanocoils has been studied in detail [102]. The three-phonon Umklapp scattering rates and the associated phonon relaxation were considered in a wide temperature range, and a certain reduction in the thermal conductivity of carbon nanocoils was numerically detected as well.

As a final remark, it warrants comment on a recent finding on the quantum transport of a helicoidal graphene nanoribbon [103]. In this geometry, the twist of the nanoribbon plays the role of an effective transverse electric field in graphene. Surprisingly, this effective electric field turned out to have a different sign for the two isospin states; here, isospin is defined with regard to the two components of a Dirac spinor. As a result, this electric field reverses polarity when the isospin is changed, leading to a separation of the isospin states of the carriers on the opposing rims of the

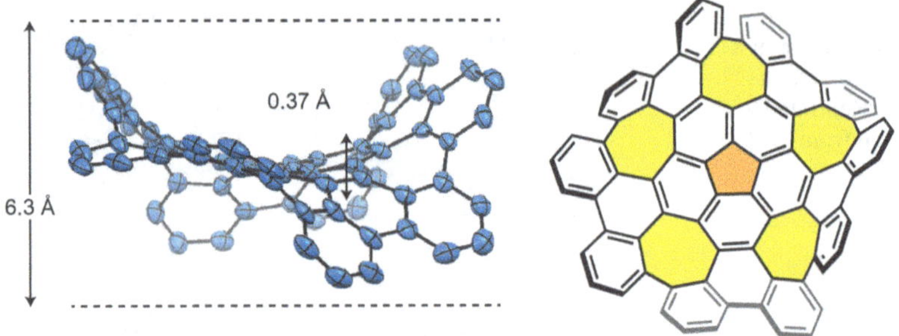

Fig. 3.18 Atomic structural model of a warped nanographene. All of the atoms shown in the image are carbons; hydrogen atoms on the perimeter are omitted for clarity. Reprinted with permission from [104]

nanoribbon. From the experimental feasibility, the isospin transitions are expected with the emission or absorption of microwave radiation which could be adjusted to be in the THz region.

3.7 State-of-the-Art Curved sp^2 Nanocarbons

3.7.1 *Nano-"Pringles"*

The final section covers several exotic sp^2 nanocarbons that were fabricated/predicted in the last few years. The first to be noted is the so-called nano-"Pringle" [104], i.e., a piece of warped nanographene comprising 80 carbon atoms joined together in a network of 26 rings. Figure 3.18 shows the atomic model of a nano-Pringle. Key to the formation of the grossly warped structure is the addition of five heptagons into the hexagonal lattice. Nano-Pringles turned out to show high solubility in common organic solvents and a widened energy gap between the highest occupied molecular orbital (HOMO) and the lowest unoccupied molecular orbital (LUMO) [104].

Interestingly, this unique structure was found to have physical and chemical properties distinct from other all-carbon families. For instance, the first-principles calculations revealed that the introduction of topological defects significantly changes the optical absorption spectra of nano-Pringles [105, 106]. In particular, the interaction between the topological defects was found to enhance the excitonic effect due to the lattice symmetry breaking, thereby generating the extra peaks at the lower photon energy side of the main peak.

From chemical perspective, the gas adsorption ability of nano-Pringles has drawn much attention recently. A series of numerical work has unveiled the adsorption property of small molecules, such as O_2, CO, and SO_2, on the nano-Pringles surface. For instance, adsorption of SO_2 turned out to induce charge transfer, indicating

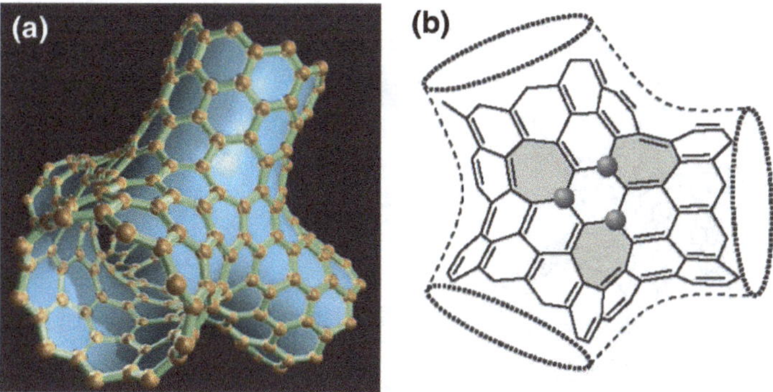

Fig. 3.19 **a** Nano-Tetrapod structure made of sp^2 carbon bondings. It is a junction of four (9, 0) nanotubes with a zigzag edge. **b** Chemical bonding near the core of the tetrapod. Trivalent carbon radicals are emphasized by gray spheres. Reprinted with permission from [109]

that the nano-Pringles can be used as an SO_2 sensor [107]. In addition, chemical adsorption of O_2 causes modulation in the HOMO-LUMO gap, implying the potential utility in optoelectronics [107]. Another salient feature of the nano-Pringles is the potential utility as the CO chemical sensors, for which the pristine graphene is not appropriate. Numerical simulation uncovered that the electronic conductivity of nano-Pringles increases with the CO concentration [108].

3.7.2 Nano- "Tetrapod"

Figure 3.19 displays the second remarkable example of exotic nanocarbons, so to say, a nano-"Tetrapod" [109]. This three-dimensional open structure is composed of a carbon network of hexagons and heptagons only. As found in Fig. 3.19, four cut-off pieces of carbon nanotubes with (9, 0) chirality are glued via several heptagons, forming a quadruped nanocarbon material. The tetrapod may serve as a building block of a nanostructured carbon foam [110].

One of the most striking features is that the nano-Tetrapod carries a net magnetic moment in the ground state. The origin of magnetism in the nano-Tetrapod is the presence of four unpaired spins in the electronic ground state. Accordingly, the ground state of the nano-Tetrapod is different either from the spin-polarized states at the zigzag edge or dangling-bond states, in which the spin polarization is attributed to the presence of under-coordinated carbon atoms. Instead, the magnetic behavior of the nano-Tetrapods stems from the trivalent carbon radicals introduced at the center of the quadruple structure. The radicals are sterically stabilized within the aromatic system of the otherwise tetravalent carbon atoms. The conclusion that unpaired spins may be introduced by carbon radicals, not only by under-coordinated carbon atoms,

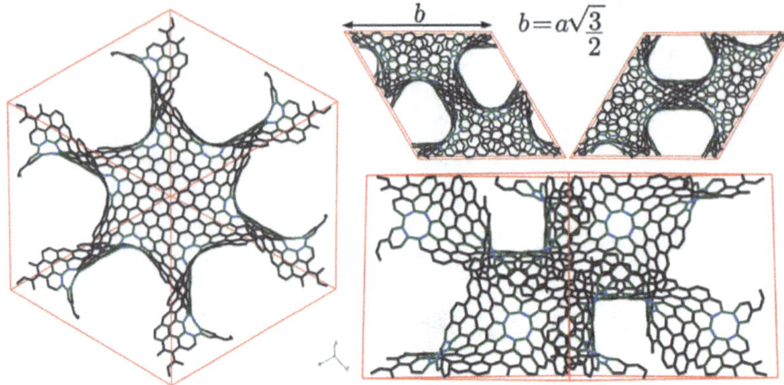

Fig. 3.20 Atomic structure of the gyroidal Schwarzite: the [111] orientation (left panel), the [100] and [010] orientation (top right panels), and the [110] orientation (bottom right panel). The arrow labeled by b indicates the lattice parameter. Reprinted with permission from [114]

will open a new avenue to design other graphitic structures with negative surface curvature.

3.7.3 Nano-"Schwarzite"

The last example of exotic sp^2 nanocarbons is carbon Schwarzite. It has been known as a negative-curvature analogue of the fullerenes since the early 1990s [111–113]. Figure 3.20 illustrates the atomic structure of a carbon Schwarzite with gyroidal type [114], in which non-hexagonal carbon rings stabilize the curved structure. Schwarzites are distinguished from the well-known sp^2 nanocarbon with 0D, 1D or 2D, by their hyperbolic geometry. Thus, such new type of geometry is expected to result in novel and fascinating properties.

In mathematics, the negative curvature surfaces of interest are called infinite periodic minimal surfaces, first studied in detail by the mathematician H. A. Schwarz in the late 19th century [115]. He pointed out that it is not possible to construct an infinite surface with a constant negative Gaussian curvature; instead, he found that patches of varying negative Gaussian curvature could be smoothly joined to provide an infinite surface with zero mean curvature, which is periodic in 3D space. Five different types of minimal surfaces were known by 1880 [116]; afterword, more than 50 types of distinct minimal surfaces have been discovered [117].

In the community of physics and materials science, Schwarzite structures have received increasing attention due to the interest in finding a graphene foam [118]. A number of periodic porous carbon structures similar to Schwarzite structures have been experimentally produced using, for example, templated synthesis [119] and liquid exfoliation techniques [120]. The enhanced control in the fabrication of

Schwarzite structures will open the way for versatile applications; hydrogen storage systems [121] and electrolyte diffusion [122] are only a few to mention. Quite recently, Schwarzite structures have been fabricated along the wall of zeolite pores; it was demonstrated that La ions embedded in zeolite pores facilitate the carbonization of ethylene or acetylene, enabling the selective formation of Schwarzite structures inside the zeolite template [123].

One interesting subject in view of quantum physics is the possible presence of pseudo-relativistic massless particles (Dirac fermions) in Schwarzite structures. The density functional calculations have revealed that a linear band crossing merging in a point for a large gyroidal Schwarzite structure [114]. Such a gapless linear energy dispersion in low-dimensional nanocarbons had a tremendous impact on conventional condensed-matter physics by imposing relativistic physics in the electronic properties of these nanosystems. Indeed, the electrons in graphene (2D) and carbon nanotubes (1D) behave like Dirac fermions as described by the crossing of linearly dispersive electronic bands, also called the Dirac cone. In this context, Schwarzites may be a remarkable playground to investigate relativistic physics of these exotic fermions, whereas their essential properties with respect to structural stability, mechanical, and electronic aspects should be largely investigated [124, 125].

3.8 Perspective

The synthesis or discovery of new materials, for example, fullerenes (1996 Nobel Prize in Chemistry), nanotubes, and graphene (2010 Nobel Prize in Physics) has been hitherto done on the basis of (1) knowledge (literature), (2) experience, and (3) intuition so far. These factors are still important, but it is wondered that we have failed to find the other fascinating materials close to them thus synthesized/discovered previously. In addition, it will take an unexpected time to discover the next new fascinating materials, because of "serendipity" based on (2) and (3) as shown in top (a) of Fig. 3.21.

Is it impossible that we can synthesize/discover new novel materials beyond "serendipity"? One possible way is to combine modern geometry with materials science. It is well known historically that Mathematics has been extensively used as a good tool in Physics and Chemistry (for example, differential and integral, point group, vector and tensor, etc.), but we have not seriously considered the correlation between geometric/topological quantity and physical/chemical quantity (properties) up to now, because there has been a large gap between mathematical simple model (point and line) and real materials (different atoms and different interactions between them).

As introduced in this chapter, geometry/topology can provide not only new forms but also new properties of nanocarbon materials. If geometric/topological quantity is correlated with physical/chemical quantity (properties), we can add new index "mathematics" to materials world consisting of two indexes "physics and chemistry" as shown in bottom (b) of Fig. 3.21. Thus we can synthesize/discover new

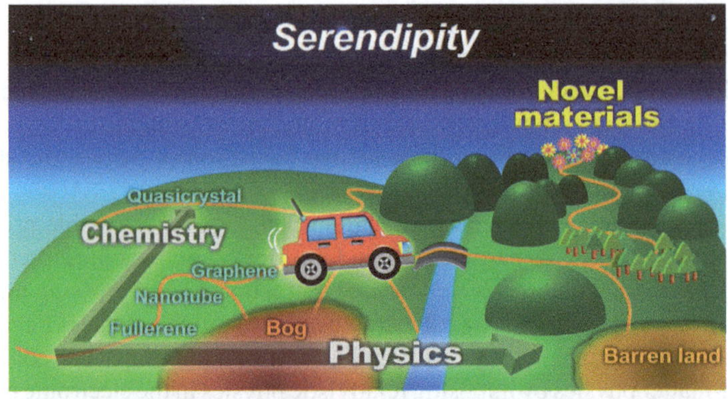

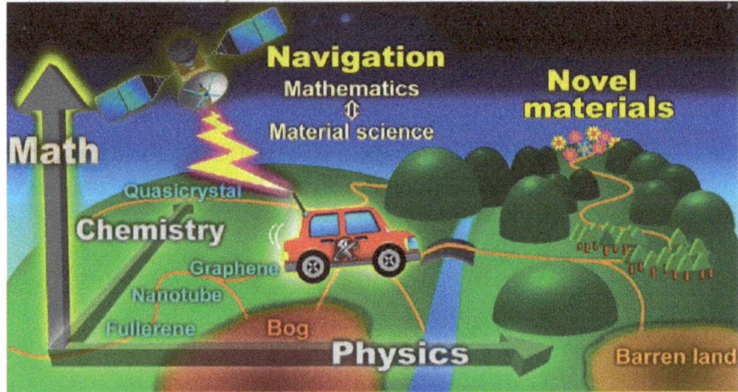

Fig. 3.21 Paradigm shift in materials science using Mathematics. Top: Conventional ways to discover novel materials. Bottom: Mathematics-navigated ways to discover novel materials

materials beyond "serendipity". More recently, materials informatics [combination of AI (artificial intelligence) with data-base of materials] has been focused from a viewpoint of materials development. It is a concern that they can find the optimal solution promptly for a given material, but may hardly find new or unexpected solution as far as using the data-base.

Acknowledgements The authors acknowledge stimulating and helpful discussions with Yoshitaka Umeno, Hideo Yoshioka, Shota Ono, Yusuke Noda, and Kaoru Ohno. This work was supported by JSPS KAKENHI Grant Numbers JP 25390147 and 15H03888.

References

1. S. Gupta, A. Saxena, J. Appl. Phys. **109**, 074316 (2011)
2. S. Gupta, A. Saxena, MRS Bull. **39**, 265–279 (2014)
3. X. Chen, G. Wu, Y. Jiang, Y. Wang, X. Chen, Analyst **136**, 4631–4640 (2011)
4. S. Iijima, Nature **354**, 56–58 (1991)

5. H. Shima, M. Sato, *Elastic and Plastic Deformation of Carbon Nanotubes* (Pan Stanford Publishing, Singapore, 2013)
6. K.S. Novoselov, A.K. Geim, S.V. Morozov, D. Jiang, Y. Zhang, S.V. Dubonos, I.V. Grigorieva, A.A. Firsov, Science **306**, 666–669 (2004)
7. M.T. Lusk, L.D. Carr, Phys. Rev. Lett. **100**, 175503 (2008)
8. H. Shima, *Functional Analysis for Physics and Engineering: An Introduction* (CRC Press, 2016)
9. J. Onoe, T. Nakayama, M. Aono, T. Hara, Appl. Phys. Lett. **82**, 595–597 (2003)
10. S. Ihara, S. Itoh, Phys. Rev. B **48**, 5643–5648 (1993)
11. A.L. Mackay, H. Terrones, Nature **352**, 762 (1991)
12. H. Terrones, A.L. Mackay, Carbon **30**, 1251–1260 (1992)
13. H. Terrones, M. Terrones, New J. Phys. **5**, 126 (2003)
14. A. Cortijo, M.A.H. Vozmediano, Nucl. Phys. B **763**, 293–308 (2007)
15. O.V. Yazyev, S.G. Louie, Phys. Rev. B **81**, 195420 (2010)
16. M.V. Diudea, Phys. Chem. Chem. Phys. **7**, 3626–3633 (2005)
17. T. Zhang, X. Li, H. Gao, Ext. Mech. Lett. **1**, 3–8 (2014)
18. J.M. Carlsson, M. Scheffler, Phys. Rev. Lett. **96**, 046806 (2006)
19. J. Kang, J. Bang, B. Ryu, K.J. Chang, Phys. Rev. B **77**, 115453 (2008)
20. D.V. Kolesnikov, V.A. Osipov, JETP Lett. **87**, 419–422 (2008)
21. Y. Noda, S. Ono, K. Ohno, Phys. Chem. Chem. Phys. **16**, 7102–7107 (2014)
22. O.V. Yazyev, S.G. Louie, Nat. Mater. **9**, 806–809 (2010)
23. A.V. Zhukov, R. Bouffanais, N.N. Konobeeva, M.B. Belonenko, JETP Lett. **97**, 400–403 (2013)
24. T. Choudhari, N. Deo, EPL (Europhys. Lett.) **108**, 57006 (2014)
25. Y. Zhu, S. Murali, M.D. Stoller, K.J. Ganesh, W. Cai, P.J. Ferreira, A. Pirkle, R.M. Wallace, K.A. Cy-chosz, M. Thommes, D. Su, E.A. Stach, R.S. Ruoff, Science **332**, 1537–1541 (2011)
26. D. Odkhuu, D.H. Jung, H. Lee, S.S. Han, S.H. Choi, R.S. Ruoff, N. Park, Carbon **66**, 39–47 (2014)
27. S. Zhang, J. Zhou, Q. Wang, X. Chen, Y. Kawazoe, P. Jena, Proc. Natl. Acad. Sci. USA **112**, 2372–2377 (2015)
28. M. Terrones, F. Banhart, N. Grobert, J.C. Charlier, H. Terrones, P.M. Ajayan, Phys. Rev. Lett. **89**, 075505 (2002)
29. M. Menon, A.N. Andriotis, D. Sri-vastava, I. Ponomareva, L.A. Chernozatonskii, Phys. Rev. Lett. **91**, 145501 (2003)
30. M. Yoon, S. Han, G. Kim, S.B. Lee, S. Berber, E. Osawa, J. Ihm, M. Terrones, F. Banhart, J.C. Charlier, N. Grobert, H. Terrones, P.M. Ajayan, D. Tománek, Phys. Rev. Lett. **92**, 075504 (2004)
31. B.I. Dunlap, Phys. Rev. B **46**, 1933–1936 (1992)
32. J. Zhang, D. Jiang, Compos. Sci. Technol. **71**, 466–470 (2011)
33. L. Liu, L. Zhang, J. Lua, Appl. Phys. Lett. **101**, 161907 (2012)
34. Y. Suda and H. Shima, Metal Powder Rep. **72**, 317–321 (2017). https://doi.org/10.1016/j.mprp.2016.06.005
35. J.W. Jang, C.E. Lee, S.C. Lyu, T.J. Lee, C.J. Lee, Appl. Phys. Lett. **84**, 2877–2879 (2004)
36. S. Shanmugam, A. Gedanken, J. Phys. Chem. B **110**, 2037–2044 (2006)
37. L. Gan, R. Lv, H. Du, B. Li, F. Kang, Carbon **47**, 1833–1840 (2009)
38. T.W. Ebbesen, Acc. Chem. Res. **31**, 558–566 (1998)
39. M. Terrones, ACS Nano **4**, 1775–1781 (2010)
40. Y. Saito, K. Hata, T. Murata, Jpn. J. Appl. Phys. **39**, L271–L272 (2000)
41. Y. Nishiura, M. Kotani, *Mathematical Challenges in a New Phase of Materials Science* (Springer, 2014)
42. J.R. Sanchez-Valencia, T. Dienel, O. Gröning, I. Shorubalko, A. Mueller, M. Jansen, K. Amsharov, P. Ruffieux, R. Fasel, Nature **512**, 61–64 (2014)
43. A.J. Stone, D.J. Wales, Chem. Phys. Lett. **128**, 501–503 (1986)

44. K.S. Novoselov, A.K. Geim, S.V. Morozov, D. Jiang, M.I. Katsnelson, I.V. Grigorieva, S.V. Dubonos, A.A. Firsov, Nature **438**, 197 (2005)
45. X. Peng, R. Ahuja, Nano Lett. **8**, 4464 (2008)
46. A.I. Podlivaev, L.A. Openov, Phys. Lett. A **379**, 1757 (2015)
47. J. Ma, D. Alfè, A. Michaelides, E. Wang, Phys. Rev. B **80**, 033407 (2009)
48. K. Balasubramanian, M. Burghard, Small **1**, 180–192 (2005)
49. J. Zhao, H. Park, J. Han, J.P. Lu, J. Phys. Chem. B **108**, 4227 (2004)
50. M. Ouyang, J.L. Huang, C.L. Cheung, C.M. Lieber, Science **291**, 97–100 (2001)
51. A. Hashimoto, K. Suenaga, A. Gloter, K. Urita, S. Iijima, Nature **430**, 870–873 (2004)
52. M. Ishigami, H.J. Choi, S. Aloni, S.G. Louie, M.L. Cohen, A. Zettl, Phys. Rev. Lett. **93**, 196803 (2004)
53. K. Suenaga, H. Wakabayashi, M. Koshino, Y. Sato, K. Urita, S. Iijima, Nat. Nanotechnol. **2**, 358–360 (2007)
54. M. Yudasaka, H. Kataura, T. Ichihashi, L.C. Qin, S. Kar, S. Iijima, Nano Lett. **1**, 487–489 (2001)
55. J.C. Meyer, C. Kisielowski, R. Erni, M.D. Rossell, M.F. Crommie, A. Zettl, Nano Lett. **8**, 3582–3586 (2008)
56. S.T. Skowron, V.O. Koroteev, M. Baldoni, S. Lopatin, A. Zurutuza, A. Chuvilin, E. Besley, Carbon **105**, 176–182 (2016)
57. B.I. Yakobson, Appl. Phys. Lett. **72**, 918–920 (1998)
58. M.B. Nardelli, B.I. Yakobson, J. Bernholc, Phys. Rev. B **57**, R4277–R4280 (1998)
59. H. Mori, S. Ogata, J. Li, S. Akita, Y. Nakayama, Phys. Rev. B **74**, 165418 (2006)
60. Y. Nakayama, Jpn. J. Appl. Phys. **46**, 5005–5014 (2007)
61. T. Zhang, X. Li, H. Gao, J. Mech. Phys. Solid. **67**, 2–13 (2014)
62. H. Masuda, J. Onoe, H. Yasuda, Carbon **81**, 842–846 (2015)
63. H. Masuda, H. Yasuda, J. Onoe, Carbon **96**, 316–319 (2016)
64. H. Ueno, S. Osawa, E. Osawa, K. Takeuchi, Fuller. Sci. Technol. **6**, 319–338 (1998)
65. S. Ono, Y. Toda, J. Onoe, Phys. Rev. B **90**, 155435 (2014)
66. H. Shima, Geometry-property relation in corrugated nanocarbon cylinders, in *Modeling of Carbon Nanotubes, Graphene and Their Composites* (Springer, 2014)
67. Y. Toda, S. Ryuzaki, J. Onoe, Appl. Phys. Lett. **92**, 094102 (2008)
68. J. Onoe, A. Takashima, Y. Toda, Appl. Phys. Lett. **97**, 241911 (2010)
69. J. Onoe, A. Takashima, S. Ono, H. Shima, T. Nishii, J. Phys.: Condens. Matter **24**, 175405 (2012)
70. J. Voit, Rep. Prog. Phys. **57**, 977 (1994)
71. H. Ishii, H. Kataura, H. Shiozawa, H. Yoshioka, H. Otsubo, Y. Takayama, T. Miyahara, S. Suzuki, Y. Achiba, M. Nakatake, T. Narimura, M. Higashiguchi, K. Shimada, H. Namatame, M. Taniguchi, Nature **426**, 540–544 (2003)
72. H. Yoshioka, H. Shima, Y. Noda, S. Ono, K. Ohno, Phys. Rev. B **93**, 165431 (2016)
73. J. Onoe, T. Ito, H. Shima, H. Yoshioka, S. Kimura, EPL (Europhys. Lett.) **98**, 27001 (2012)
74. Y. Noda, S. Ono, K. Ohno, J. Phys. Chem. A **119**, 3048–3055 (2015)
75. T.A. Beu, J. Onoe, A. Hida, Phys. Rev. B **72**, 155416 (2005)
76. R.C.T. da Costa, Phys. Rev. A **23**, 1982 (1981)
77. H. Shima, H. Yoshioka, J. Onoe, Phys. Rev. B **79**, 201401 (2009)
78. M. Kociak, A.Y. Kasumov, S. Guéron, B. Reulet, I.I. Khodos, Y.B. Gorbatov, V.T. Volkov, L. Vaccarini, H. Bouchiat, Phys. Rev. Lett. **86**, 2416 (2001)
79. J.Y. Park, S. Rosenblatt, Y. Yaish, V. Sazonova, H. Üstünel, S. Braig, T.A. Arias, P.W. Brouwer, P.L. McEuen, Nano Lett. **4**, 517 (2004)
80. P. Zhou, K.-A. Wang, P.C. Eklund, G. Dresselhaus, M.S. Dresselhaus, Phys. Rev. B **48**, 8412–8417 (1993)
81. T. Hertel, R. Fasel, G. Moos, Appl. Phys. A **75**, 449 (2002)
82. I. Chatzakis, Appl. Phys. Lett. **103**, 043110 (2013)
83. S. Ono, Y. Toda, J. Onoe, Phys. Rev. B **90**, 155435 (2014)

84. H. Shima, Y. Suda, Mechanics of helical carbon nanomaterials, in *Advanced Computational Nanomechanics* (Wiley, 2015)
85. X. Chen, S. Zhang, D.A. Dikin, W. Ding, R.S. Ruoff, L. Pan, Y. Nakayama, Nano Lett. **3**, 1299–1304 (2003)
86. A. Volodin, D. Buntinx, M. Ahlskog, A. Fonseca, J.B. Nagy, C.V. Haesendonck, Nano Lett. **4**, 1775–1779 (2004)
87. V. Gayathri, N.R. Devi, R. Geetha, Int. J. Hydro. Energ. **35**, 1313–1320 (2010)
88. L.Z. Liu, H.L. Gao, J.J. Zhao, J.P. Lu, Nanoscale Res. Lett. **5**, 478–483 (2010)
89. S. Ihara, S. Itoh, J. Kitakami, Phys. Rev. B **47**, 12908–12911 (1993)
90. V. Ivanov, J.B. Nagy, Ph Lambin, A. Lucas, X.B. Zhang, X.F. Zhang, D. Bemaerts, G.V. Tendeloo, S. Amelinckx, J. van Landuyt, Chem. Phys. Lett. **223**, 329–335 (1994)
91. K.T. Lau, Composites B **37**, 437–448 (2006)
92. L.P. Biró, S.D. Lazarescu, P.A. Thiry, A. Fonseca, J.B. Nagy, A.A. Lucas, P. Lambin, Europhys. Lett. **50**, 494–500 (2000)
93. S.L. Lim, Y. Suda, K. Takimoto, Y. Ishii, K. Maruyama, H. Tanoue, H. Takikawa, H. Ue, K. Shimizu, Y. Umeda, Jpn. J. Appl. Phys. **52**, 11NL04 (2013)
94. J. Goldstone, R.L. Jaffe, Phys. Rev. B **45**, 14100 (1992)
95. H. Taira, H. Shima, J. Phys. Condens. Matt. **22**, 075301 (2010)
96. G. Cuoghi, A. Bertoni, A. Sacchetti, Phys. Rev. B **83**, 245439 (2011)
97. G.H. Liang, Y.L. Wang, L. Du, H. Jiang, G.Z. Kang, H.S. Zong, Physica E **83**, 246–255 (2016)
98. C. Ortix, Phys. Rev. B **91**, 245412 (2015)
99. P. Gentile, M. Cuoco, C. Ortix, Phys. Rev. Lett. **115**, 256801 (2015)
100. W. Lu, Sci. Technol. Adv. Mater. **6**, 809–813 (2005)
101. J. Zhao, J. Wu, J.W. Jiang, L. Lu, Z. Zhang, T. Rabczuk, Appl. Phys. Lett. **103**, 233511 (2013)
102. Z.P. Popović, M. Damnjanović, I. Milošević, Carbon **77**, 281–288 (2014)
103. V. Atanasov, A. Saxena, Phys. Rev. B **92**, 035440 (2015)
104. K. Kawasumi, Q. Zhang, Y. Segawa, L.T. Scott, K. Itami, Nat. Chem. **5**, 739 (2013)
105. Y. Noguchi, O. Sugino, J. Chem. Phys. **142**, 064313 (2015)
106. X. Wang, S. Yu, Z. Lou, Q. Zeng, M. Yang, Phys. Chem. Chem. Phys. **17**, 17864–17871 (2015)
107. Y. Dai, Z. Li, J. Yang, Carbon **100**, 428–434 (2016)
108. S. Jameh-Bozorghi, H. Soleymanabadi, Phys. Lett. A **381**, 646–651 (2017)
109. N. Park, M. Yoon, S. Berber, J. Ihm, E. Osawa, David Tomanek, Phys. Rev. Lett. **91**, 237204 (2003)
110. A.V. Rode, E.G. Gamaly, B. Lu-ther-Davies, Appl. Phys. A **70**, 135 (2000)
111. A.L. Mackay, H. Terrones, Nature **352**, 762 (1991)
112. T. Lenosky, X. Gonze, M. Teter, E. Vert, Nature **355**, 333 (1992)
113. M. O'Keeffe, G.B. Adams, O.F. Sankey, Phys. Rev. Lett. **68**, 2325 (1992)
114. A. Lherbier, H. Terrones, J.-C. Charlier, Phys. Rev. B **90**, 125434 (2014)
115. H.A. Schwarz, *Gesammelte Mathematische Abhandlungen* (Springer, 1890)
116. E.R. Neovius, *Bestimmung Zweier Spezieller Periodische Minimalflächen* (J. C. Frenkel & Sohn, Hel-sinki, 1883)
117. W. Fischer, E. Koch, Acta Cryst. A **45**, 166,169, 485, 558, 726 (1989)
118. Y. Wu, N. Yi, L. Huang, T. Zhang, S. Fang, H. Chang, N. Li, J. Oh, J.A. Lee, M. Kozlov, A.C. Chipara, H. Terrones, P. Xiao, G. Long, Y. Huang, F. Zhang, L. Zhang, X. Lepró, C. Haines, M.D. Lima, N.P. Lopez, L.P. Rajukumar, A.L. Elias, S. Feng, S.J. Kim, N.T. Narayanan, P.M. Ajayan, M. Terrones, A. Aliev, P. Chu, Z. Zhang, R.H. Baughman, Y. Chen, Nat. Commun. **6**, 6141 (2015)
119. H. Nishihara, T. Kyotani, Adv. Mater. **24**, 4473–4498 (2012)
120. J.G. Werner, T.N. Hoheisel, U. Wiesner, ACS Nano **8**, 731–743 (2014)
121. P. Kowalczyk, R. Hołyst, M. Terro-nes, H. Terrones, Phys. Chem. Chem. Phys. **9**, 1786–1792 (2007)
122. A. Nicolaï, J. Monti, C. Daniels, V. Meunier, J. Phys. Chem. C **199**, 2896–2903 (2015)

123. K. Kim, T. Lee, Y. Kwon, Y. Seo, J. Song, J.K. Park, H. Lee, J.Y. Park, H. Ihee, S.J. Cho, R. Ryoo, Nature **535**, 131–135 (2016)
124. J.R. Owens, C. Daniels, A. Nicolaï, H. Terrones, V. Meunier, Carbon **96**, 998–1007 (2016)
125. D.C. Miller, M. Terrones, H. Terrones, Carbon **96**, 1191–1199 (2016)

Chapter 4
Topology by Design in Magnetic Nano-materials: Artificial Spin Ice

Cristiano Nisoli

Abstract Artificial Spin Ices are two dimensional arrays of magnetic, interacting nano-structures whose geometry can be chosen at will, and whose elementary degrees of freedom can be characterized directly. They were introduced at first to study frustration in a controllable setting, to mimic the behavior of spin ice rare earth pyrochlores, but at more useful temperature and field ranges and with direct characterization, and to provide practical implementation to celebrated, exactly solvable models of statistical mechanics previously devised to gain an understanding of degenerate ensembles with residual entropy. With the evolution of nano–fabrication and of experimental protocols it is now possible to characterize the material in real-time, real-space, and to realize virtually any geometry, for direct control over the collective dynamics. This has recently opened a path toward the deliberate design of novel, exotic states, not found in natural materials, and often characterized by topological properties. Without any pretense of exhaustiveness, we will provide an introduction to the material, the early works, and then, by reporting on more recent results, we will proceed to describe the new direction, which includes the design of desired topological states and their implications to kinetics.

4.1 Introduction

From quasi-particles, fractionalization, pattern formation, to the nano-machinery of life in DNA replication and transcription, or to the coherent behavior of a flock, an ant colony, or a human group, emergent phenomena are generated by the collective dynamics of surprisingly simple interacting building blocks. Indeed, much of the more recent research in condensed matter pertains to the modeling of unusual emergent behaviors, typically from correlated building blocks in natural materials, either at the quantum or classical level [1]. A few years ago [2] we proposed a different approach: design, rather than simply deduce, collective behaviors, through the

C. Nisoli (✉)
Theoretical Division, Los Alamos National Laboratory, Los Alamos, NM 87545, USA
e-mail: cristiano@lanl.gov

S. Gupta and A. Saxena (eds.), *The Role of Topology in Materials*,
Springer Series in Solid-State Sciences 189,
https://doi.org/10.1007/978-3-319-76596-9_4

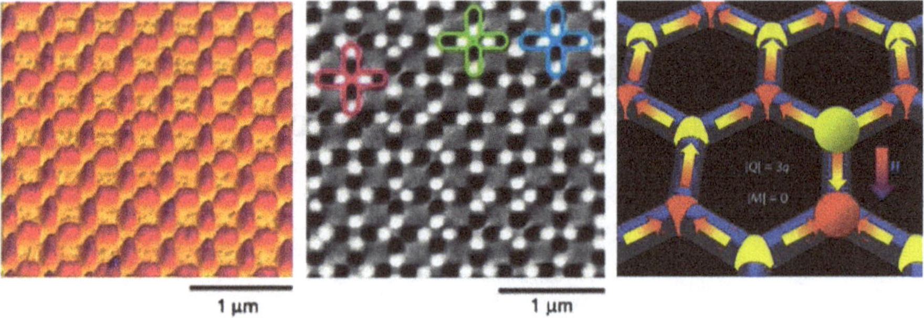

Fig. 4.1 Artificial spin ice in its most common geometries. Left: Atomic force microscopy image of square ice showing its structure (figures from [2]). Center: a magnetic force microscopy image of square ice, showing the orientation of the islands' magnetic moments (north poles in black, south poles in white); Type-I (pink) Type-II (blue) and Type-III (green) vertices are highlighted (see the text and Fig. 4.7 for a definition). Right: schematics of honeycomb spin ice

interaction of simple, artificial building blocks whose interaction could lead to exotic states not seen in natural materials.

Arrays of elongated, mutually interacting, single-domain, magnetic nano-islands arranged along a variety of different geometries, (Fig. 4.1) were ideal candidates. The magnetic state of each island could be described by a classical Ising spin, and advances in lithography allowed their nano-fabrication in virtually any geometry. The advantage of such approach is twofold: (1) the low energy dynamics, which underlies possible exotic states, is dictated by geometry, which here is open to design; (2) characterization methods—Magnetic Force Microscopy (MFM), PhotoElectron Emission Microscopy (PEEM), Transmission Electron Microscopy (TEM), Surface Magneto-Optic Kerr Effect (MOKE), Lorentz Microscopy—allow direct visualization of the magnetic degrees of freedom for unprecedented validation. Nano-scale is an good choice: the size of the building blocks, which are shape-anisotropic, elongated nano-islands (typically, NiFe alloys $200 \times 80 \times 5 - 30\,\mathrm{nm}^3$ patterned by nano-lithography on a non-magnetic Si substrate), has to be inferior to the typical magnetic domain, to provide single domains with magnetization directed along the principal axis, that can be interpreted as switchable spins.

These Artificial Spin Ices (ASI) were employed at first to study frustration in a controllable setting, to mimic the behavior of spin ice rare earth pyrochlores, but at more useful temperature and field ranges and with direct characterization, and to provide practical implementation to celebrated, exactly solvable models of statistical mechanics previously devised to gain an understanding of degenerate ensembles. Soon, a growing number of groups has extended the use of ASI [3], to investigate topological defects and dynamics of magnetic charges and spin fragmentation [4–12], information encoding [13, 14], in and out of equilibrium thermodynamics [15–24], avalanches [25, 26], direct realizations of the Ising system [27–29], magnetoresistance and the Hall effect [30, 31], critical slowing down [32], dislocations [33], spin wave excitations [34], and memory effects [35, 36]. Meanwhile similar strategies [37–42] have found realization in trapped colloids [43–45], vortices in nano-patterned superconductors [46, 47] and even at the macroscale [48]. With the evolution of nano-fabrication and of experimental protocols it is now possible to characterize the material in real-time, real-space [49–53], and to realize virtually any geometry, for direct control over the collective dynamics. This has recently opened a path toward the deliberate design of novel, exotic states [54–58] not found in natural materials [59, 60].

Frustration is a fundamental ingredient in design: it controls the interplay of length and energy scales, dictating the emergent dynamical properties that lie at the boundaries between order and disorder, and leading to a lively, quasi-disordered ensemble called *ice manifold*, to be exploited in the design of exotic behaviors.

Correlated spin systems have of course a long history in Physics. In classical statistical mechanics, the Ising model [61] paved the way to our understanding of long-range order from symmetry breaking as a second order phase transition, universality classes and scaling [62], and finally the renormalization group [63] with implications reaching well beyond condensed mater systems [64]. However, frustrated spin systems often do not order, generally resulting in quasi-disordered manifolds governed by some geometric or topological rule. Often their collective dynamics lends itself to emergent descriptions that are only partially reminiscent of the constitutive spin structure. The situation is somehow similar to everyday life, where frustration results from a set of constraints that cannot be all satisfied at the same time, leading to a manifold of compromises among which the choice is most often equivalent and can be influenced by a small bias. Thus, obstructed optimization provides high susceptibility that can generate the complex social dynamics we witness everyday. These analogies between social settings and frustrated materials are not merely philosophical: ideas borrowed from the frustrated spin ice physics have been exploited in social networks to describe wealth allocation [65].

Much as in life, frustration is understood in Physics as a set of constraints that cannot be all satisfied. Typically the constraint is the optimization of an energy, usually the pairwise interaction between elementary degrees of freedom. This too leads to a degenerate manifold which preserves non-zero entropy density at low temperatures, in apparent violation of the third law of thermodynamics.

We will see how frustration is exploited in the design of artificial spin ices. Initially, the aim was pure exploratory science, with the goal to understand frustration and disorder in a controllable environment that could be characterized directly. These materials could mimic the frustrated ice rule that defines the exotic manifold of rare earth pyrochlores (see below), yet at room temperature rather than at the Kelvin scale. They could also provide the first realization of the celebrated exactly solvable models of statistical mechanics, such as the antiferromagnetic Ising system on a triangular lattice described above, or the various vertex models introduced and/or solved by Lieb, Wu and Baxter between the late 60s and early 80s [66–69]. As both experimental protocols and theoretical understanding evolved, however, it became clear how the material could open new paths in a material-by-design effort: instead of finding, more or less serendipitously, natural materials of interesting or novel behavior, one could think about a bottom-up approach, where a suitable design could produce desired exotic properties.

In this chapter we will start with a brief description of fundamental concepts and then earlier realizations, fleshing out the basics pertaining to their experimental protocols, nano-fabrication, and characterization (for details we refer to references or to the following reviews [3, 70, 71]). Geometric frustration should really be called topological, as it is essentially a topological property, yet it is based on interaction, which is instead geometric, and generally not topologically invariant: we will discuss the two issues in parallel, and see how new approaches and new designs, based on a different level of frustration, so called vertex-frustration, can indeed allow access to bona fide topological states.

4.2 Frustration, Topology, Ice, and Spin Ice

The concept of geometric frustration in its broader mathematical form involves a geometric system, a manifold of degrees of freedom and a set of prescriptions on how they should arrange with respect to each other. The system is frustrated if there are loops along which not all these prescriptions can be satisfied (Fig. 4.1). Clearly the concept is very general and extends beyond Physics. One recognizes topology immediately in the nature of such definition: any homotopy, that is any continuous transformation that does not tear those loops, will lead to a system of the same frustration.

In Physics, in general (a) these "prescriptions" correspond to the optimization of a certain energy, and (b) that energy is usually a pairwise interaction between binary degrees of freedom.

An early example is the famous antiferromagnetic Ising model on a triangular lattice [72], a system of binary spins interacting antiferromagnetically on a triangular lattice (Fig. 4.2). There the interaction among nearest neighbor spins cannot be satisfied simultaneously on a triangular plaquette, leading to a disordered manifold. The disorder is, however, non-trivial, and its entropy per spin is not merely $s = k_B \ln(2) \simeq 0.6931 k_B$, because rules apply, due to frustration: of all the energy links, only one per plaquette is frustrated, and it relieves the frustration on adjacent plaquettes. Indeed its entropy per spin at $T = 0$ is $s \simeq 0.3383 k_B$, different from zero, in violation of the third law of thermodynamics, and about half of the entropy of a completely random configuration.

However, as most realistic interactions in Physics are geometric (for instance, the dipolar interaction between magnets is anisotropic) they immediately break the topological structure of frustration in a real system. We will discuss later how renouncing the point (b) opens the way to great freedom of design in artificial spin ices.

Perhaps the first famous occurrence of frustration in the history of Physics pertained to water ice. In the 1930s Giaque and Ashley [73, 74] performed a series of carefully conducted calorimetric experiments, deduced the entropy of water ice at very low temperature, and found that it was not zero. The answer to this mystery

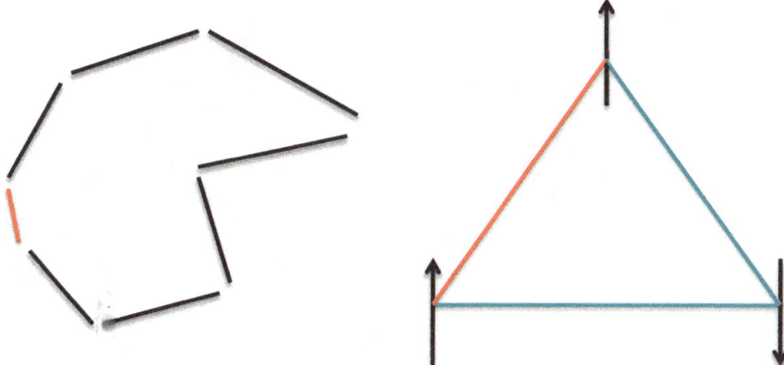

Fig. 4.2 Geometric frustration can be understood schematically as a set of prescriptions that cannot be satisfied simultaneously around certain loops. The red link on the figure on the left represents an "unhappy link" in a generally frustrated system. More specifically, for an Ising antiferromagnet (right) the loop in question is a loop of interactions among nearest neighbors. On a triangular lattice, triangular loops are frustrated, as one of the three links (red) must be unhappy

would be provided by Linus Pauling a few years later [75]. Ice comes in many crystalline forms, but all imply oxygen atoms residing at the center of tetrahedra, sharing four hydrogen atoms with four nearest neighbor oxygen atoms (Fig. 4.3). Two of such hydrogens will be covalently bond, and two will realize an hydrogen bond: two are "in", two are "out" of the tetrahedron. This is the so called ice-rule previously introduced by Bernal and Fowler [76]. Each tetrahedron has 6 admissible configurations out of the $2^4 = 16$ ideally possible, and the collective degeneracy grows exponentially in the number of tetrahedra N as W^N, leading to a non-zero entropy per tetrahedron $s = k_B \ln W$ for this disordered manifold. In what can be considered as one of the most precise and felicitous back-of-the envelope estimate in the history of statistical mechanics, Pauling counted such degeneracy as $W = 3/2$, remarkably close to both the experimental value and to the numerical value ($W = 1.50685 \pm 0.00015$ [77]).

These works pointed to the reality of exotic disordered states in the most common and vital substance on earth. Decades later, they also motivated the introduction by Lieb, Wu, and Rys of simplified models of mathematical physics, known as vertex models, which could in many cases be solved exactly [66–68, 79, 80]. Those are two dimensional models of in-plane spins impinging on vertices, where different energies are associated to different vertex-configurations, and whose statistical mechanics is usually solved via transfer matrix methods.

Ice-like systems have then received renewed interest in the 1990s, when unusual behaviors were discovered in the low temperature regime of rare earth titanates such as $Ho_2Ti_2O_7$ whose magnetic moments exhibit a net ferromagnetic interaction between nearest neighbor spins, yet no ordering at low temperature, suggesting strong frustration. Similar to protons in water ice, the magnetic moments of these materials reside on a lattice of corner-sharing tetrahedra, and they are constrained to point either directly toward or away from the center of a tetrahedron (Fig. 4.3). The resulting

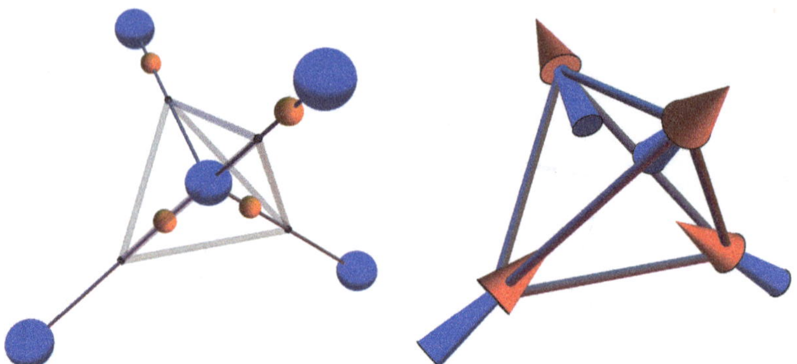

Fig. 4.3 The ice rule. Left: In water ice oxygen atoms sit at the center of tetrahedra, connected to each other by a hydrogen atom. Two of such protons are close (covalently bonded) to the oxygen at the center, two are further away, close to two of the four neighboring oxygens. Right: One might replace this picture with spins pointing in or out depending on whether the proton is close or far away. Then two spins point in, two point out. This corresponds to the disposition of magnetic moments on pyrochlore spin ices, rare earth titanates whose magnetic ensemble does not order at low temperature, because of frustration, and, much like water ice, provides non-zero low temperature entropy density. (Figures from [78])

ferromagnetic interaction favors a 2-in/2out ice-rule. The similarity noted by Harris et al. [81] was confirmed experimentally by Ramirez et al. [82].

Besides providing important model systems with novel field-induced phase transitions and unusual forms of glassiness, and an early and practical example of a classical topological state, spin ices harbor a new fractionalization phenomenon in their low energy dynamics: emergent magnetic monopoles [78, 83]. To facilitate understanding consider the two-dimensional schematics of Fig. 4.4, which represent a disordered ice manifold, an ensemble of spins where all the vertices obey the ice rule. The reader will notice that it is impossible to explore the manifold by single spin flips, without breaking the ice-rule. Only by flipping proper loops of spins we can obtain a new configuration within the ice-manifold. This is already a hint of the topological nature of the state.

If we flip one spin only, we create two defects (3-in/1-out and vice versa). We can separate those defects by further flips, and we have two deconfined magnetic monopoles, and one can prove through multipole expansion that their interaction is Coulomb [78]. Of course these monopoles are in effect simply the opposite ends of a long, floppy dipole, in red in figure, called the Dirac string; however, owing to the disorder of the manifold, the system is no longer reminiscent of the Dirac string connecting the monopoles. Thus, excitations over the ice manifold can be described by a fractionalization of the spins into individual, separable magnetic charges which interact via a Coulomb law.

In view of the more complex geometries that we will discuss later, let us generalize the notion of ice manifold and ice rule for a general lattice, or graph, or network [65], whose edges are spins impinging in vertices of various coordination z. Then we say

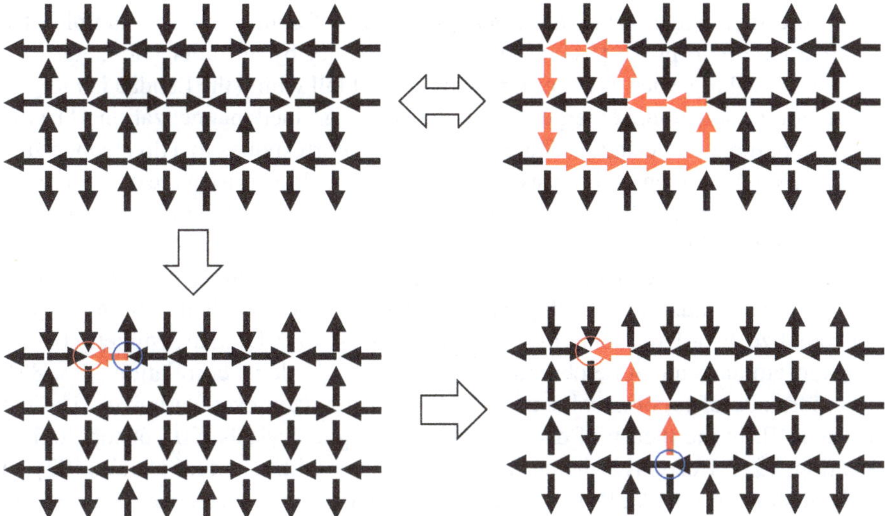

Fig. 4.4 The ice manifold (top left), an ensemble of spins obeying the ice rule (in each vertex two arrows point in, two out). One can obtain another realization of the ensemble only by flipping a proper loop of spins. Flipping a single spin creates a couple of magnetic monopoles of opposite charge (positive is red, negative is blue). The monopoles can be separated by further spin flips (creating a "Dirac string", shown in red), interact via Coulomb interaction, but are topologically protected, as they can only be created and annihilated in pairs

that a vertex of coordination z with n spins pointing toward it has *topological charge* $q = 2n - z$, corresponding to the difference between spins pointing in and out. In general, we call spin ice systems those in which $|q|$ is minimized locally at each vertex (typically, but not necessarily, by nearest neighbor spin-spin interaction). For a lattice of even coordination, such as the square ice or pyrochlore ice introduced before, the ice manifold is characterized by zero charge on each vertex. However, for lattices of odd coordination there cannot be any charge cancellation, and thus in the ice manifold each vertex will have charge $q = \pm 1$ in equal fraction, as the total charge of a system of dipoles must always be zero. That is the case of Kagome ice, which we will discuss in the next section.

These magnetic charges are topologically protected monopoles in square or pyrochlore ice: their magnetic charge is indeed also a topological charge and one can see from Fig. 4.3 that charges can only be created and annihilated in opposite pairs. One could indeed create a single monopole in an open system but that would simply imply pushing the second one at the boundaries. If we, however, placed the ensemble of Fig. 4.3 on a torus, thus without boundaries, then clearly there would not be a net monopole charge in the bulk. Even in an open system, the total net charge will be proportional to the flux of the magnetic moment through the boundaries, and as the latter is bound by the net magnetization of the spins, one finds that the density of net charge must scale with the reciprocal length of the system. These two-dimensional considerations extend to the three-dimensional spin ice.

In more theoretical terms, the topological state of spin ice is an example of a so called Coulomb phase [84], a phase described not by an order parameter or a symmetry breaking, such as the ordered phases that fall within the Landau paradigm, but rather by a solenoidal emergent field (the coarse grained magnetization $\vec{M}$). One can say that the ice rule and its charge cancellation corresponds to a divergence-free condition $\vec{\nabla} \cdot \vec{M} = 0$ on each vertex. Then, a monopole in $\vec{x}_0$ is a source of divergence $\vec{\nabla} \cdot \vec{M} = q_m \delta(\vec{x} - \vec{x}_0)$ of charge $q_m = Qq$ where $Q = M/L$ (the magnetic moment M of the spin divided by its length L), and q is the topological charge of the vertex defined above. This is considered as an example of *classical topological order* [85].

Indeed, while quantum topological order has provided a valuable framework to conceptualize disordered states of spin liquids that escape a Landau symmetry breaking paradigm and cannot be obviously characterized by local correlations [86, 87], the importance of topological states had been recognized even earlier in classical physics [88]: in the theory of dislocations [89], liquid crystals [90], or topological transitions [91]. Recently, whether in direct analogy with quantum physics [92], in purely abstract terms [93, 94], or motivated by real systems such as pyrochlore spin ices [84, 85, 95], a consistent notion of *classical* topological order in *discrete* systems has been proposed, to conceptualize (i) a degenerate, locally disordered manifold (ii) described by a topologically non-trivial, emergent field (iii) whose topological defects (in spin ice, magnetic monopoles [78, 83]) coincide with excitations above the manifold.

Topological protection implies that states *within* the manifold can be linked only via collective changes of entire loops of a discrete degree of freedom. Thus any realistic low-energy dynamics happens necessarily *above the manifold*, through creation, motion, and annihilation of pairs of protected topological excitations. Typically, their constrained and discrete kinetics leads to ergodicity breaking, fractionalization and thus various forms of glassy behaviors [93, 96]. We will see later that such order can also be found in novel, non-trivial geometries of artificial spin ice characterized by vertex-frustration, such as Shakti spin ice.

4.3 Simple Artificial Spin Ices

After introducing the main concepts, we warm up to the field of artificial spin ice by summarizing briefly the early work on classical geometries, based on the square and honeycomb lattices. Further, more general details about fabrication and characterization will be discussed in the context of these early realizations.

4.3.1 Kagome Spin Ice

Even before artificial spin ice realizations [16, 97, 98] (Fig. 4.5) honeycomb structures have been extensively studied theoretically as they describe the two-dimensional

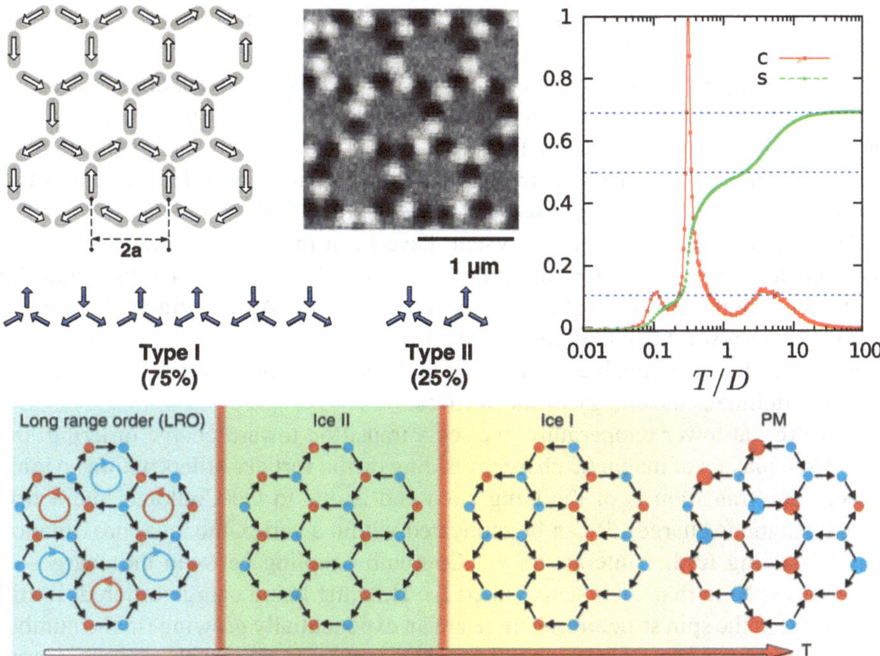

Fig. 4.5 Top, from left to right: Schematics and MFM image of the hexagonal arrays with the 8 vertices of the honeycomb/Kagome artificial spin ice. White arrows show the vertex Ice I state, and the percentages indicate the vertex multiplicity. Type-I vertices have lower energy than Type-II and correspond to the generalized ice rule. Temperature dependence (top right) of the specific heat c and entropy per spin s of the Kagome spin ice obtained by [55]. The dashed lines show values of entropy per spin $s = 0.693$ (Ising paramagnet), 0.501 (Ice I), and 0.108 (charge-ordered spin ice, or Ice II). Bottom: the four phases of Kagome ice ordered by increasing temperature. Figures are adapted from [16, 32, 55]

behavior of the three dimensional spin ice pyrochlores under a magnetic field aligned along a particular crystalline axis. A honeycomb ice is often called the Kagome spin ice, as the spins reside on the edges of a honeycomb lattice, which is a Kagome lattice, the honeycomb dual lattice. In the context of artificial spin ice, Kagome represented the only simple geometry with a degenerate ice manifold. Indeed, as we will see in the next section, square ice has a frustrated yet perfectly ordered, antiferromagnetic ground state.

Before proceeding with Kagome ice, some more general details on artificial spin ice materials are in order, starting with the energetics involved. In general, magnetic, elongated nano-islands can be described as nano-spins, binary degrees of freedom describing their magnetization along their principal axis. This is, however, already an approximation of the magnetic texture of the nano-structure: indeed both direct characterization and micromagnetic simulations show potentially significant relaxation of the magnetization field at the tips of the islands, due to the local field of the surrounding islands.

A further approximation, which seems to work surprisingly well, implies describing the inter-island interaction via a vertex model. There we assign energies to the various vertex configurations as in Fig. 4.5. The nano-islands being magnetic dipoles, one expects this approach to eventually break down. It does indeed in enticing ways, revealing inner, low entropy phases within the ice manifold.

The equilibrium phases of the system have been investigated numerically [99, 100] via Metropolis Monte-Carlo simulations with full dipolar interaction. Figure 4.5 shows that at high temperature the system is paramagnetic. As temperature is reduced, we cross over toward a disordered ice-manifold, called Ice I, where each vertex has charge ± 1. This is as much as a vertex model approximation would explain, as the ice-rule minimizes the energy of the vertices.

However, at lower temperature, we see a transition toward charge ordering: the disordered plasma of magnetic charges residing on the vertices orders within an ionic crystal. The transition is of the Ising class and is due to the Coulomb interaction among magnetic charges. It can be replicated within a vertex-model approximation only by adding further interaction via Coulomb coupling between the charges of the vertices. Note that such state, called Ice II, while being charge ordered, is still disordered in the spin structure, as there are an exponentially growing (in the number of spins) number of possible spin configurations that correspond to the charge ordered state. Finally, further lowering the temperature, another transition leads to an ordered state, where order is brought in by the long range effects of the dipolar interaction.

These states were variously investigated experimentally. Ice I proved easy to reach. Indeed, even non-thermal methods were able to reach it [16, 97, 98]. Those methods pertains to thicker islands that are thus not superparamagnetic at room temperature (that is, do not flip their magnetization under thermal fluctuations). These islands are therefore coercive enough that MFM would provide a non-destructive characterization at room temperature. The AC demagnetization [17] of such samples is sufficient to reach the ice manifold, a fact which already points to its lack of topological protection.

The facility with which such state could be reached is telling. Indeed, while magnetic charges are topologically protected in pyrochlore ices [85], as we saw above, they are not bona fide topological numbers in the ice manifold of Kagome ice. There, vertices of odd coordination can gain and lose charge freely from the surrounding, disordered, and overall neutral plasma of magnetic charges. Consequently, the ice-manifold can be explored *from within* by consecutive single-spin flips, without any need for collective loop-moves such as those shown in Fig. 4.4.

Instead, the charge-ordered state, or Ice II, cannot be explored by individual spin flips. A glimpse of the Ice II phase shown in Fig. 4.5 should convince that any spin flip within the manifold will lead out of the manifold, as it will locally destroy the charge order.

Signatures of the Ice II state were first suspected after AC demagnetization [11]. They were subsequently investigated via thermal methods capable of providing a bona fide thermal spin ensemble [101, 102]. These are of three kinds: annealing at

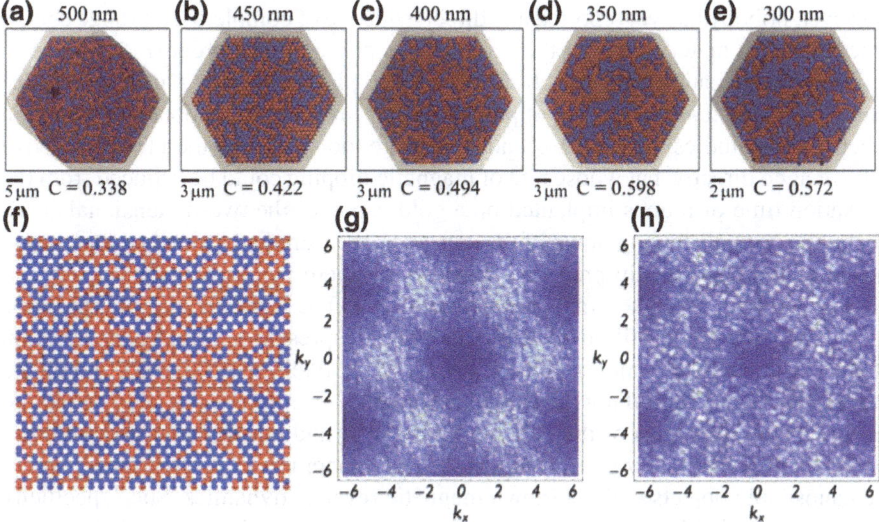

Fig. 4.6 Magnetic charge ordering in Kagome ice. **a–e** Charge domain maps obtained via Lorenz TEM relative to the annealing of Fe-Pd alloy artificial Kagome ices of different edge length (from 500 to 300 nm) show increasing size of the ionic crystallites of charges as the lattice constant decreases, and thus the mutual interaction among magnetic charges increases. C is the charge-charge correlation parameter ($C = 1$ for a fully charge ordered state). Images adapted from [101]. **f** Charge map obtained via MFM after annealing of permalloy Kagome ice of lattice constant 260 nm showing incipient domains of charge-ordering and **g** its static structure factor showing incipient peaks corresponding to crystalline order. **h** Static structure factor for lattice constant 490 nm, showing no incipient peaks

higher than room temperature followed by characterization at room temperature [52, 101, 102]; thermalization with real-time, real-space characterization [51, 53, 57]; and thermalization without real-space characterization [32, 49]. In the first, the material is not superparamagnetic at room temperature, but it is heated slightly above the Curie temperature of the nano-islands (which can vary, depending on the size and chemical composition of the nano-structure, from about 600 °C for permalloy down to about 100 °C for Fe-Pd alloys) and then annealed down into a frozen state, usually characterized via MFM. In the second method the nano-islands are chosen to be thin enough (usually thickness of 2–3 nm) to be superparamagnetic at room temperature or below, and thus need to be characterized via PEEM, at a proper beam source. In the third, various averaged quantities are extracted, such as the average flip rate of spins, through muon spectroscopy [32]—while spin noise spectroscopy [103, 104] should also be a viable method.

Figure 4.6 shows the results of thermal annealing on artificial hexagonal ice made of permalloy [102], which demonstrate formation of crystallites of magnetic charges, due to the Coulomb interaction between the charges themselves. Control over the size of those ionic crystallites has also been obtained by employing an alloy of iron and palladium, rather than permalloy, which has a much lower Curie temperature [101].

However, nobody has yet reported any direct evidence of complete long range charge order in such a material, nor of the zero entropy phase of spin order (Fig. 4.5).

Indirect indications that such low entropy phases within the Kagome ice manifold—or at least some kind of phases—can be reached were obtained via muon spectroscopy studies, involving islands that were too small and therefore too active to be imaged directly, but whose rate of magnetic flipping could be deduced from the relaxation time of muons implanted on a gold cap over the two-dimensional array. There, the critical slowing down of the spins was measured and found to correspond to that of the numerically predicted transitions, where parameters for the numerical simulations were taken from the material [32]. These results, the first to probe deep inside the ice manifold of Kagome lattice, represent a strong corroboration of the existence of a complex phase diagram, most likely the theoretical predicted one. However, we should not forget that topological or ordered states can be hard to reach via spin dynamics of the Glauber kind [105], and that indeed the actual spin dynamics might be rather more complex than a Glauber model. Indeed these "spins" are nanoscopic objects with their own magnetic reversal dynamics. Such specificity might bias certain kinetic pathways, leading to non-equilibration or ergodicity breaking even in Ising models that are not susceptible to these phenomena: thus the phases whose critical slowing down was experimentally revealed could be only reminiscent of the one predicted at equilibrium, which of course adds to their potential interest.

4.3.2 Square Ice

With the exception of the work of Tanaka et al. [97], early works concentrated on the square geometry (Fig. 4.1) [2, 15, 16]. Square ice also represented the benchmark on which to test demagnetization and annealing methods which lead to experimental protocols for thermal ensembles [20, 49, 52].

It is important to understand that square artificial spin ice does not resemble the square ice of Lieb [66], or the degenerate square ice described in Fig. 4.4, firstly because it admits topological defects in the form of magnetic monopoles absent in the six-vertex model, but most importantly because it is not degenerate. In this sense it shares similarities with the Rys F-model [80] but those should not be overstated, as the (physically unnatural) absence of monopoles in the latter leads to an infinitely continuous transition to antiferromagnetic ordering [67], whereas in the former the transition is of second order.

Figure 4.7 shows the energetic hierarchy of vertices with 90° angles (including those of coordination $z = 3, 2$ to be discussed later). Because of the anisotropy of the dipolar interaction, nearest neighbor perpendicular islands interact more strongly than collinear ones, leading to lifting of degeneracy within the ice manifold. The system, if modeled at the vertex level, can be described as a J_1, J_2 antiferromagnetic Ising model on a square lattice, with a transition to antiferromagnetic ordering, which indeed has been obtained experimentally via thermal annealing, as shown in Fig. 4.7.

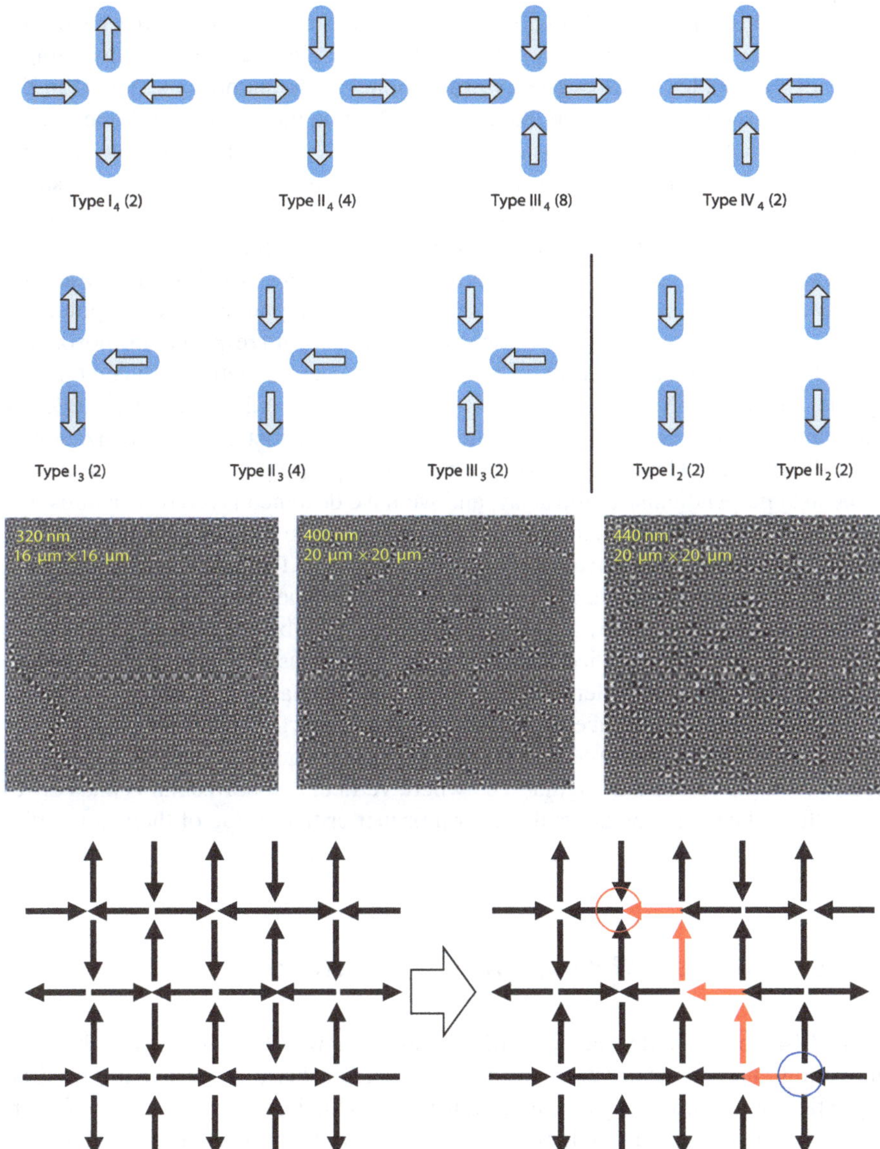

Fig. 4.7 Top: Vertex-configurations for 90° angles of coordination $z = 4, 3, 2$ (degeneracy in brackets) listed in order of increasing energy. Middle: MFM images of thermally annealed square ice at different lattice constants showing an ordered domain crossed by a Dirac string (for the specimen at 320 nm) and a multi-domain ensemble separated by domain walls of monopoles and diract strings (at 400 and 440 nm); note also the frozen in monopole pairs (figures adapted from [102]). Bottom: the lowest energy state of square ice as an antiferromagnetic tiling of Type-I$_4$ vertices; creating and separating a monopole pair entails a Dirac string (red) of Type-II$_4$ vertices, that are energetically more costly than the Type-I$_4$, leading to the linear confinement of the pair

Within the ordered state of square ice potentially interesting transitions have been proposed [106–108]: when the system is not degenerate, creating and separating a couple of monopoles requires energy proportional to the number of Type-II in the Dirac string (see Fig. 4.7). Much like quarks or Nambu monopoles [109] these pairs are linearly confined, and the tensile strength of their Dirac string drives the ordering as the temperature is reduced. However, one can imagine that a topological transition corresponding to monopole deconfinement might take place under proper conditions when the energy of the Dirac string is offset by its fluctuating entropy.

Square ice, however, can be made properly degenerate, which means described by a spin ensemble such as the one of Fig. 4.4, revealing an emergent topological Coulomb phase. One way is to raise the vertical islands with respect to the horizontal ones [99], a method that has recently been pursued experimentally [110] demonstrating a degenerate manifold whose static structure factor coincides with the numerically computed one for a six-vertex model, thus providing the first artificial realization of a two dimensional Coulomb phase. We have also proposed to iterate such design on the axis perpendicular to the array, and we have designed layered structures that are geometrically different but topologically equivalent to three dimensional spin ice pyrochlores [111]. Those have not found realization yet. The only three dimensional realization of artificial spin ice was obtained by filling the voids of an artificial opal film with Cobalt [112, 113], a promising approach to bring to room temperature some of the features of spin ice pyrochlores. Of course, as always with three dimensional realizations, the challenge there lies not only in nano-fabrication, but also in characterization, as real-space methods are generally surface methods.

Finally, another way to produce a Coulomb phase in square ice has been presented recently, and involves "rectangular ice" where vertical and horizontal islands differ in length, and degeneracy is obtained for a proper critical value of their ratio [107, 114].

4.4 Exotic States Through Vertex-Frustration

Until 2014, the only degenerate artificial spin ice was Kagome. As both nanofabrication and characterization protocols evolved, it became clear that the initial inspiration of the entire project—to design exotic behaviors in the geometry of interacting, binary degrees of freedom—could become viable, if not for one problem: in real systems, the frustration of the pairwise interaction is wedded to the geometry. What this means is explained in Fig. 4.8 where brickwork spin ice and Kagome spin ice are shown to lead to completely different ground states, one disordered, the other ordered, despite the two geometries being topologically equivalent. Indeed, the dipolar interaction is not topologically invariant, but instead depends very much on the mutual arrangements of the dipoles.

To overcome this limitation and gain freedom in the design of new materials capable of various states and unusual behaviors, the first step is to decouple frustration

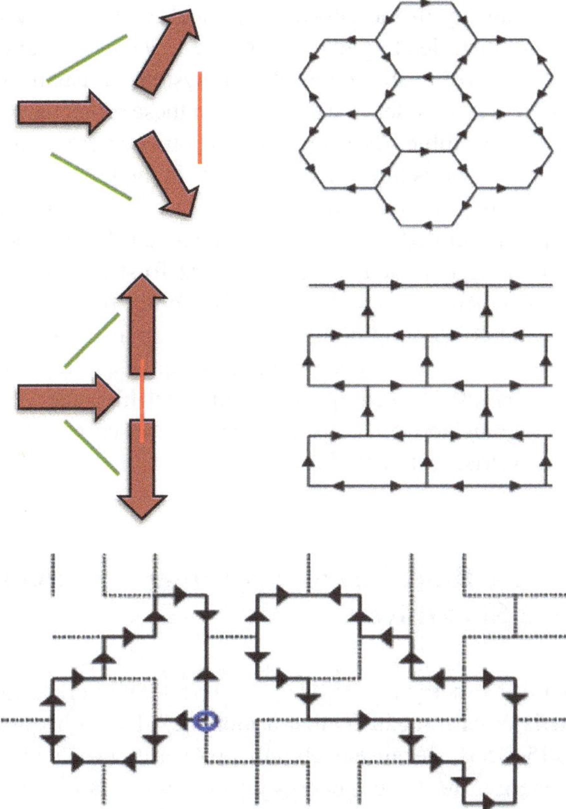

Fig. 4.8 Geometry versus topology. Topologically equivalent geometries leads to completely different spin ensembles, due to the anisotropy of the dipolar interaction. The honeycomb spin ice (top right) is topologically equivalent to the ladder spin ice (middle, right) yet the nearest neighbor interactions lead to an ordered ground state in the latter (see also Fig. 4.7 for the energetic hierarchy of the vertices) and a disordered manifold in the former. Pairwise interactions are frustrated in both systems, however in the honeycomb lattice all the spins interacting in the vertex have the same mutual angle (top left) and thus any of the three interactions can be frustrated, whereas it is energetically favorable to frustrate the interaction between parallel spins in the ladder lattice (middle left). At the bottom is an example of vertex-frustration, where the allocation of vertices of lowest energy is frustrated, leading to "unhappy vertices" (blue circles) on certain loops, instead of unhappy energy links (red lines above)

from geometry. As the pairwise interaction is anisotropic, something else will have to be frustrated. A possibility is the vertex itself.

Consider a geometry made of 90° vertices of coordination $z = 4, 3, 2$ (Fig. 4.7). Each vertex has a unique configuration of minimal energy (up to a flip of all the spins). Imagine now arranging them in such a way that, however, not all vertices can be assigned to the lowest energy configuration [54]. This will lead to "unhappy vertices" (UV), that is, topologically protected local excitations (Fig. 4.8). In proper geometries, the degeneracy of the allocation of such vertices grows exponentially with the size of the system, leading to a degenerate low energy manifold [54, 58].

Crucial here is that within this manifold the system is usually captured by an emergent description that considers the allocation of these protected local excitations, rather than by its spin ensemble. As a consequence, other emergent properties appear that are in general not obvious nor indeed apparent in the local spin structure.

While vertex models [79] were introduced to describe frustrated systems, they were themselves not frustrated. They simply subsumed the degeneracy of a frustrated system within a degenerate energetics. Vertex-frustrated geometries can thus be considered the first frustrated vertex models. Vertex-frustration is of course a nearest-neighbor level concept, although it induces topological states that are collective. However, the real materials being made of dipoles, other phases are present within their vertex-frustrated low energy manifold, much like inner phases are present in the diagram of Kagome above. Let us now see how this comes about in three such geometries: Shakti, Tetris, and Santa Fe.

4.5 Emergent Ice Rule, Charge Screening, and Topological Protection: Shakti Ice

Consider the Shakti geometry in Fig. 4.9 [55]. Each minimal, rectangular loop of Shakti is frustrated. What it means is that it must be affected by an odd number of unhappy vertices [54, 55]. Because each unhappy vertex aways affects two nearby loops and costs energy, the lowest energy configuration is realized when nearby loops are dimerized by a single unhappy vertex [54]. If one considers the geometry, one finds (Fig. 4.9c) that each plaquette made by two rectangular loops will host two unhappy vertices in 4 possible locations, much like the ice rule in water ice prescribes that 2 hydrogenatoms are within the tetrahedron containing each oxygen atom (Fig. 4.3), in 2 of the 4 possible allocations. In both cases the same ice-rule applies, but here in emergent form: not in terms of the original spins, but in terms of allocation of unhappy vertices. Thus, the lowest energy manifold, at the nearest neighbor vertex description employed here, corresponds then to an emergent six-vertex model. This has been shown experimentally [56].

Nonetheless, as we had cautioned before, this nearest neighbor description defines the ice-manifold, within which intervene other non-trivial phenomena, due to the long range nature of the interaction. A particularly interesting one regards the screening of magnetic charge. Shakti has multiple coordination, therefore while in its low energy state all the vertices of coordination $z = 4$ are in the ice rule, they are surrounded by vertices of coordination $z = 3$ which always have a magnetic charge ± 1 (in natural units, previously defined), and are disordered. When a vertex of coordination $z = 4$ hosts a magnetic monopole, the overall neutral plasma of charge around it rearranges to screen it, as shown in Fig. 4.10 [56].

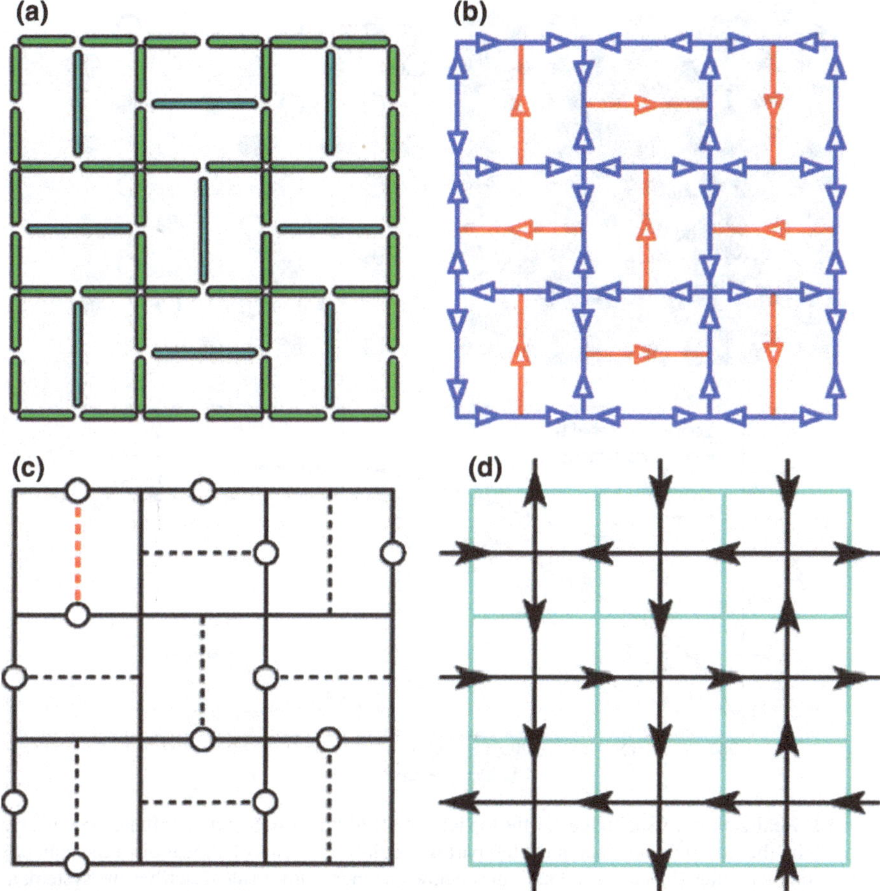

Fig. 4.9 Theory of Shakti spin ice. The structure of the system (**a**) is such that its lowest energy spin ensemble **b** is disordered. A look at the spin structure in **b** does not seem particularly insightful. However, if we translate that spin map unto the allocation of locally excited vertices, denoted by circles in **c** we then see that each plaquette will host two and only two unhappy vertices in four possible positions. This is equivalent to a six vertex model (**d**) where pseudo-spins are assigned to each plaquettes and point toward (away from) the unhappy vertices in plaquette of vertical (horizontal) long island. Figures adapted from [55]

It is important to understand that magnetic monopoles are *not* proper topological charges for Shakti, as they are not protected. Each $z = 4$ vertex being surrounded by a sea of charges, it can gain or lose charge to and from it. However, the Shakti state is a bona fide topological phase, which means that some other topological charge should be identified in it.

That the manifold has topological protection can be immediately suspected by noting that a single spin flip takes out of the manifold, and only a proper loop of collective spin flips realizes change *within* the manifold. This can be understood

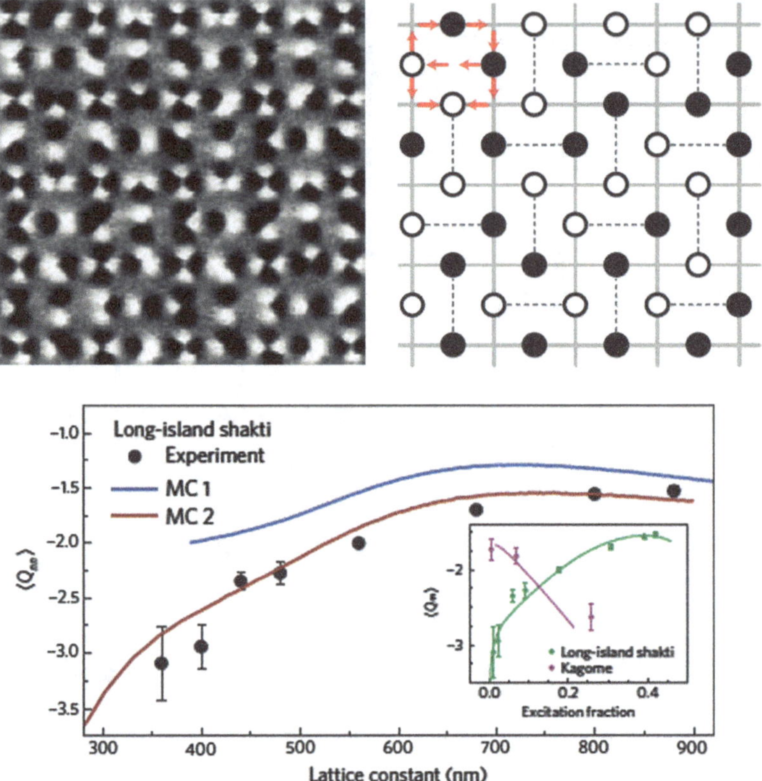

Fig. 4.10 Realizations of Shakti Ice. On the top left, an MFM image of Shakti ice after annealing. On the top right, the experimental data of the left part is translated in terms of allocation of the unhappy vertices, on plaquettes (black dots). One can see how an emergent ice rule describes the system, as each plaquette can have only two of four slots for unhappy vertices occupied. Bottom: screening of monopoles from magnetic charges $\langle Q_{nn} \rangle$ denotes the average magnetic charge surrounding a magnetic monopole on a $z = 4$ vertex, at the nearest neighbor level. Figures adapted from [56]

easily from Fig. 4.9, as all spins impinging in a $z = 3$ vertex also impinge into $z = 4, 2$ vertices, which are in their lowest energy in the manifold. Flipping such spin will thus necessarily cause excitations.

To identify the topological structure, we go back to the properties of the low energy state. We saw that because each UV affects two nearby plaquettes (Fig. 4.9) and costs energy, the lowest energy configuration is realized when nearby plaquettes are "dimerized" by a single UV [54]. The ice manifold of Shakti is thus described by a dimer cover model on the lattice connecting the rectangular plaquettes, which is topologically equivalent to a square lattice (Fig. 4.11) (from now on called "dimer lattice"), and which can be solved exactly [115].

The following is then standard: a discrete, emergent vector field $\vec{E}$ can be introduced, perpendicular to each edge, of length 1 (o 3) if the edge is unoccupied (or

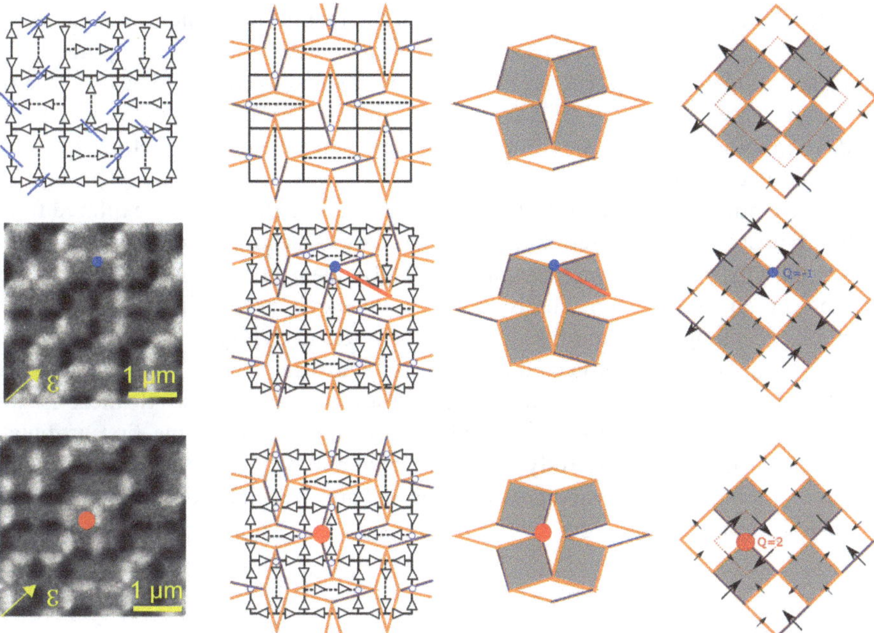

Fig. 4.11 Top: Shakti manifold as a dimer cover model. From left to right: Disordered spin ensemble for the ground state of Shakti ice manifold. The manifold is completely described by the allocation of the UVs (circles) which affect two nearby rectangular plaquettes (connected by the blue segments). Thus, an unhappy vertex is a dimer (blue segments) connecting frustrated plaquettes, and the ground state is a complete dimer-cover model on the (Ochre color) lattice with nodes in the center of rectangular plaquettes, topologically equivalent to a square lattice. There we introduce the emergent vector field $\vec{E}$, as in the text. The circulation of the vector field along any closed loop is zero. Middle and Bottom: The Shakti's low-energy manifold. XMCD image of Shakti spin ice, for a spin ensemble with one excitation (red and blue dots) and the corresponding emergent dimer cover representation. Now excitations appear as multiple occupancy and/or diagonal dimers (Type II_2s). $\vec{E}$ is no longer irrotational and its circulation defines the topological charge as $q = \frac{1}{4}\oint_\gamma \vec{E} \cdot d\vec{l}$. (Image copyright: Yuyang Lao.)

occupied) by a dimer, and direction entering (exiting) a gray square of Fig. 4.11 from top or bottom, and exiting (entering) it from the sides. The "line integral" $\int_\gamma \vec{E} \cdot d\vec{l}$ for such a discrete vector field along a directed line γ crossing the edges is the sum of the vectors along the line with sign taken along the line's direction. For a complete cover the emergent field is irrotational ($\oint_\gamma \vec{E} \cdot d\vec{l} = 0$) leading to the definition of a "height function" [93] h such that $\vec{E} = \vec{\nabla} h$ and thus demonstrating the topological state.

Beyond the standard dimer model, this picture can incorporate the low-energy excitations of Shakti ice as scramblings of the cover. As Fig. 4.11 shows, above the ground state a frustrated plaquette (i.e. a node of the dimer lattice) can be dimerized three times instead of one (over-dimerization) by UVs, or also diagonally by a

Type-II_4 or a Type-II_2 vertex. In the presence of such scramblings the emergent vector field $\vec{E}$ is not irrotational anymore. Indeed its circulation around any topologically equivalent loop encircling a scrambling defines the quantized topological charge of the defect as $q = \frac{1}{4}\oint_\gamma \vec{E} \cdot d\vec{l}$ (Fig. 4.11). Thus, the excitations of the Shakti ice manifold are topological charges, turning the discrete scalar field h that defines its order into a multivalued phase.

We have now the full picture: a topological phase, which cannot be explored from within, but only via a discrete kinetics of excitations whose topological charge is conserved. This picture is emergent, and not at all evident from, or indeed reminiscent of, the original spin structure. It also has consequences for the kinetics, in terms of ergodicity breaking, non-equilibration, and glassiness, as it is typical of a topological state with topologically protected excitations, that cannot be reabsorbed into the manifold individually, and which evolves via a discrete kinetics. All these issues are still to be investigated in full, as Shakti ice might provide the first artificial, controllable, modifiable and fully characterizable magnetic system which provides non-topographic vistas of ergodicity breaking and non-equilibration as consequences of a classical topological order.

4.6 Dimensionality Reduction: Tetris Ice

While Shakti spin ice provides a topologically protected low-energy manifold, no such protection is present in the ground state of Tetris ice (Fig. 4.12), which can be explored by consecutive spin flips. As Fig. 4.12 shows, the lattice can be decomposed into T-shaped "tetris" pieces and it has a principal axis of symmetry.

The geometry can be considered as layered one-dimensional systems. On the blue islands in Fig. 4.12 there cannot be any Type-II_3 unhappy vertex [54], and therefore the blue portion of the lattice, which we call backbone, must be ordered at the lowest energy. The unhappy vertices must reside on the red portions, which we call staircase, and which therefore remain disordered at low temperature. As temperature is lowered, we have thus a dimensional reduction of an alternating ordered-disordered one-dimensional system, which was indeed confirmed experimentally [57].

This dimensional reduction is also apparent in the kinetics. Tetris was the first of the new geometries to be characterized in real-time, real-space and from the supplementary information of [57] it is possible to watch clips of its kinetics as the temperature is lowered or raised. Starting at high temperature, all the spins flip at about the same rate. As the temperature is lowered, ordered domains begin to form in correspondence with the backbones, where eventually the spins become static, while the spins on the staircases continue to fluctuate.

While Tetris spin ice's lowest energy state described above has been confirmed experimentally, it follows from a nearest neighbor approximation. The profile of low energy excitations, however, has not been yet studied in any systematic way, and promises interesting new effects. For instance, as one-dimensional systems, the backbones can never order completely, and will always host excitations above the

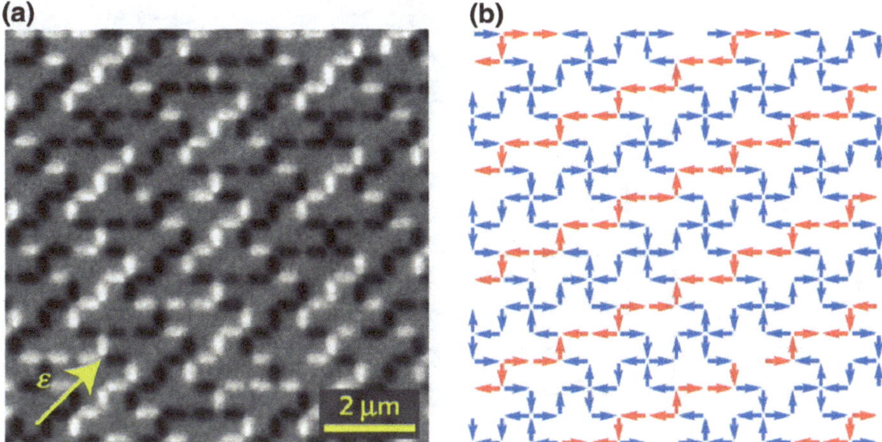

Fig. 4.12 Tetris Ice. **a** XMCD-PEEM image of a 600 nm Tetris lattice. The black/white contrast indicates whether the magnetization of an island has a component parallel or antiparallel to the polarization of the incident X-ray, which is indicated by the yellow arrow. **b** Map of the moment configurations showing ordered backbones (blue) and disordered staircases (red). (Images from [57].)

low-energy manifold. Of course Tetris is in fact a two-dimensional system, which decomposes into one-dimensional ones only in the lowest energy configuration. Slightly above such a manifold, one expects correlations among excitations that belong to different backbones. Such correlations must be controlled both by the magnetic interaction between these defects—as Tetris is, after all, a system of dipoles that can interact at long-range—but also through entropic interactions, since the backbones are separated by disordered staircases of non-zero density of entropy. None of the above issues has yet been studied theoretically, and they might indeed provide a useful setting to explore the onset of phase decoupling into lower-dimensional states, a broader problem relevant to liquid crystal phases [116] or weakly coupled sliding phases [117, 118].

4.7 Polymers of Topologically Protected Excitations: Santa Fe Ice

We end this vista on how novel and unusual spin ice geometries influence topology with Santa Fe of Fig. 4.13, which was inspired by a terra cotta floor in the homonymous New Mexican capital—incidentally, the oldest in the United States. While Shakti and Tetris are maximally frustrated, which means that any minimal loop inside the geometry needs to be affected by an unhappy vertex, in Santa Fe only the dashed loops in the figure are frustrated and they are surrounded by unfrustrated

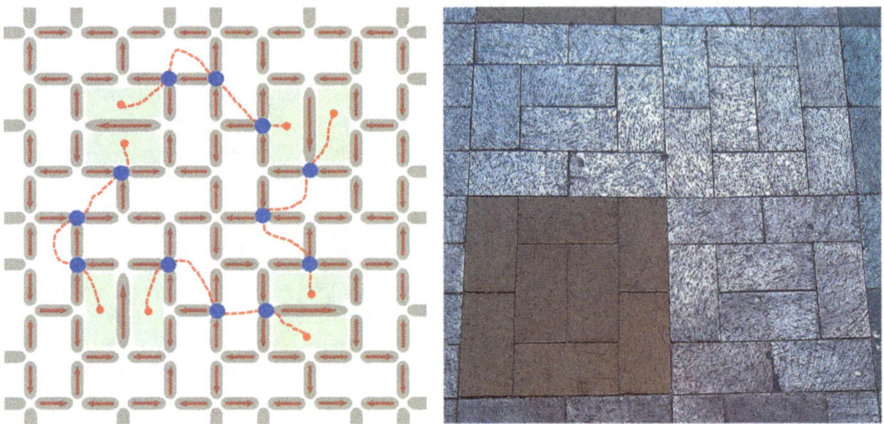

Fig. 4.13 The Santa Fe Ice can support both frustrated (shaded, green) and unfrustrated loops. There, "polymers" of unhappy vertices (blue dots) thread through unfrustrated loops to connect frustrated ones. On the right, the brick floor at the convention center in Santa Fe, New Mexico, USA

ones. It is an inviolable topological constraint that at any energy frustrated loops can be affected by only an odd number of excitations, and unfrustrated ones by only an even number (or none).

An unhappy vertex on a frustrated loop of Santa Fe lattice affects a nearby unfrustrated one. However, an unfrustrated loop can only be affected by an even number of defects, and thus there will be a second unhappy vertex on it, affecting in turn a nearby unfrustrated loop, et cetera. It follows that magnetic "polymers", whose "monomers" are local protected excitations, must begin from and end into frustrated loops.

As each monomer costs energy, the lowest energy configuration of the magnetic ensemble will correspond to the shortest possible polymers, which are made of three monomers, each connecting nearby frustrated loops as in Fig. 4.13. The entropy of such a state can easily be computed exactly. As each polymer dimerizes two frustrated loops, the degeneracy is given by the dimer cover model on the square lattice whose node is made of nearby frustrated loops, times the number of ways in which polymers can be chosen once their pinned ends are fixed. Thus the ice manifold decomposes into the direct product of two states: the dimer-cover manifold, which selects which loops are joined by the polymers, and the degeneracy of the polymers themselves.

At low temperature the kinetics will reduce to the fluctuations of the magnetic polymers without changing the pinning location of their ends, and thus without changing the dimer cover picture. Thus, the low energy manifold can be explored from within, but only in part. The kinetics within the manifold remains local and the polymer's fluctuations are uncorrelated. As the temperature rises the polymers lengthen to include more than three monomers. At that point they can bump into each other, fuse in a cross, and then separate in different ways. This transition can lead to a different dimerization, as the new polymers, emerging from "collisions" of the old ones, are now pinned to different ending point. Thus the dimer-cover ensemble is explored via this mechanism of polymer colliding, fusing together and then breaking again into different ones. This of course involves excitations over the ice manifold, further demonstrating the partial topological protection that pertains only to the dimer-cover sector of the low-energy manifold.

4.8 Conclusions

We have argued that by assembling together interacting, elementary building blocks, here Ising spins in the form of single domain, magnetic nano-islands, we can invert a tendency that has dominated condensed matter physics for half a century. Instead of finding serendipitously exotic states and behavior in nature, and then model them via higher level, emergent hamiltonians, one can devise materials—magnetic materials in this case–that do not exist in nature by developing first the model of their collective dynamics. Then, advances in nano-lithography and chemical synthesis can allow for their realization, while new thermal protocols afford characterization, often in real-time and real-space, for unprecedented direct validation of the theoretical expectations.

We have shown that while the field began with simple geometries, reminiscent of natural materials, current advances make it possible to realize dedicated geometries of completely different properties and behaviors. These allow now to access rather sophisticated topological states, as their emergent description loses reminiscence of the original degrees of freedom, the underlying spin structure. This approach opens a new path in the material-by-design effort, on which unusual topological states can be deliberately designed.

References

1. P.W. Anderson et al., More is different. Science **177**(4047), 393–396 (1972)
2. R.F. Wang, C. Nisoli, R.S. Freitas, J. Li, W. McConville, B.J. Cooley, M.S. Lund, N. Samarth, C. Leighton, V.H. Crespi, P. Schiffer, Artificial 'spin ice' in a geometrically frustrated lattice of nanoscale ferromagnetic islands. Nature **439**(7074), 303–6 (2006)
3. C. Nisoli, R. Moessner, P. Schiffer, Colloquium: artificial spin ice: designing and imaging magnetic frustration. Rev. Mod. Phys. **85**(4), 1473 (2013)
4. E. Mengotti, L.J. Heyderman, A.F. Rodríguez, F. Nolting, R.V. Hügli, H.-B. Braun, Real-space observation of emergent magnetic monopoles and associated Dirac strings in artificial kagome spin ice. Nat. Phys. **7**(1), 68–74 (2010)
5. S. Ladak, D.E. Read, G.K. Perkins, L.F. Cohen, W.R. Branford, Direct observation of magnetic monopole defects in an artificial spin-ice system. Nat. Phys. **6**, 359–363 (2010)
6. S. Ladak, D. Read, T. Tyliszczak, W.R. Branford, L.F. Cohen, Monopole defects and magnetic coulomb blockade. New J. Phys. **13**(2), 023023 (2011)
7. K. Zeissler, S.K. Walton, S. Ladak, D.E. Read, T. Tyliszczak, L.F. Cohen, W.R. Branford, The non-random walk of chiral magnetic charge carriers in artificial spin ice. Sci. Rep. **3**, 1252 (2013)

8. C. Phatak, A.K. Petford-Long, O. Heinonen, M. Tanase, M. De Graef, Nanoscale structure of the magnetic induction at monopole defects in artificial spin-ice lattices. Phys. Rev. B **83**(17), 174431 (2011)
9. S. Ladak, D.E. Read, W.R. Branford, L.F. Cohen, Direct observation and control of magnetic monopole defects in an artificial spin-ice material. New J. Phys. **13**(6), 063032 (2011)
10. S.D. Pollard, V. Volkov, Y. Zhu, Propagation of magnetic charge monopoles and dirac flux strings in an artificial spin-ice lattice. Phys. Rev. B **85**(18), 180402 (2012)
11. N. Rougemaille, F. Montaigne, B. Canals, A. Duluard, D. Lacour, M. Hehn, R. Belkhou, O. Fruchart, S. El Moussaoui , A. Bendounan et al., Artificial kagome arrays of nanomagnets: a frozen dipolar spin ice. Phys. Rev. Lett. **106**(5), 057209 (2011)
12. B. Canals, I.-A. Chioar, V.-D. Nguyen, M. Hehn, D. Lacour, F. Montaigne, A. Locatelli, T.O. Menteş, B. Santos Burgos, N. Rougemaille, Fragmentation of magnetism in artificial kagome dipolar spin ice. Nat. Commun. **7** (2016)
13. P.E. Lammert, X. Ke, J. Li, C. Nisoli, D.M. Garand, V.H. Crespi, P. Schiffer, Direct entropy determination and application to artificial spin ice. Nat. Phys. **6**(10), 786–789 (2010)
14. Y.-L. Wang, Z.-L. Xiao, A. Snezhko, J. Xu, L.E. Ocola, R. Divan, J.E. Pearson, G.W. Crabtree, W.-K. Kwok, Rewritable artificial magnetic charge ice. Science **352**(6288), 962–966 (2016)
15. C. Nisoli, R. Wang, J. Li, W. McConville, P. Lammert, P. Schiffer, V. Crespi, Ground state lost but degeneracy found: the effective thermodynamics of artificial spin ice. Phys. Rev. Lett. **98**(21), 217203 (2007)
16. C. Nisoli, J. Li, X. Ke, D. Garand, P. Schiffer, V.H. Crespi, Effective temperature in an interacting vertex system: theory and experiment on artificial spin ice. Phys. Rev. Lett. **105**(4), 047205 (2010)
17. X. Ke, J. Li, C. Nisoli, P.E. Lammert, W. McConville, R. Wang, V.H. Crespi, P. Schiffer, Energy minimization and ac demagnetization in a nanomagnet array. Phys. Rev. Lett. **101**(3), 037205 (2008)
18. L.F. Cugliandolo, Artificial spin-ice and vertex models. J. Stat. Phys. 1–16 (2017)
19. D. Levis, L.F. Cugliandolo, L. Foini, M. Tarzia, Thermal phase transitions in artificial spin ice. Phys. Rev. Lett. **110**(20), 207206 (2013)
20. J.P. Morgan, A. Stein, S. Langridge, C.H. Marrows, Thermal ground-state ordering and elementary excitations in artificial magnetic square ice. Nat. Phys. **7**(1), 75–79 (2010)
21. Z. Budrikis, J.P. Morgan, J. Akerman, A. Stein, P. Politi, S. Langridge, C.H. Marrows, R.L. Stamps, Disorder strength and field-driven ground state domain formation in artificial spin ice: experiment, simulation, and theory. Phys. Rev. Lett. **109**(3), 037203 (2012)
22. Z. Budrikis, P. Politi, R.L. Stamps, Diversity enabling equilibration: disorder and the ground state in artificial spin ice. Phys. Rev. Lett. **107**(21), 217204 (2011)
23. P.E. Lammert, V.H. Crespi, C. Nisoli, Gibbsianizing nonequilibrium dynamics of artificial spin ice and other spin systems. New J. Phys. **14**(4), 045009 (2012)
24. C. Nisoli, On thermalization of magnetic nano-arrays at fabrication. New J. Phys. **14**(3), 035017 (2012)
25. R.V. Hügli, G. Duff, B. O'Conchuir, E. Mengotti, L.J. Heyderman, A.F. Rodríguez, F. Nolting, H.B. Braun, Emergent magnetic monopoles, disorder, and avalanches in artificial kagome spin ice. J. Appl. Phys. **111**(7), 07E103 (2012)
26. P. Mellado, O. Petrova, Y. Shen, O. Tchernyshyov, Dynamics of magnetic charges in artificial spin ice. Phys. Rev. Lett. **105**(18), 187206 (2010)
27. S. Zhang, J. Li, I. Gilbert, J. Bartell, M.J. Erickson, Y. Pan, P.E. Lammert, C. Nisoli, K.K. Kohli, R. Misra et al., Perpendicular magnetization and generic realization of the ising model in artificial spin ice. Phys. Rev. Lett. **109**(8), 087201 (2012)
28. U.B. Arnalds, J. Chico, H. Stopfel, V. Kapaklis, O. Bärenbold, M.A. Verschuuren, U. Wolff, V. Neu, A. Bergman, B. Hjörvarsson, A new look on the two-dimensional ising model: thermal artificial spins. New J. Phys. **18**(2), 023008 (2016)
29. C. Nisoli, Nano-ising. New J. Phys. **18**(2), 021007 (2016)
30. W.R. Branford, S. Ladak, D.E. Read, K. Zeissler, L.F. Cohen, Emerging chirality in artificial spin ice. Science **335**(6076), 1597–1600 (2012)

31. B.L. Le, J. Park, J. Sklenar, G.-W. Chern, C. Nisoli, J.D. Watts, M. Manno, D.W. Rench, N. Samarth, C. Leighton, P. Schiffer, Understanding magnetotransport signatures in networks of connected permalloy nanowires. Phys. Rev. B **95**, 060405 (2017)
32. L. Anghinolfi, H. Luetkens, J. Perron, M.G. Flokstra, O. Sendetskyi, A. Suter, T. Prokscha, P.M. Derlet, S.L. Lee, L.J. Heyderman. Thermodynamic phase transitions in a frustrated magnetic metamaterial. Nat. Commun. **6** (2015)
33. J. Drisko, T. Marsh, J. Cumings, Topological frustration of artificial spin ice. Nat. Commun. **8** (2017)
34. S. Gliga, A. Kákay, R. Hertel, O.G. Heinonen, Spectral analysis of topological defects in an artificial spin-ice lattice. Phys. Rev. Lett. **110**(11), 117205 (2013)
35. I. Gilbert, G.-W. Chern, B. Fore, Y. Lao, S. Zhang, C. Nisoli, P. Schiffer, Direct visualization of memory effects in artificial spin ice. Phys. Rev. B **92**(10), 104417 (2015)
36. A. Libál, C. Reichhardt, C.J. Olson Reichhardt, Hysteresis and return-point memory in colloidal artificial spin ice systems. Phys. Rev. E **86**(2), 021406 (2012)
37. A. Libál, C. Reichhardt, C.J. Olson Reichhardt, Realizing colloidal artificial ice on arrays of optical traps. Phys. Rev. Lett. **97**(22), 228302 (2006)
38. A. Libál, C.J. Olson Reichhardt, C. Reichhardt, Creating artificial ice states using vortices in nanostructured superconductors. Phys. Rev. Lett. **102**(23), 237004 (2009)
39. A. Libal, C. Nisoli, C. Reichhardt, C.J. Reichhardt, Dynamic control of topological defects in artificial colloidal ice (2016), arXiv:1609.02129
40. C.J. Olson Reichhardt, A. Libal, C. Reichhardt, Multi-step ordering in kagome and square artificial spin ice. New J. Phys. **14**(2), 025006 (2012)
41. D. Ray, C.J. Olson Reichhardt, B. Jankó, C Reichhardt, Strongly enhanced pinning of magnetic vortices in type-ii superconductors by conformal crystal arrays. Phys. Rev. Lett. **110**(26), 267001 (2013)
42. C. Nisoli, Dumping topological charges on neighbors: ice manifolds for colloids and vortices. New J. Phys. **16**(11), 113049 (2014)
43. A. Ortiz-Ambriz, P. Tierno, Engineering of frustration in colloidal artificial ices realized on microfeatured grooved lattices. Nat. Commun. **7** (2016)
44. P. Tierno, Geometric frustration of colloidal dimers on a honeycomb magnetic lattice. Phys. Rev. Lett. **116**(3), 038303 (2016)
45. J. Loehr, A. Ortiz-Ambriz, P. Tierno, Defect dynamics in artificial colloidal ice: real-time observation, manipulation, and logic gate. Phys. Rev. Lett. **117**(16), 168001 (2016)
46. M.L. Latimer, G.R. Berdiyorov, Z.L. Xiao, F.M. Peeters, W.K. Kwok, Realization of artificial ice systems for magnetic vortices in a superconducting moge thin film with patterned nanostructures. Phys. Rev. Lett. **111**, 067001 (2013)
47. J. Trastoy, M. Malnou, C. Ulysse, R. Bernard, N. Bergeal, G. Faini, J. Lesueur, J. Briatico, J.E. Villegas, Freezing and melting of vortex ice (2013), arXiv:1307.2881
48. P. Mellado, A. Concha, L. Mahadevan, Macroscopic magnetic frustration. Phys. Rev. Lett. **109**(25), 257203 (2012)
49. V. Kapaklis, U.B. Arnalds, A. Harman-Clarke, E.Th. Papaioannou, M. Karimipour, P. Korelis, A. Taroni, P.C.W. Holdsworth, S.T. Bramwell, B. Hjörvarsson, Melting artificial spin ice. New J. Phys. **14**(3), 035009 (2012)
50. U.B. Arnalds, A. Farhan, R.V. Chopdekar, V. Kapaklis, A. Balan, E.Th. Papaioannou, M. Ahlberg, F. Nolting, L.J. Heyderman, B. Hjörvarsson, Thermalized ground state of artificial kagome spin ice building blocks. Appl. Phys. Lett. **101**(11), 112404 (2012)
51. A. Farhan, P.M. Derlet, A. Kleibert, A. Balan, R.V. Chopdekar, M. Wyss, L. Anghinolfi, F. Nolting, L.J. Heyderman, Exploring hyper-cubic energy landscapes in thermally active finite artificial spin-ice systems. Nat. Phys. (2013)
52. J.M. Porro, A. Bedoya-Pinto, A. Berger, P. Vavassori, Exploring thermally induced states in square artificial spin-ice arrays. New J. Phys. **15**(5), 055012 (2013)
53. V. Kapaklis, U.B. Arnalds, A. Farhan, R.V. Chopdekar, A. Balan, A. Scholl, L.J. Heyderman, B. Hjörvarsson, Thermal fluctuations in artificial spin ice. Nat. Nanotechnol. **9**(7), 514–519 (2014)

54. M.J. Morrison, T.R. Nelson, C. Nisoli, Unhappy vertices in artificial spin ice: new degeneracies from vertex frustration. New J. Phys. **15**(4), 045009 (2013)
55. G.-W. Chern, M.J. Morrison, C. Nisoli, Degeneracy and criticality from emergent frustration in artificial spin ice. Phys. Rev. Lett. **111**, 177201 (2013)
56. I. Gilbert, G.-W. Chern, S. Zhang, L. O'Brien, B. Fore, C. Nisoli, P. Schiffer, Emergent ice rule and magnetic charge screening from vertex frustration in artificial spin ice. Nat. Phys. **10**(9), 670–675 (2014)
57. I. Gilbert, Y. Lao, I. Carrasquillo, L. O'Brien, J.D. Watts, M. Manno, C. Leighton, A. Scholl, C. Nisoli, P. Schiffer, Emergent reduced dimensionality by vertex frustration in artificial spin ice. Nat. Phys. **12**(2), 162–165 (2016)
58. R.L. Stamps, Artificial spin ice: the unhappy wanderer. Nat. Phys. **10**(9), 623–624 (2014)
59. I. Gilbert, C. Nisoli, P. Schiffer, Frustration by design. Phys. Today **69**(7), 54–59 (2016)
60. C. Nisoli, V. Kapaklis, P. Schiffer, Deliberate exotic magnetism via frustration and topology. Nat. Phys. **13**(3), 200–203 (2017)
61. E. Ising, Beitrag zur theorie des ferromagnetismus. Zeitschrift für Phys. A Hadron. Nucl. **31**(1), 253–258 (1925)
62. L.P. Kadanoff, Scaling laws for ising models near tc, *From Order to Chaos: Essays: Critical, Chaotic and Otherwise* (World Scientific, Singapore, 1993), pp. 165–174
63. K.G. Wilson, Renormalization group and critical phenomena. i. renormalization group and the kadanoff scaling picture. Phys. Rev. B **4**(9), 3174 (1971)
64. K.G. Wilson, J. Kogut, The renormalization group and the ? expansion. Phys. Rep. **12**(2), 75–199 (1974)
65. B. Mahault, A. Saxena, C. Nisoli, Emergent inequality and self-organized social classes in a network of power and frustration. PloS One **12**(2), e0171832 (2017)
66. E.H. Lieb, Residual entropy of square ice. Phys. Rev. **162**(1): 162 (1967)
67. E.H. Lieb, Exact solution of the f model of an antiferroelectric. Phys. Rev. Lett. **18**(24), 1046 (1967)
68. F.Y. Wu, Critical behavior of two-dimensional hydrogen-bonded antiferroelectrics. Phys. Rev. Lett. **22**, 1174–1176 (1969)
69. R.J. Baxter, Corner transfer matrices. Phys. A **106**(1), 18–27 (1981)
70. C. Marrows, Experimental studies of artificial spin ice (2016), arXiv:1611.00744
71. L.J. Heyderman, R.L. Stamps, Artificial ferroic systems: novel functionality from structure, interactions and dynamics. J. Phys. Condens. Matter **25**(36), 363201 (2013)
72. G.H. Wannier, Antiferromagnetism. The triangular ising net. Phys. Rev. **79**(2), 357 (1950)
73. W.F. Giauque, M.F. Ashley, Molecular rotation in ice at 10 k. free energy of formation and entropy of water. Phys. Rev. **43**(1), 81 (1933)
74. W.F. Giauque, J.W. Stout, The entropy of water and the third law of thermodynamics. The heat capacity of ice from 15 to 273 k. J. Am. Chem. Soc. **58**(7), 1144–1150 (1936)
75. L. Pauling, The structure and entropy of ice and of other crystals with some randomness of atomic arrangement. J. Am. Chem. Soc. **57**(12), 2680–2684 (1935)
76. J.D. Bernal, R.H. Fowler, A theory of water and ionic solution, with particular reference to hydrogen and hydroxyl ions. J. Chem. Phys. **1**(8), 515–548 (1933)
77. J.F. Nagle, Lattice statistics of hydrogen bonded crystals. i. the residual entropy of ice. J. Math. Phys. **7**(8), 1484–1491 (1966)
78. C. Castelnovo, R. Moessner, S.L. Sondhi, Magnetic monopoles in spin ice. Nature **451**(7174), 42–5 (2008)
79. R.J. Baxter, *Exactly Solved Models in Statistical Mechanics* (Academic, New York, 1982)
80. F. Rys, Ueber ein zweidimensionales klassisches Konfigurationsmodell, Ph.D. thesis, 1963
81. M.J. Harris, S.T. Bramwell, D.F. McMorrow, T.H. Zeiske, K.W. Godfrey, Geometrical frustration in the ferromagnetic pyrochlore ho 2 ti 2 o 7. Phys. Rev. Lett. **79**(13), 2554 (1997)
82. A.P. Ramirez, A. Hayashi, R.J. Cava, R. Siddharthan, B.S. Shastry, Zero-point entropy in 'spin ice'. Nature **399**, 333–335 (1999)
83. I.A. Ryzhkin, On magnetic relaxation in rare earth metal perchlorate metals. Zhurnal Ehksperimental'noj i Teoreticheskoj Fiziki **128**(3), 559–566 (2005)

84. C.L. Henley, The "coulomb phase" in frustrated systems. Annu. Rev. Condens. Matter Phys. **1**(1), 179–210 (2010)
85. C. Castelnovo, R. Moessner, S.L. Sondhi, Spin ice, fractionalization, and topological order. Annu. Rev. Condens. Matter Phys. **3**(1), 35–55 (2012)
86. X.-G. Wen, Vacuum degeneracy of chiral spin states in compactified space. Phys. Rev. B **40**(10), 7387 (1989)
87. X.-G. Wen, Quantum orders and symmetric spin liquids. Phys. Rev. B **65**(16), 165113 (2002)
88. P.M. Chaikin, T.C. Lubensky, *Principles of Condensed Matter Physics* (Cambridge university press, Cambridge, 2000)
89. V. Volterra, Sur l'équilibre des corps élastiques multiplement connexes, *Annales scientifiques de l'École Normale Supérieure*, vol. 24 (Elsevier, 1907), pp. 401–517
90. M.V. Kurik, O.D. Lavrentovich, Defects in liquid crystals: homotopy theory and experimental studies. Physics-Uspekhi **31**(3), 196–224 (1988)
91. J.M. Kosterlitz, D.J. Thouless, Ordering, metastability and phase transitions in two-dimensional systems. J. Phys. C Solid State Phys. **6**(7), 1181 (1973)
92. C. Castelnovo, C. Chamon, Topological order and topological entropy in classical systems. Phys. Rev. B **76**(17), 174416 (2007)
93. C.L. Henley, Classical height models with topological order. J. Phys. Condens. Matter **23**(16), 164212 (2011)
94. R.Z. Lamberty, S. Papanikolaou, C.L. Henley, Classical topological order in abelian and non-abelian generalized height models. Phys. Rev. Lett. **111**(24), 245701 (2013)
95. L.D.C. Jaubert, M.J. Harris, T. Fennell, R.G. Melko, S.T. Bramwell, P.C.W. Holdsworth, Topological-sector fluctuations and curie-law crossover in spin ice. Phys. Rev. X **3**(1), 011014 (2013)
96. C. Castelnovo, R. Moessner, S.L. Sondhi, Thermal quenches in spin ice. Phys. Rev. Lett. **104**(10), 107201 (2010)
97. M. Tanaka, E. Saitoh, H. Miyajima, T. Yamaoka, Y. Iye, Magnetic interactions in a ferromagnetic honeycomb nanoscale network. Phys. Rev. B **73**(5), 052411 (2006)
98. Y. Qi, T. Brintlinger, J. Cumings, Direct observation of the ice rule in an artificial kagome spin ice. Phys. Rev. B **77**(9), 094418 (2008)
99. G. Möller, R. Moessner, Magnetic multipole analysis of kagome and artificial spin-ice dipolar arrays. Phys. Rev. B **80**(14), 140409 (2009)
100. G.-W. Chern, P. Mellado, O. Tchernyshyov, Two-stage ordering of spins in dipolar spin ice on the kagome lattice. Phys. Rev. Lett. **106**, 207202 (2011)
101. J. Drisko, S. Daunheimer, J. Cumings, Fepd 3 as a material for studying thermally active artificial spin ice systems. Phys. Rev. B **91**(22), 224406 (2015)
102. S. Zhang, I. Gilbert, C. Nisoli, G.-W. Chern, M.J. Erickson, L. O'Brien, C. Leighton, P.E. Lammert, V.H. Crespi, P. Schiffer, Crystallites of magnetic charges in artificial spin ice. Nature **500**(7464), 553–557 (2013)
103. N.A. Sinitsyn, Y.V. Pershin, The theory of spin noise spectroscopy: a review. Rep. Prog. Phys. **79**(10), 106501 (2016)
104. S.A. Crooker, D.G. Rickel, A.V. Balatsky, D.L. Smith, Spectroscopy of spontaneous spin noise as a probe of spin dynamics and magnetic resonance. Nature **431**(7004), 49–52 (2004)
105. R.J. Glauber, Time-dependent statistics of the ising model. J. Math. Phys. **4**(2), 294–307 (1963)
106. L.A.S. Mól, W.A. Moura-Melo, A.R. Pereira, Conditions for free magnetic monopoles in nanoscale square arrays of dipolar spin ice. Phys. Rev. B **82**(5), 054434 (2010)
107. F.S. Nascimento, L.A.S. Ml, W.A. Moura-Melo, A.R. Pereira, From confinement to deconfinement of magnetic monopoles in artificial rectangular spin ices. New J. Phys. **14**(11), 115019 (2012)
108. L.A. Mól, R.L. Silva, R.C. Silva, A.R. Pereira, W.A. Moura-Melo, B.V. Costa, Magnetic monopole and string excitations in two-dimensional spin ice. J. Appl. Phys. **106**(6), 063913 (2009)
109. Y. Nambu, Strings, monopoles, and gauge fields. Phys. Rev. D **10**(12), 4262 (1974)

110. Y. Perrin, B. Canals, N. Rougemaille, Extensive degeneracy, coulomb phase and magnetic monopoles in artificial square ice. Nature **540**(7633), 410–413 (2016)
111. G.-W. Chern, C. Reichhardt, C. Nisoli, Realizing three-dimensional artificial spin ice by stacking planar nano-arrays. Appl. Phys. Lett. **104**(1), 013101 (2014)
112. A.A. Mistonov, N.A. Grigoryeva, A.V. Chumakova, H. Eckerlebe, N.A. Sapoletova, K.S. Napolskii, A.A. Eliseev, D. Menzel, S.V. Grigoriev, Three-dimensional artificial spin ice in nanostructured co on an inverse opal-like lattice. Phys. Rev. B **87**(22), 220408 (2013)
113. A.A. Mistonov, I.S. Shishkin, I.S. Dubitskiy, N.A. Grigoryeva, H. Eckerlebe, S.V. Grigoriev, Ice rule for a ferromagnetic nanosite network on the face-centered cubic lattice. J. Exp. Theor. Phys. **120**(5), 844–850 (2015)
114. I.R.B. Ribeiro, F.S. Nascimento, S.O. Ferreira, W.A. Moura-Melo, C.A.R. Costa, J. Borme, P.P. Freitas, G.M. Wysin, C.I.L. de Araujo, A.R. Pereira, Realization of rectangular artificial spin ice and direct observation of high energy topology (2017), arXiv:1704.07373
115. P.W. Kasteleyn, The statistics of dimers on a lattice: I. the number of dimer arrangements on a quadratic lattice. Physica **27**(12), 1209–1225 (1961)
116. P.G. De Gennes, G. Sarma, Tentative model for the smectic b phase. Phys. Lett. A **38**(4), 219–220 (1972)
117. C.S. O'Hern, T.C. Lubensky, J. Toner, Sliding phases in xy models, crystals, and cationic lipid-dna complexes. Phys. Rev. Lett. **83**(14), 2745 (1999)
118. S.L. Sondhi, K. Yang, Sliding phases via magnetic fields. Phys. Rev. B **63**(5), 054430 (2001)

Chapter 5
Topologically Non-trivial Magnetic Skyrmions in Confined Geometries

Haifeng Du and Mingliang Tian

Abstract Magnetic skyrmion is a small magnetic whirl that possesses non-trivial topology and behaves like a particle. The specially twisted spin arrangement within skyrmion gives rise to topological stability and low critical current to drive its motion, both of them benefit the potential technological application in memory devices. In this chapter, we briefly introduce the notation of topology of magnetic skyrmions and recent progress ranging from the hard disk storage in conventional magnetic memory device to racetrack memory by using magnetic skyrmions. The related magnetic phases in skyrmion materials based on the Dzyaloshinskii-Moriya (DM) interactions are discussed in detail. Furthermore, experimental achievements for the formation and stability of highly geometry-confined skyrmions are outlined, where the topological effects are fully embodied.

5.1 Introduction

Magnetic memory devices mainly magnetic hard disk drives (HDD) have been the primary repository of digital data for more than half a century. Data is recorded in a thin ferromagnetic film with the binary data bits 0 and 1 represented by the direction of magnetization of a small domain (Fig. 5.1). Data read from the disk is accomplished by transferring the magnetization of the small magnetic domain into electric signals via the giant magnetoresistance (GMR) effect, while writing data is accomplished by using the magnetic field to control the direction of the small magnetic domain. In spite of its extreme success in memory device, the decreased size of the magnetic domains in HDD may lead to the loss of their magnetic state due to the thermally

H. Du · M. Tian (✉)
The Anhui Province Key Laboratory of Condensed Matter Physics at Extreme Conditions, and High Magnetic Field Laboratory, Chinese Academy of Science (CAS), Hefei 230031, Anhui, China
e-mail: tianml@hmfl.ac.cn

H. Du · M. Tian
School of Physics and Materials Science, Anhui University, Hefei 230601, Anhui, China

S. Gupta and A. Saxena (eds.), *The Role of Topology in Materials*,
Springer Series in Solid-State Sciences 189,
https://doi.org/10.1007/978-3-319-76596-9_5

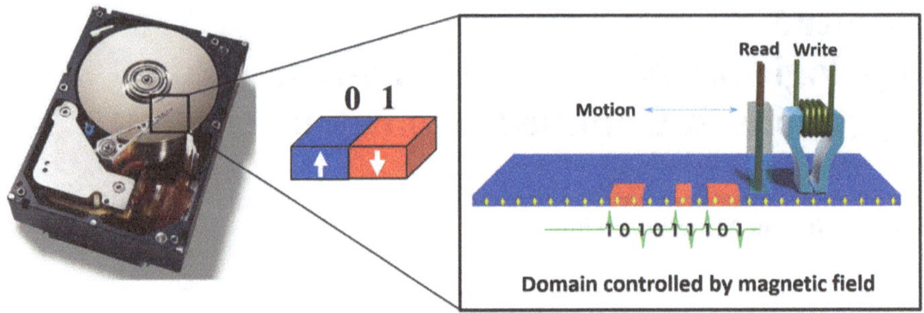

Fig. 5.1 Hard disk drive and an overview of how it works

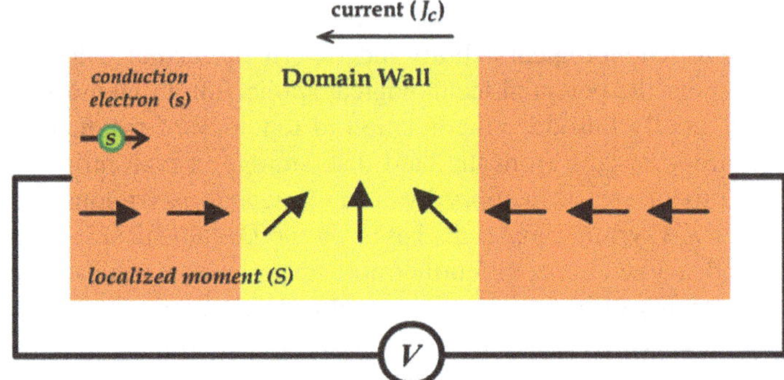

Fig. 5.2 Spin transfer torque effect in which the orientation of localized magnetic moments can be controlled by using a spin-polarized current

induced magnetic instability, commonly known as the superparamagnetic limit [1]. Meanwhile, the movable mechanical engines consume large energy and thus set a limitation on their writing/reading speed.

Therefore, a number of alternative magnetic memory architectures towards achieving high reliability, performance and capacity data storage have been proposed. A typical one is based on controllable manipulation of magnetic domain walls (DW) by using electrical current [2]. It is well known that electrons carry not only charges but also spins. In a non-ferromagnetic material, the spins of electrons are generally randomly oriented and do not show spin polarization. However, when a ferromagnetic (FM) component is incorporated into a device, a spin-polarized current is produced when a current passes through the FM component with fixed magnetization direction, i.e. fixed layer. If this spin-polarized current is further passed into a second thinner magnetic layer (the "free layer"), its orientation of the magnetization can be changed due to the transformation of angular momentum of electrons to this FM layer. This effect is the so-called spin transfer torque (STT) (Fig. 5.2a), and can be used to flip the orientation of the magnetic domains. This means that domain walls can be moved by spin polarized current.

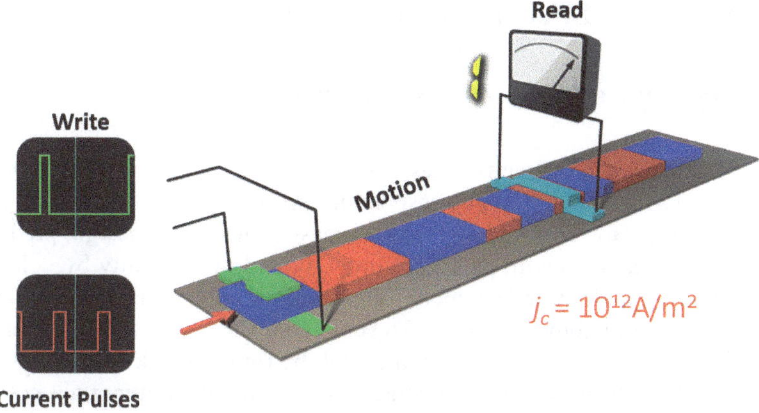

Fig. 5.3 Racetrack memory and an overview of how it works

Based on this effect, a memory device was then proposed by IBM and called racetrack memory (RM) that actually inherited the bubble memory concept [3]. In the RM (Fig. 5.3), the information bit (0 or 1) is encoded in magnetic domains separated by domain walls (DWs) that can move by means of a spin polarized current. A read/write device is fixed somewhere to polarize or measure the magnetization in a given domain. By controlling current pulses in the device, the domain walls can be moved at speeds of hundreds of miles per hour and then can be stopped precisely at the position needed, allowing massive amounts of stored information to be accessed in less than a billionth of a second. The advantage of data access speed makes the racetrack memory so intriguing. Projections are that the racetrack memory can read or write a bit of information in less than 1–10 ns, depending on the length of the racetrack, while a hard drive performs the same operation in 3,000,000 ns. However, because the RM based on STT effect requires very high current density ($j_c \sim 10^{12}A/m^2$) to move the domain walls, together with many other physical and material problems such as the pinning effects, the Walker limit, the conventional DW-based RM turned out too complicated to be competitive compared to HDDs. Increasing efforts and resources dedicated to the development of the racetrack memory are then highly required.

In magnets without inversion symmetry, so called chiral magnets, relatively weak spin-orbit coupling leads to the formation of smooth twisted magnetic structure with a long period. Recently, a new local magnetic state, named magnetic skyrmion was discovered in chiral magnets. It has a series of advantages that might allow the device to be more robust and more efficiently manipulated with currents. Here, we will give a brief introduction to the magnetic skyrmions. In Sect. 5.2, we review the topological effect related to magnetic skyrmions. In Sect. 5.3, we concentrate on the formation mechanism of magnetic skyrmions. In the last section, we turn to some special properties of geometrically-confined skyrmions.

5.2 Topological Effect in Magnetic Skyrmions

5.2.1 Topology in Magnetic Materials

Traditionally, study on magnetism has mainly followed the route by exploring its microscopic magnetic structure, corresponding properties, and then to realize certain functionality. However, there is now increasing interest to design, predict and fabricate novel materials with particular property and functionality by delineating the active role of topology and geometry. The topology class is mathematically characterized by the homotopy theory [4]. The emergent state possesses a property that is protected in a symmetry sense, and then can be rigorously defined by its corresponding mathematical topological characteristics. One of the important notions in the homotopy theory is the topological defect, which is a stable configuration of matter formed in the very early universe and characterized by a homotopy class. Such configurations are in the original, symmetric or old phase, but they persist after a phase transition to the asymmetric or new phase is completed. There are a number of possible types of defects in a magnetic system. To distinguish these topological defects belonging to one and the same homotopy class, it is convenient to define a global parameter, named as winding (also skyrmion) number, Q. Such a parameter is provided by the degree of a mapping $f\colon M \to N$, where M and N are orientable and compact manifolds, respectively. In the case of both manifolds belonging to n-spheres, the winding number of the mapping, $\deg f$, counts how many times M is wrapped around N under the map f. In two dimensional case, the easy-plane spin varies as $\boldsymbol{m}(\tau) = (\cos \Phi(\tau), \sin \Phi(\tau))$, where $0 \le \tau \le 2\pi$ parameterizes the loop and continuity requires $\boldsymbol{m}(0) = \boldsymbol{m}(2\pi)$, the winding number can be expressed as

$$Q = 1/2\pi \int_0^{2\pi} \partial_\tau \Phi d\tau \tag{5.1}$$

In fact, before the discovery of magnetic skyrmions, this topology method has been used to analyze the complex switching processes observed in ferromagnetic nanoparticles [5]. The switching process involving the creation, propagation, and annihilation of domain walls with complex internal structure. But, from the topological viewpoint, these complex domain walls are composite objects made of only two or more elementary defects: vortices with integer winding numbers ($Q = \pm 1$) and edge defects with fractional winding numbers ($Q = \pm 1/2$). The simplest domain walls are composed of two edge defects with opposite winding numbers. Creation and annihilation of the defects are constrained by conservation of a topological charge. The elementary topological bulk and edge defects will significantly affect the interaction and domain trajectory in the nanowires. For example, head-to-head and tail-to-tail magnetic domain walls in the nanowire are two fundamental domains that behave as free magnetic monopoles carrying a single magnetic charge. They attract one another since adjacent walls always carry opposite charges, leading to annihilation of the two domain walls. However, the topological edge defects that have the same winding number suppress annihilation of the walls because of a short-range repulsive interaction [6].

In three-dimensional case, Q is the number of spins winding around the unit ring after going through the magnetic ring. It can be done by wrapping the spin vector, $\boldsymbol{m}$ around a unit sphere, where the Q is defined as follows:

$$Q = 1/4\pi \int \boldsymbol{m} \cdot \left(\partial_x \boldsymbol{m} \times \partial_y \boldsymbol{m}\right) dxdy \quad (5.2)$$

According to these definitions, the winding number Q of each configuration also characterizes the number of magnetic monopoles therein.

5.2.2 Topological Stability of Magnetic Skyrmions and Emergent Magnetic Monopoles

In magnetic materials, a notable example of a topologically stable object is the skyrmion [7–10]. It is a swirl-like spin texture, in which the magnetic moments point in all directions wrapping a unit sphere (Fig. 5.4a). It is then easy to obtain a unit topological charge (Fig. 5.4c) according to (5.3), while it is zero for conventional spin textures like the ferromagnetic state. It explained the topological stability of magnetic skyrmions. Two objects or configurations are topologically different if they belong to a different homotopy group if there exists no continuous transformation of one configuration into the other without cutting or gluing. The same applies to skyrmions: skyrmion lattices in chiral magnets can be regarded as macroscopic lattices formed by topological entities with particle-like properties where the particle-like character of the skyrmions is reflected in the integer winding number of their magnetization.

It must be noted that a non-trivial topology does not imply in itself energetic stability. There is in fact unnecessary relation between topology and energetic stability. Topological stability referred to a system going from one topological state to another is always accompanied by a discontinuity in the continuous field and is only a mathematical concept. For instance, a torus going into a sphere, a rupture must be created on some part of the torus's surface. In this case, the torus would be mathematically described as topologically stable. By contrast, energy stability is always connected to real physical systems and is described by the free energy, which is required to create the rupture and is always finite. That is to say, the mathematical concept of topology is only used to describe a physical system. The attributes of the system including the energy stability depend on the system's physical parameters. To build an efficient link between topological and energy stability, a non-zero phenomenological field rigidity, which is used to account for the finite energy needed to rupture the field's topology, should be introduced. By calculating a breakdown energy-density of the field, a topology-related energy barrier is introduced. For the magnetic skyrmions,

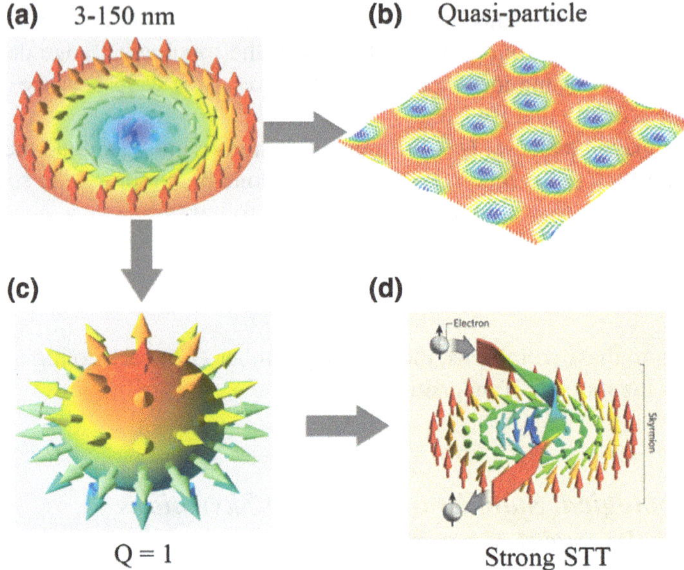

Fig. 5.4 Magnetic skyrmion and its properties. **a** The spin configuration of single skyrmion. **b** skyrmion lattice. **c** Unit topological charge of skyrmion due to the nontrivial spin arrangement. **d** Strong spin transfer torque effect (STT)

such a topological barrier is extracted by evaluating the energy routine during the transition of the skyrmion number or topological charge in the dynamical process of a skyrmion creation or annihilation. In fact, the barrier height linked to the topological charge has been theoretically calculated [11].

Leaving the energy stability out of consideration, transition between different topological classes can lead to some exotic topological state. For the topology-protected magnetic skyrmions, they cannot be destroyed or created by smooth deformation of other trivial spin textures such as the ferromagnetic state, helical or conical phase. Then what will be happen when the topological winding in a skyrmion can be destroyed. By using a magnetic force microscope (MFM) to map out the distribution and shape of the skyrmions on the surface of the bulk chiral magnets $Fe_{0.5}Co_{0.5}Si$ [12], the field evolution of magnetic states in the sample surface is revealed. The skyrmions disappeared as the magnetic field was reduced to zero and were seen to decay by coalescing with their neighbors to form lines on the surface. At zero magnetic field, the lines created a tiger-stripe pattern. But, there is no individual skyrmion within the bulk of the sample. Computer simulations well reproduced the experimental observation that a similar coalescence occurs beneath the surface. It is then well established that the formation and destruction of magnetic skyrmions is driven by the creation and motion of singular defects. The singular points can be identified with emergent monopoles.

5.2.3 Topological and Skyrmion Hall Effect

As a result of their unique spin topology, magnetic skyrmions provide a platform to study many intriguing real-space topological transport phenomena. A well-known example is the Hall effect, which may show a peculiar behavior arising from non-trivial spin arrangements. The ordinary Hall effect of a conductor in a perpendicular magnetic field originates from the Lorentz force acting on the charge carriers. It is usually measured as a voltage transverse to the current. In ferromagnetic conductors, an additional anomalous Hall effect (AHE)—for which the theories usually assume a collinear spin structure—arises from magnetization and spin–orbit interaction and is present even in zero magnetic field. However, in a non-coplanar spin configuration the spin chirality of three magnetic moments spanning a triangle can induce a finite Berry phase and an associated fictitious magnetic field. This field generates an AHE even without the spin–orbit interaction, the so-called topological Hall effect (THE) [13, 14]. Thus, the Berry phase reflects the chirality and winding number of the knots. The topological Hall effect arises besides the normal Hall effect. The topological Hall resistivity is proportional to the density of emerging magnetic field, which is the average density of magnetic field that is associated with the flux quanta Φ_0 contained in the skyrmion of diameter R. The strength of emergent magnetic field can be as large as $\langle \boldsymbol{b}_z \rangle \approx 100$ T given a skyrmion of 10 nm in diameter, which thus provides a unique platform to study the high magnetic field response of electrons. Meanwhile, based on the well-known Faraday's law, the motion of a magnetic skyrmion thus produces a time-dependent electric field $\langle \boldsymbol{E}_t \rangle$ that leads to the emergent electromagnetic induction. These effects have been observed in bulk chiral magnets including MnSi, GeMn [15].

Another interesting phenomenon is the skyrmion Hall effect [16, 17]. Because magnetic skyrmion behaves like a particle with a unit topological charge $Q = \pm 1$. Similar to the transverse deflection of charged particles as a result of Lorentz force, the motion of magnetic skyrmion exhibits a well-defined transverse component as a result of topological Magnus force. The skyrmion hall effect is directly related to the device applications. Since skyrmions can be shifted by electrical currents and feel a repulsive force from the edges of the magnetic track as well as from single defects in the wire, they can move relatively undisturbed through the track. This is a highly desired property for racetrack devices, which are supposed to consist of static read- and write-heads, while the magnetic bits are shifted in the track. However, another important aspect of skyrmion dynamics originates from skyrmion Hall effect is that the skyrmions do not only move parallel to the applied current, but also perpendicular to it. This leads to an angle between the skyrmion direction of motion and the current flow called the skyrmion Hall angle. As a result, the skyrmions should move under this constant angle until they start getting repelled by the edge of the material and then keep a constant distance to it. Recent investigation has proved that the billion-fold reproducible displacement of skyrmions is indeed possible and can be achieved with high velocities. Furthermore, it turned out that the skyrmion Hall angle depends on the velocity of the skyrmions, which means that the components of the motion

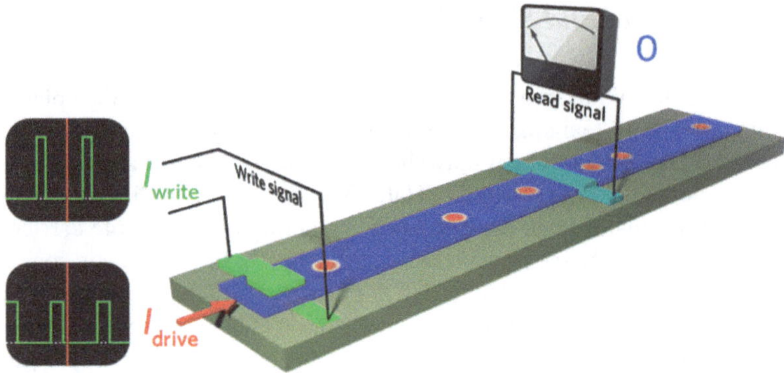

Fig. 5.5 Skyrmion-based Racetrack memory

parallel or perpendicular to the current flow do not scale equally with the velocity of the skyrmions. This is not predicted in the conventional theoretical description of skyrmions. Part of the solution of this unexpected behavior could be the deformation of the skyrmion spin structure, calling for more theoretical effort to fully understand the properties of skyrmion in confined geometries.

5.2.4 Skyrmion-Based Racetrack Memory (RM)

The peculiar twists of the magnetization within the skyrmion give rise to a nontrivial topology so that the spin current is able to efficiently couple with skyrmion. This process further links to the topological Hall effect and efficient spin-transfer torque effect. As a consequence, the critical current density to drive skyrmions is 4 or 5 orders of magnitude lower than that needed to move the conventional ferromagnetic domain walls. Moreover, the single skyrmion possesses much smaller size than a conventional domain. The typical value of the size is on the order of 3–150 nm depending on the intrinsic parameters of skyrmion materials. These properties including small size, high stability and mobility all benefit to build future skyrmion-based RM [18]. Recent investigations have demonstrated the current-induced creation and motion as well as electric detecting of individual skyrmions in confined geometries even at room temperature [19]. These advances raise great expectations for realizing skyrmion-based RM (Fig. 5.5).

5.3 Origin of Magnetic Skyrmion

5.3.1 Magnetic Phase Diagram in Chiral Magnets

Within the framework of micromagnetism that is the continuum theory of magnetic moments, the magnetic microstructure is determined by the couplings among local magnetic moments. Among them, the common one is the exchange interaction. It is a purely quantum phenomenon, which has no analogy in the classical world. The magnetic moments and spontaneous magnetization are realized by the exchange interaction between electrons. In a simple two-electron picture, exchange gives rise to ferromagnetic or antiferromagnetic coupling depending on the sign of exchange constant J. In the discrete model, the most general expression for two sites exchange energy between neighboring magnetic moments $\mathbf{S}_i$ and $\mathbf{S}_j$ is:

$$E_e = -J\mathbf{S}_i \cdot \mathbf{S}_j \tag{5.3}$$

Although the exchange interaction keeps spins aligned, it does not align them in a particular direction. This is accomplished by the magnetic anisotropy. Without magnetic anisotropy, the spins in a magnet randomly change direction in response to thermal fluctuations and the magnet is superparamagnetic. There are several kinds of magnetic anisotropy, the most common of which is magnetocrystalline anisotropy. This is a dependence of the energy on the direction of magnetization relative to the crystallographic lattice. Another common source of anisotropy can be induced by internal strains or interface. Single-domain magnets also can have a shape anisotropy due to the magnetostatic effects of the sample shape. This also belongs to stray field energy, which is connected with the magnetic field generated by the magnetic body itself. It arises because each magnetic moment in a ferromagnetic sample represents a magnetic dipole and therefore contributes to the total magnetic field inside the sample.

A magnetic moment will try to reduce its energy by aligning itself parallel to an external magnetic field. The energy that describes the interaction of a magnetic moment with an applied field $\mathbf{B}$ is called Zeeman energy:

$$E_h = -\mathbf{S}_i \cdot \mathbf{B} \tag{5.4}$$

Zeeman energy as well as anisotropies are local energy terms because their energy contributions are determined only by the local values of magnetization vector.

The magnetic structures are determined by the competition among different energies. These naturally lead to non-collinear spin textures such as magnetic domain walls, magnetic vortices and magnetic bubbles. Unlike the controllable manipulation of magnetic domains using magnetic field in the HDD, these non-collinear spin textures may provide new opportunities for the spintronic devices because spins of conduction electrons can efficiently couple with the localized spin textures via a spin angular momentum transfer, the so-called spin transfer torque (STT) mechanism.

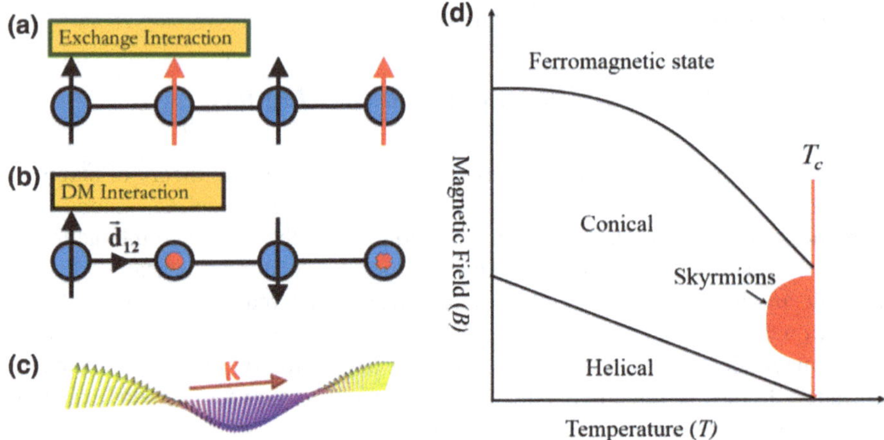

Fig. 5.6 Magnetic phases in B20 chiral magnets. The competition between ferromagnetic exchange in (**a**) and non-collinear DM interactions in (**b**) leads to a helical ground state in (**c**), which unpins under the application of a high magnetic field to form the conical phase with its wave vector along the direction of magnetic field, and then condenses into a skyrmion crystal with a hexagonal arrangement if temperature is close to the Curie temperature T_C. Images are taken from [24]

The formation and stabilization of magnetic skyrmions often require some non-collinear magnetic interactions. A typical one is the Dzyaloshinskii-Moriya (DM) interaction [20]. In addition to this coupling, it has also been addressed that magnetic dipolar interaction, frustrated exchange interaction [21], and four-spin exchange interaction [22] are all able to create skyrmions. Among them, the DM-induced skyrmions have evident advantages including small and tunable size [23], and extra stability even in highly confined geometries [24]. Due to the importance of DM couplings, it is valuable to introduce its origin before entering the non-collinear magnetic structures in chiral magnets. DM interactions come from the symmetry-breaking in the crystal lattice or interface and surface. It is written as [25]:

$$E_D = \mathbf{D}_{ij} \cdot \left(\mathbf{S}_i \times \mathbf{S}_j\right) \tag{5.5}$$

Vectors $\mathbf{D}_{ij}$ are the nearest neighbor DM coupling constants and originate from the symmetry-breaking in the crystal lattice. It is obvious that spins tend to be perpendicular to each other under the DM interaction according to (5.5) (Fig. 5.6b). Competition between ferromagnetic and DM interaction leads to a helical ground state (Fig. 5.6c). Since DM interaction is coupled to the lattice, it is chiral. The corresponding magnetic materials are also called chiral magnets. Including the above-mentioned Heisenberg exchange and Zeeman energies, the total energy of a B20 helimagnet with the magnetization **S** in the continuum limit can be written as

$$E = J\,(\nabla \mathbf{S})^2 + D\mathbf{S} \cdot (\nabla \times \mathbf{S}) - \mathbf{B} \cdot \mathbf{S} \tag{5.6}$$

The emergent magnetic structures in helical magnets can be well explained within the above-mentioned physical model concerning the DM, ferromagnetic exchange, and external magnetic interaction. The spin helix is created by the completion between DM and ferromagnetic exchange interactions. The wave-vector **k** is fixed depending on the high symmetry crystal axis (Fig. 5.6a–c). Under the action of a magnetic field **B**, a conical phase with **k** || **B** is energetically favorable when the magnetic field is of moderate strength and the temperature (*T*) is lower than the Curie temperature T_C. A higher field will further transfer the conical phase into a ferromagnetic state. Because the rotation sense of both helical and conical states orients along only one direction, they are called single-twist magnetic structure. By contrast, magnetic skyrmion is a vortex-like magnetic structure with the magnetization within the skymion rotating in two directions. Skyrmion thus belongs to double-twist modulated magnetic configuration. In bulk materials, the formation of the skyrmion state is commonly explained by the thermal fluctuation effect, where skyrmions occupy only a small temperature-magnetic field *(T-B)* region in the magnetic phase diagram. The magnetic phase diagram in Fig. 5.6d represents a common behavior in B20 helical magnets.

On the contrary, highly stable magnetic skyrmions can be realized by the reduced dimensionality of helical magnets. It has been confirmed experimentally that thin plates of B20 magnets with its thickness around the featured skyrmion size can hold a skyrmion in significantly extended *T-B* phase region even to the zero temperature [26]. The extended skyrmion state in low dimensional B20 magnets has been explained by several different mechanisms. The spatial confinement effects is the first one [27]. Numerical calculation showed that a type of 3D skyrmion, instead of conventional 2D case, appears if the thickness of the film is below a threshold. A 3D skyrmion is characterized by a superposition of double-twist rotation of magnetic skyrmions in the perpendicular plane and conical modulations along the skyrmion axis. The 3D skyrmion is proposed to be energetically favorable and then thermodynamically stable in a broad *T-B* range. The high stability of magnetic skyrmions in 2D materials such as mechanically thinned flakes and quasi-2D MnSi nanostripes can be well explained by this mechanism [28]. Beyond 3D-modulated configuration, highly stable skyrmions can also exist in magnetic materials with the uniaxial magnetic anisotropy [29]. In this case, if the external magnetic field is applied along the direction of easy axis of the system, the energy of conical phase would be significantly increased under a combined action of magnetic field and uniaxial anisotropy. Skyrmion state is thus energetically favorable in large regions of *T-B* space. But, it should be noticed that the skyrmion size and crystal lattice constant are also depend on the strength of uniaxial anisotropy. However, there is no clear experimental evidence to identify this mechanism. For a thin film with uniaxial anisotropy, if the external magnetic field is applied in the film plane, extended elliptic skyrmion gratings are proposed to be energetically favorable if the hard axis of the system is perpendicular to the film plane. In the highly confined helical magnets, the experimental observations, as discussed below, cannot be explained by independently using the three mechanisms.

5.3.2 Mechanism of DM Interaction

Magnetic skyrmions were first discovered in metallic alloy MnSi by neutron scattering [8], and later confirmed in semiconductor $Fe_{0.5}Co_{0.5}Si$ by Lorentz Transmission Electron Microscopy (TEM) [9]. Both materials belong to the same lattice class of B20, a non-centrosymmetric binary structure, in the Strukturbericht Symbol. Subsequent investigation proved that the emergence of skyrmion is a common feature in other B20 magnets such as FeGe and Cu_2OSeO_3 [28, 31]. The continuing discovery of more skyrmion materials is unambiguously an important and urgent task for the development of skyrmion science. To guide the search for more skyrmion materials, it is instructive to classify all possible materials with DM interaction by symmetry. The pioneering phenomenological theory was introduced by Bogdanov et al. [32], where the technique of symmetric tensor is employed [33, 34] and the classification of point groups hosting skyrmions is achieved [35].

In detail, the Fourier component of the Heisenberg exchange is given by $E_{Hei}(\mathbf{k}) = J\mathbf{k}^2 |\mathbf{S}_\mathbf{k}|^2$, where $\mathbf{S}_\mathbf{k}$ is the Fourier component of the spin $\mathbf{S}_\mathbf{k} = \frac{1}{V}\int d\mathbf{r}\mathbf{S}(\mathbf{r})\exp(i\mathbf{k}\cdot\mathbf{r})$. Under rotations, $\mathbf{S}$ transforms in the same way as $\mathbf{k}$. As a result, $E_{\mathrm{Hei}}(\mathbf{k})$ is rotationally invariant. Furthermore, the inversion and mirror symmetries are also respected. Therefore this is a generic quadratic term for all ferromagnets. On the other hand, the DM interaction provides a linear term in momentum; $E_{\mathrm{DM}}(\mathbf{k}) = iD\mathbf{S}_\mathbf{k}\cdot(\mathbf{k}\times\mathbf{S}_{-\mathbf{k}})$. Although the rotational symmetry is still preserved, the inversion symmetry is apparently broken, which is the well-known origin of the DM interactions. However a long overlooked fact is the broken mirror symmetry in this DM interaction. This comes from the fact that $\mathbf{S}_\mathbf{k}\cdot(\mathbf{k}\times\mathbf{S}_{-\mathbf{k}})$ is a pseudoscalar. Under any improper rotation such as mirror reflection in the lattice, this term flips sign, and should be ruled out in the energy. This term is a unique feature of lattices, such as T23 group, with only pure rotations. Now a question arises whether this form is the only allowed term linear in $\mathbf{k}$ for any other material.

To answer this question, we can rewrite any k-linear term as a tensor product

$$E_{\mathrm{DM}}(\mathbf{k}) = i d_{ijm} k^i S^j_{-\mathbf{k}} S^m_\mathbf{k}, \tag{5.7}$$

where d_{ijm} is a third order tensor that can be constructed from symmetry analysis. Any symmetry operation R can be represented as a 3×3 matrix in natural basis (x, y, z). Under such operation, vector $\mathbf{k}$ transforms as $k_i \rightarrow k_j R_{ji}$, while the pseudovectors $\mathbf{S}_{\pm\mathbf{k}}$ transform as $S_i \rightarrow |R|\, S_j R_{ji}$, where $|R|$ is the determinant of R matrix. If R is an improper rotation, $|R| = -1$. Once R is a symmetry operation, energy should be invariant under such rotation, therefore the tensor d_{ijm} must satisfy the Neumann's principle:

$$d_{ijm} = R_{ip} R_{jq} R_{mr} d_{pqr} \tag{5.8}$$

In practice, one does not need to go through all symmetry operations in order to determine the d tensor. Most operations can be written as products of some independent matrices, called generating matrices [36], within the same point group. For T23 point group, the generating matrices are $C_2$and C_3 rotations. The Neumann's principle thus leads to the constraint that $d_{xyz} = d_{yzx} = d_{zxy}$, and $d_{xzy} = d_{yxz} = d_{zyx}$. One can symmetrize these parameters by $d_{xyz} = S + D$ and $d_{xzy} = S - D$. However because the whole Hamiltonian can be reorganized as $\sum_{\mathbf{k}} H_{DM}(\mathbf{k}) = \sum_{\mathbf{k}} i\left(d_{ijm} - d_{imj}\right) k^i S^j_{-\mathbf{k}} S^m_{\mathbf{k}}$, the symmetric component S does not contribute. The resulting Hamiltonian is thus $\sum_{\mathbf{k}} i D \varepsilon_{ijm} k^i S^j_{-\mathbf{k}} S^m_{\mathbf{k}}$, which reproduces the DM interaction in (5.6).

The same method applies to any other lattice. If $S_{-\mathbf{k}}$ and $S_{\mathbf{k}}$ have the same index, its contribution to the Hamiltonian vanishes when completing the summation over momenta. The relevant terms are 6 components of d_{xyz} permutations, and other 12 components of d_{xxy}, d_{xyx} and their permutations. For future convenience, these components are symmetrized as $d_{xyz} = \alpha_S + \alpha_A$, $d_{yxz} = \alpha_S - \alpha_A$, $d_{yzx} = \beta_S + \beta_A$, $d_{xzy} = \beta_S - \beta_A$, $d_{zxy} = \gamma_S + \gamma_A$, $d_{zyx} = \gamma_S - \gamma_A$, and $\xi_{\alpha\beta} = d_{\alpha\alpha\beta} - d_{\alpha\beta\alpha}$. As a result, the total Hamiltonian is given by

$$E = \int d^3\mathbf{r} \Big[J(\nabla\mathbf{S})^2 - \mathbf{B}\cdot\mathbf{S} + \gamma_A \mathbf{S}\cdot(\partial_{\hat{z}} \times \mathbf{S}) + \frac{1}{2}(\alpha_S + \alpha_A - \beta_S + \beta_A)\,\mathbf{S}\cdot(\partial_{\hat{x}} \times S)$$
$$+ \frac{1}{2}(-\alpha_S + \alpha_A + \beta_S + \beta_A)\,\mathbf{S}\cdot\left(\partial_{\hat{y}} \times S\right) + \xi_{\alpha\beta} S^\beta \partial_\alpha S^\alpha \Big] \quad (5.9)$$

where $\partial_{\hat{r}}$ is the directional derivative along $\mathbf{r}$ direction. These linear order derivative terms are the DM interactions. Under low magnetic field $\mathbf{B}$, a spin helix is thus formed along certain directions given the competition between anisotropic DM interaction and the Heisenberg exchange. The last term in this energy characterizes Neel type helices and skyrmions, where the magnetic moments are coplanar with the propagation direction, while the rest of the DM terms belong to the Bloch type, where moments and propagation direction are always perpendicular to each other. A complete list of point groups contributing to nonzero DM interactions is summarized in the following table (Table 5.1).

The inversion symmetry is lost in any of the classes. In addition, the mirror symmetry is also missing in a large portion of the allowed point groups, especially in the cubic crystal system. Higher order spin anisotropies can be constructed in the same way. Once elevated, they might change the ground state, such as anti-vortex state in S_4 and D_{2d} once easy axis is established [34, 37]. The B20 compounds, harboring typical Bloch skyrmions, are located in class B-IV. The most important message delivered from this derivation is that class B-V, the O group, is described by exactly the same Hamiltonian as the B20 compounds in class-IV. The spin anisotropies are also the same in these two classes. Therefore the spin physics observed in B20 compounds are also persistent in the O group. A promising family of magnetic skyrmions, A_2Mo_3N with A=Fe, Co, Rh, and their alloys is found [35]. Interestingly, the pure Co_2Mo_3N shows antiferromagnetic behavior and an unexpected superconducting phase in Rh_2Mo_3N at a critical temperature 4.4 K is also found. Searching skyrmion materials in inversion-asymmetric magnets is a cutting edge topic in magnetism [38].

Table 5.1 Constraints of nonzero d_{ijk} parameters for all possible point groups to create DM interactions. Data is taken from [35]

Class	Constraints	Point groups
Block type $\xi_{\alpha\beta} = 0$		
B-I	No constraints	D_2
B-II	$\alpha_S = \beta_S = \gamma_S = 0$	D_4, D_3, D_6
B-III	$\alpha_A = \beta_A = \gamma_A = 0$	D_{2d}
B-IV	$\alpha_S = \beta_S = \gamma_S$, $\alpha_A = \beta_A = \gamma_A$	T
B-V	$\alpha_S = \beta_S = \gamma_S = 0$, $\alpha_A = \beta_A = \gamma_A$	O
Neel type $\alpha = \beta = \gamma = 0$		
N-I	$\xi_{112} = \xi_{221} = \xi_{331} = \xi_{332} = 0$	C_{2v}
N-II	$\xi_{113} = \xi_{223}, \xi_{112} = \xi_{221} = \xi_{331} = \xi_{332} = 0$	C_{3v}, C_{4v}, C_{6v}
Mixed type		
M-I	No constraints	C_1
M-II	$\alpha = \beta = \alpha = 0$, $\xi_{113} = \xi_{223} = 0$	C_{1h}
M-III	$\alpha_A = \beta_A = \gamma_A = 0$ $\xi_{112} = \xi_{221} = \xi_{331} = \xi_{332} = 0, \xi_{113} = -\xi_{223}$	S_4
M-IV	$\alpha_S = \beta_S = \gamma_S = 0$ $\xi_{112} = \xi_{221} = \xi_{331} = \xi_{332} = 0, \xi_{113} = -\xi_{223}$	C_3, C_4, C_6

5.4 Magnetic Skyrmions in Confined Geometries

As discussed in the first section that the skyrmion-based device is based on the controllable manipulation of individual skyrmions in nanostructured chiral magnets (see Fig. 5.5), particular attention in the field of skyrmions is now focusing on isolated skyrmions in confined geometries. This section will discuss the fabrication of nanostructured samples, real space observation of skyrmions, stability and high flexibility of highly geometrically-confined skyrmions.

5.4.1 Sample Fabrication Techniques

Nanostructured samples can be fabricated by a top-down method from the bulk by using the dual-beam system (Helios Nanolab, 600i FEI, see Fig. 5.7a), i.e. the focused ion beam (FIB) and scanning electron microscope (SEM), combined with a

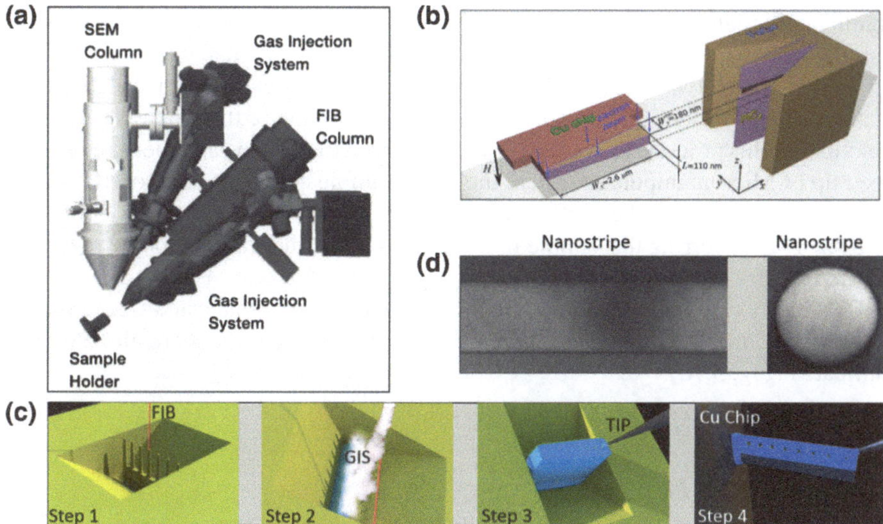

Fig. 5.7 **a** FEI Helios Nanolab 600i System. **b** A schematic of fabricating a wedged-shape FeGe nanostrip that is chosen to study highly geometrically-confined effect on the skyrmion morphology by using off-axis electron holography and Lorentz TEM, as discussed below. The detailed sample fabrication is shown above. In the TEM measurements, the magnetic field induced by controlling the strength of current in object lens is always along the electron beam direction. Both the dimensions of sample and the coordinate system are marked. **c** Schematic procedure for fabricating nanodisk by using FIB-SEM system. **d** TEM images for the final nanostrip and nanodisk. Images are taken from [24, 49]

Gas Injection System (GIS), and Micromanipulator (Omniprobe 200+, Oxford) [39]. Dual Beams system consists of a high-resolution SEM column with a fine-probe ion source (FIB). These instruments allow the preparation of samples from specific areas of a sample as well as nano-machining. The OmniProbe AutoProbe™ 200 in situ sample lift-out system allows the preparation of site specific TEM samples without the need for support films. Figure 5.7b represents briefly the fabrication process of wedged nanostrip for the transmission electron microscopy (TEM) observation. In detail, the whole process is schematically depicted in Fig. 5.7c through Fig. 5.4 steps.

Step 1: Carving a row of FeGe columns on the surface of a FeGe bulk by using FIB with 30 kV voltage. The process is similar to the procedure to fabricate a TEM specimen [40]. The diameter is designed to be tens to hundreds of nanometers depending on the requirements. In this process, a thin amorphous layer with a thickness of 20 nm will be produced because of the damage of high energy Gallium ions. The amorphous surface is further reduced to 2–4 nm by using low energy Gallium ions with 2–5 kV voltage.

Step 2: Coating the FeGe columns with an amorphous PtC_x film. This process is accomplished by combining the Gas Injection System (GIS) and FIB-SEM dual beams. It should be noticed that this amorphous PtC_x is not a magnetic material so that it will not affect the magnetic properties of FeGe sample. The coating layers can significantly reduce the Fresnel fringes around the sample edges to make it possible to image the magnetic structure of nanostructured sample by using Lorentz TEM.

Step 3: Transferring the sample into a silicon surface by using an advanced lift-out method with the help of Omni probe 200+ Micromanipulator that possesses rotation function. In this process, the thin flake is first carved to U-shape and then stuck to the need tip of Micromanipulator. Then, the FeGe sample coating with two PtC_x layers is released from the needle to the surface of the silicon.

Step 4: Transferring the sample into Cu-Chip for further fabrication by recycling the steps 1–3.

By slightly changing the procedure including the sample shape and size, a variety of TEM specimens such as nanostrip, nanodisk and wedged films are all able to be fabricated (Fig. 5.7d).

5.4.2 Lorentz TEM

Imaging magnetic skyrmions in real space is highly desired to understand the basic properties of individual skyrmions in confined geometries. The first real space observation of magnetic skyrmions is accompanied by Lorentz TEM [9]. Since then, almost every magnetic imaging technique has been used to investigate the static and dynamical behaviors (Table 5.2). For example, magnetic force microscopy takes advantage of both the short-range interatomic and long-range magnetostatic interactions between the sharp probe and the magnetic material surface. It has been used to help unveiling the magnetic monopole [12]. Magneto-optical effects originate from the interaction between polarized light and the magnetization and have been used to image the formation of individual skyrmions via geometrical confinement [16]. Pump-probe X-ray transmission electron microscope has high temporal resolution and has been used to studying the inertia effect of skyrmions. A large effective mass of magnetic skyrmion was obtained and attributed to the intrinsic breathing mode that acts as a collective source for inertia [41].

Given the size of individual magnetic skyrmions is commonly 3–150 nm, imaging the complex magnetic structure in nanostructured elements require spatial resolution and sensitivity. TEM-based magnetic imaging technique has then obvious advantages. Traditionally, there are two methods to obtain the magnetic signal in the TEM. They are Lorentz TEM and electron holograph (EH), respectively, as shown in Fig. 5.8.

Lorentz TEM relies on the fact that a high energy beam electron will be deflected by the magnetic induction inside and around the magnetic sample [42]. There are two modes in Lorentz microscopy: the Fresnel mode, in which domain walls and magnetization ripples are observed, and the Foucault mode, where domains are imaged. For the Fresnel mode, the minimum observed contrast indicates the in-focus position of the sample. In this position, the domain walls do not appear. Thus out-of-focus conditions by defocusing the objective lens are used to obtain the magnetic contrast of domain walls (see the schematic electron path through the sample in Fig. 5.8a).

Table 5.2 Magnetic imaging techniques

Name	Resolution		Magnetic field	Temperature	Remarks
	Spatial	Time			
Lorentz TEM	2 nm	ms	*–0.2–2* T	*5–300* K	*Detect magnetic moments In-plane components (Thin films)*
MFM	10 nm	s	–16–16 T[a]	2–300 K[a]	Detect the leakage magnetic field
SP-STM	Atomic	s	–9–9 T[a]	Low	Magnetic moments atomic smooth in sample surface
X-ray Holography	20 nm	ns	Integrated[a]	Unknown	Magnetic moments ultra-fast dynamics
SMOKE	300 nm	ns	–9–9 T[a]	2–300 K[a]	High sensitivity Surface detection
PEEM	5 nm	s	0 T	Room	High sensitivity only surface detection (1–2 nm)

[a]It indicates the parameters can be conveniently adjusted

To illustrate the principle of Fresnel mode, we only consider a sample composed by two domains with a 180° domain wall in the following.

An electron moving through a region of space with an electrostatic field and a magnetic induction field B experiences the Lorentz force F: $F = e(\mathbf{v} \times \mathbf{B})$, where **v** is the velocity of the electron. F acts normal to the travel direction of the electron, a deflection will occur. Note only components of the magnetic induction normal to the electron beam give rise to a deflection. The deflection angle is linked to the in-plane magnetization and the thickness of the sample. The Lorentz force from the magnetic specimen acts on the electrons passing through the specimen and splits each of the diffraction spots into two (as shown schematically in the Fig. 5.8a). One split spot contains information from domains with magnetization lying in one direction, and the other spot contains information from the antiparallel domains. Under in-focus conditions, there is no magnetic contrast because these deflecting electrons

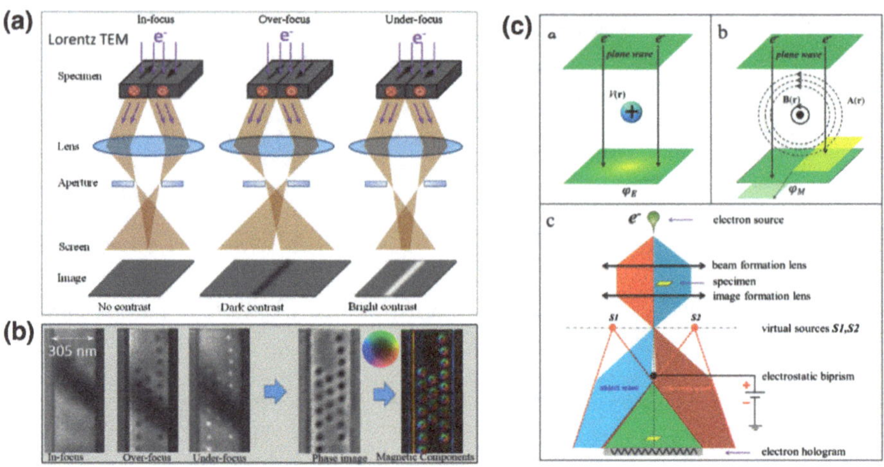

Fig. 5.8 **a** Ray diagram illustrating the Fresnel image of a magnetic thin plate composed by two 180° domain walls. **b** A real process of reconstructing the in-plane components of a FeGe nanostrip by using the TIE method. The defocus values used in these images are 196 μm $\Delta z = 196\,\mu\text{m}$. **c** Schematic diagrams showing different contributions to the electron-optical phase shift arising from local variations in electrostatic scalar potential and magnetic vector potential measured using off-axis electron holography. Electrostatic phase contribution φ_E originating from the electrostatic potential V(r) within and around the TEM specimen. Magnetic phase contribution φ_M originating from the magnetic vector potential $\boldsymbol{A}$(r) within and around the TEM specimen. Simplified ray diagram for off-axis electron holography. Essential components are the coherent electron source, electromagnetic lenses and electrostatic biprism for separation and overlap of two parts of the electron wave to form an electron hologram. The object and reference waves can be considered as originating from two virtual sources *S1* and *S2*. The electron-transparent specimen occupies approximately half of the field of view. For recording information about the magnetic properties of the specimen, the conventional TEM objective lens is normally switched off and a non-immersion Lorentz lens is used as the primary imaging lens. A pre-calibrated magnetic field can then be applied to the specimen in the electron beam direction by exciting the conventional objective lens slightly. The final electron hologram can be recorded digitally for further analysis to yield information about the projected electromagnetic potential within and around the specimen. Images are taken from [24, 49]

are finally focused in the image plane. Under out-of-focus conditions, the magnetic domain walls are imaged as alternate bright (convergent) and dark (divergent) lines depending on the over-focus conditions. The bright lines occur when the domain walls are positioned such that the magnetization on either side deflects the electrons toward the wall. This means that the inversion of the magnetic contrast in the domain wall will be observed when we transfer the over- and under-focus conditions. This is a common feature in the Fresnel mode images.

From the above discussion that the magnetic contrasts at the domain wall depend on the defocused conditions, it is thus possible to reconstruct the in-plane distribution of magnetic components around the domain walls. A commercial software package QPt has been developed to realize the purpose based on the transport-of-intensity equation (5.10) (TIE), where three magnetic images at three defocus values (under-, in-, and over-focus) were used.

$$\frac{2\pi}{\lambda}\frac{\partial I\,(x,y)}{\partial z} = -\nabla_{\mathrm{xy}}.\left[I\,(x,y)\,\nabla_{\mathrm{xy}}\phi\,(x,y)\right] \tag{5.10}$$

where $I\,(x, y)$ and $\phi\,(x, y)$ are the intensity and phase distributions of propagating wave distribution, respectively and λ is the electron wavelength. The desired magnetization **m** of the sample is further obtained by the Maxwell-Ampére equations with the relationship

$$\mathbf{m} \times \mathbf{n} = -\frac{\hbar}{et}\phi\,(x,y) \tag{5.11}$$

where e, $\hbar$ and t are the electron charge, the reduced Planck constant and the thickness of the sample, respectively. **n** is the unit vector along the beam direction. The intensity gradient $\partial I/\partial z$ can be approximately expressed as $\Delta I/\Delta z$, because the defocus step Δz is usually much less than focal length. Here, a representative process to map the in-plane magnetic components of skyrmions in a nanostrip is illustrated in Fig. 5.8b.

While Lorentz TEM provides a high spatial resolution to image small skyrmions, it is still very hard to observe small samples. This occurs because of the Fresnel fringe effect. Under out-of-focus conditions, the abrupt change of the projected electromagnetic potentials or sample thickness in the sample will inevitably lead to Fresnel contrast, which is more obvious around the sample edge [43]. In this sense, the ability of directly observing the fine magnetic structure by using Lorentz TEM in Fresnel model assumes that the thickness variation or projected electrostatic potentials are negligibly small as compared with the magnetic induction contributions [44]. This is obviously not true around the specimen edges, the abrupt change in thickness leads to significant variation in Fresnel fringes, smearing out the real magnetic contrast. The analysis of magnetic information at the edge is thus extremely difficult. Previous Lorentz TEM investigations on FeGe thin plates have illustrated the artificial contrast extends above ~100 nm[26]. Concerning the helical period of FeGe studied here, it is ~70 nm. This is sufficient to completely eradicate or severely distort the real domain structure of the edge of interest. By contrast, if the nanostructured samples such as nanodisk and nanostrip are coated by amorphous PtC_x, the Fresnel fringes effect will be significantly decreased because of the negligible thickness variation, making it possible to observe nanostructured samples.

5.4.3 Off-Axis Electron Holography for Imaging Magnetic Contrast

According to the Ehrenberg-Siday-Aharonov-Bohm (ESAB) effect in quantum mechanics, the wave function of an electrically charged particle is affected by the electromagnetic potential through which it traverses. In TEM, the phase change of an electron wave that traverses an electron-transparent specimen (written in one dimension here for simplicity) can be expressed in the form

$$\varphi_{EM}\left(\mathrm{x}\right) = \varphi_E + \varphi_M = C_E \int V\left(x, z\right) dz - 2\pi \frac{e}{h} \iint B_{\perp}\left(x, z\right) dx dz, \qquad (5.12)$$

where x is a direction in the plane of the specimen, z is the incident electron beam direction, C_E is an interaction constant that takes a value of 6.53×10^6 rad·V·m^{-1} at a microscope accelerating voltage of 300 kV, V is the electrostatic potential within and around the specimen and $B_{\perp}$ is the in-plane component of the magnetic induction within and around the specimen. The total recorded phase φ_{EM} is the sum of the electrostatic contribution to the phase φ_E originating from the electrostatic potential $V\left(r\right)$ (see the left-top panel in Fig. 5.8c) and the magnetic contribution to the phase φ_M (see the left-bottom panel in Fig. 5.8c).

One of the most widely used techniques for recording the total phase φ_{EM} within and around a specimen directly is the TEM mode of off-axis electron holography. The technique requires the use of a highly coherent field emission gun (FEG) electron source to examine a specimen, in which the region of interest is positioned so that it occupies approximately half of the field of view. The application of a voltage to an electron biprism results in overlap of part of the electron wave that has passed through vacuum alone with part of the same electron wave that has passed through the specimen, as shown schematically in right panel in Fig. 5.8c. If the electron source is sufficiently coherent, then an interference fringe pattern (an electron hologram) is formed in the overlap region, in addition to an image of the specimen. The amplitude and the phase of the specimen wave are encoded in the intensity and the position, respectively, of the interference fringes. For studies of magnetic materials, a Lorentz lens (a high-strength minilens) allows the microscope to be operated at high magnification with the objective lens switched off and the sample located in magnetic-field-free conditions. An external magnetic field can then be applied to the specimen either by using a magnetizing specimen holder or by exciting the conventional microscope objective lens to a pre-calibrated value.

It should be noted that the total phase change φ_{EM} is generally a sum of the electrostatic contribution to the phase φ_E arising from local variations in specimen thickness and composition and the magnetic contribution to the phase φ_M arising from the magnetic vector potential associated with the specimen. Since the magnetic phase information is of primary interest, φ_M has to be separated from φ_E, especially close to the edge of the nanostructure, where the specimen thickness and composition change rapidly. This can be obtained by substracting two phase shifts at two temperatures that are below and above the Curie temperature of magnetic material, respectively bacause the electrostatic contribution to the phase φ_E can be recorded as the temperature is above the Curie temperature.

5.4.4 Edge-Mediated Skyrmion Phase and Field-Driven Cascade Phase Diagram

With the advance in the nanostructured sample fabrication and imaging technology, it is possible to study the creation, stability and morphology of highly geometry-confined magnetic skyrmions. These contents are in turn discussed in this and next sections and closely related to the non-trivial topology of magnetic skyrmions. It is well known that topological charge is a constant of motion in the space of continuous functions. Therefore, by changing the spin texture from a ferromagnetic or collinear state to a skyrmion state, we must change the topological charge. It is hard to be realized due to an energy barrier, but can be realized in nature by defects or edges initiating Bloch points or other discontinuous states, facilitating the switch of the topological charge. This claim is based on topologically protected property of skyrmion states. More precisely, in geometrically confined systems (such as nanowires), skyrmion configurations are not topologically protected and a transition between uniform and skyrmion state would not be facilitated via singularity (Bloch Point)—they can always be created or destructed at the boundary [45].

This behavior is actually what was observed in the nanodisk. To investigate the phase transitions between different topological classes in detail, a FeGe nanodisk with a diameter of $D \sim 270$ nm is fabricated by the above-mentioned top-to-down method. The size of the nanodisk is designed to be about three times larger than the helical period of B20 FeGe, ~70 nm. The morphology of the nanodisk is shown in Fig. 5.9a. The orientation of magnetic field **B**, pointed upward, is marked by red "⊙". The helical ground state is obviously observed at the temperature $T \sim 100$ K, as shown in Fig. 5.9b, in which the dark and bright stripes are the magnetic contrasts standing for the in-plane magnetic moments distributions. Following the above-mentioned TIE method, the in-plane magnetic configuration of the spin helix can be constructed (Fig. 5.1c). The spin helices have four turns and are distorted around the sample edge (Fig. 5.9c, e) due to the spatial confinement leading to the specific form of the boundary conditions [30, 46].

Under the action of magnetic field, the spin orientation of the helical state will be changed, where confined helices with only three turns are formed at $B \sim 142$ mT. The period of interior helical state is almost unchanged ~70 nm (Fig. 5.9d). At $B \sim 162$ mT, the helix will change into elongated skyrmions by shrinking the length of spin helix. These elongated skyrmions are also called bimerons [47] (Fig. 5.9e, n). This process is always accompanied by forming a completed vortex-like edge state according to theoretical prediction. But, it should be noted that the observed vortex-like magnetic contrast around the nanodisk edge is a mixture of real edge state and Fresnel fringe-induced artificial contrast [48]. With further increasing the magnetic field, the elongated skyrmions will reduce their size to circular shape. Interestingly, at the relatively low magnetic field, the new-formed skyrmions always stay together around the sample edge, as shown in Fig. 5.9f. This gives a strong hint that the skyrmions attract each other, and also are attracted to the edge [24, 49].

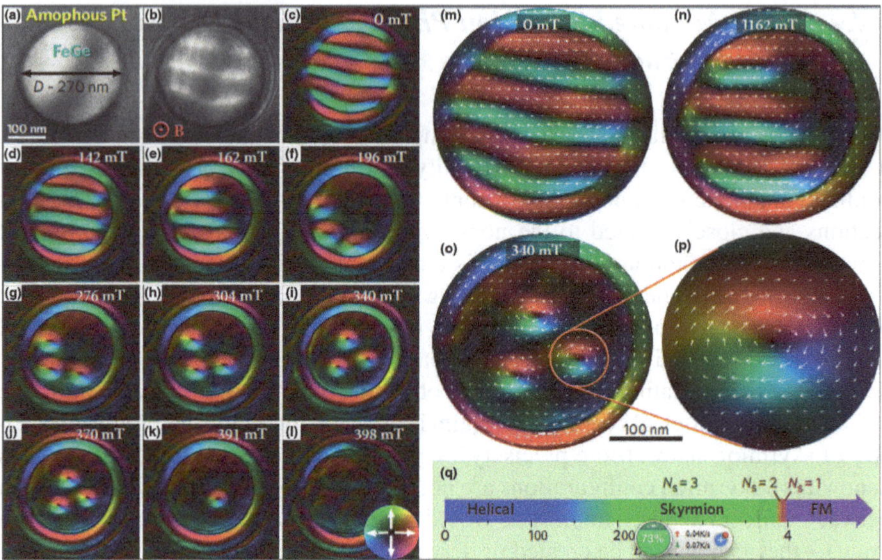

Fig. 5.9 Magnetic field-driven evolution of magnetic structure in a FeGe nanodisk. **a** TEM image of disk that was coated by non-magnetic PtCx. **b** The spin helix ground state observed by Lorentz TEM. **c–l** Magnetic structure as a function of magnetic field with its orientation pointing outward. The direction and strength of in-plane components are represented by a color wheel in panel (**l**). With the increase of the magnetic field, a spin helix transfers into magnetic skyrmions, which further move to the center of disk and disappear one by one at high magnetic field. **m–o** The enlarged images in panels (**c**), (**d**), and (**i**), respectively. **p**. The fine in-plane magnetic structure for a single skyrmion in panel (**o**) **p**. The magnetic field intervals of hosting varied magnetic structures. N_s represents skyrmion number. Images are taken from [49]

The experimental temperature discussed above is 100 K that is far below the magnetic transition temperature T_c ~280 K [20]. It has been well established that decreasing temperature does not benefit the formation of magnetic skyrmions. For example, in 2D FeGe plates with an approximate thickness of the nanodisk, the lowest temperature of hosting the skyrmion lattice is only about 200 K. Here, the temperature region of hosting magnetic skyrmions extends through the whole interval possible in the experiment. The significantly enhanced stability of skyrmion phase in the highly geometry-confined nanodisk is in sharp contrast to the above-mentioned low stability of skyrmions in bulk [50] and also the thickness-dependent stability in two dimensional films [28]. This phenomenon implies that geometric confinement or sample edge may be used to benefit the formation of magnetic skyrmions. Using the method to calculate the topological charge, we can also project the spin of helical ground state into a sphere. It is found that the surrounding region for the helical ground state with distorted edge twists is much larger than that for a pure ground state. It is actually the distorted edge twists that play a positive role to create skyrmions.

These new-formed skyrmions always stay around the sample edge. With the increase of the magnetic field, the skyrmions move into the center of the disk with an increased distance of skyrmion-skyrmion and skyrmion-edge, as shown in Fig. 5.9g–j. This common behavior is always observed in the nanodisk with varied diameter and can be readily explained by the repulsive interactions of skyrmion-skyrmion and skyrmin-edge [51].

Before the annihilation of these skyrmions at highest magnetic field, they form a triangle at a magnetic interval B $\in$ [196 mT, 370 mT], as shown in Fig. 5.9i, j. With the slight increase of the magnetic field to 391 mT, two skyrmions disappear instantaneously, as shown in Fig. 5.9k and the left one skyrmion stays in the center of the disk, leaving one skyrmion sitting nearly at the center of the disk due to the skyrmion-edge repulsive interaction (Fig. 5.9k). The single skyrmion is finally destroyed at the higher field 398 mT, as shown in Fig. 5.9l.

As mentioned above, skyrmion is a topologically non-trivial spin texture with unit topological charge. Then, the observed cluster state with varied skyrmion number N_s in the disk belongs to a different topological state. The field-driven cascade transition between them has thus a common means. Recent magnetoresistance (MR) measurements on single B20 MnSi nanowires have also demonstrated the cascade transitions [52]. Interestingly, compared with the cluster state with small number of skyrmions, both the Lorentz TEM observation and MR measurements showed that the cluster state with maximum number of skyrmions, N_s^m, had a high stability with wider magnetic field intervals at such low temperature (Fig. 5.9q). By contrast, from numerical simulations based on the calculation of equilibrium state, the magnetic intervals of hosting different skyrmion clusters are more complex and comparable at certain conditions, depending on the disk size [53].

At high temperature, skyrmions behave like balls in a box and form a closely-packed arrangement in the disk. A representative phase diagram at varied temperature and magnetic field is shown in Fig. 5.10. As the temperature $T < 190$ K, skyrmions distribute sparsely and form a cluster state, in which the N_s^m is nearly fixed. In other words, the skyrmions do not accommodate the whole disk plane with closed packed mode as expected theoretically. On the contrary, closely packed skyrmions only form at high temperature $T > 190$ K. Similar behavior was also found for nanodisks of different diameters, and provides evidence that the N_s^m at low temperatures is linearly dependent on the diameter of nanodisk.

5.4.5 High Flexibility of Geometrically-Confined Skyrmions

Mathematically, a magnetic skyrmion has a nontrivial geometrical aspect with respect to its spins. Its non-trivial topology persists under continuous deformation, indicating the tunable skyrmion morphology. For example, the elongated or shrunk skyrmions show the same unit topological charge (Fig. 5.11). But, the appearance of a non-trivial topological object should physically be to some extent energy favorable. For the skyrmion materials studied here, previous theoretical calculation and real-space imaging on bulk or two-dimensional films have identified that skymions appear in circle shape and condense into crystal lattice with a fixed lattice constant [9, 37].

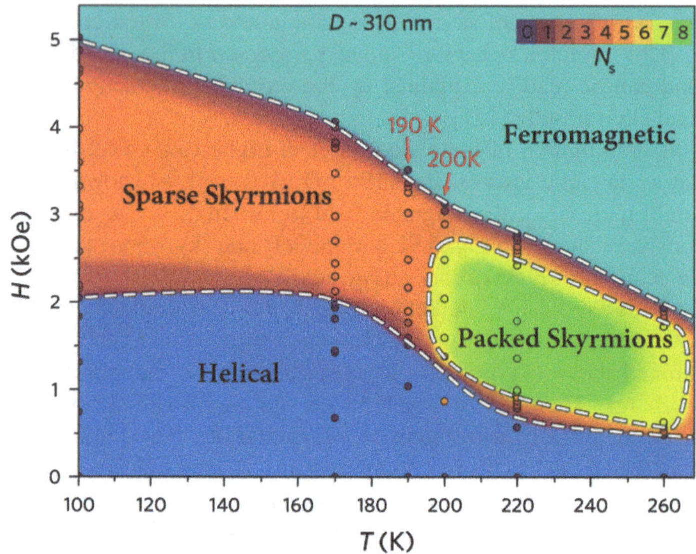

Fig. 5.10 *Magnetic* phase diagram for a FeGe nanodisk with a diameter of 310 nm. Skyrmions form a cluster state with its maximal number depending on the turn number of initial helical state at low temperature. At high temperature, skyrmions form a closely-packed arrangement. Image is taken from [49]

Accordingly, the benefits of its topological stability are not well exhibited though its topological properties (chirality and winding number) have been manifested by the emergent topological Hall effect [14, 15]. By contrast, as the material dimension is reduced to be comparable to the featured skyrmion size, the topological stability enables a skyrmion to change its shape or size to match the sample geometry. Investigating the skyrmion stability in magnetic nanostructures thus provides a model system to exploit the topological stability under the subject of energy stability.

With the common model describing B20 chiral magnets [27], numerical simulation has confirmed the hypothesis. In detail, consider a nanostripe with a width W_y that is comparable to single skyrmion size and a length W_x. Figure 5.12 shows the skyrmion morphology at a middle magnetic field in the nanostripes with varied width. A critical width W_y^c is determined according to the numerical calculations. Below this critical width, skrymions show shrunk shape. Approximately, the morphology of skyrmions can be described by elliptical shape with their semi-axes a and b along and perpendicular to a nanostripe, respectively. Above the critical width W_y^c, skyrmions show elongated shape. For a wider nanostripe, two zigzag chains of skyrmions form. It should be noticed that the numerical simulation is based on real nanostripes with certain thickness. For a single layer system, theoretical analysis has demonstrated that elliptical skyrmions would lose its stability from the viewpoint of energy stability.

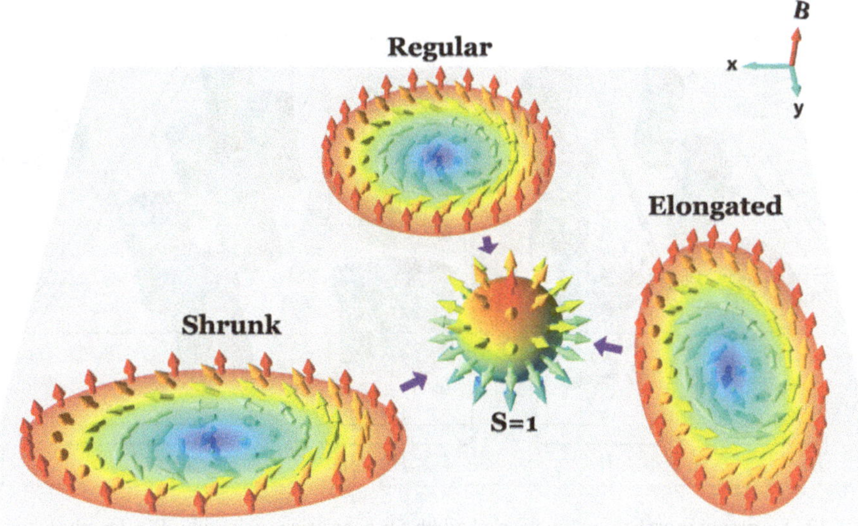

Fig. 5.11 Schematic representations of topological deformation of magnetic skyrmions by confined geometry. Three kinds of skyrmion configurations (regular, shrunk, and elongated) possess the same unit topological charge, S = 1, because the magnetic moments in these skyrmions can all cover the whole unit sphere only one time

Experimental verification of these elliptical skyrmions has also been done on a wedged-shape sample fabricated by the above-mentioned top-to-down method (Fig. 5.7b). The width of the nanostripe is designed to cross the featured skyrmions lattice constant of FeGe materials (Fig. 5.13a). Due to these ultrasmall sizes, even coating the sample with amorphous PtC_x in the Lorentz TEM cannot obtain the accurate magnetic structure, especially around the sample edge. By contrast, the off-axis EH technique, introduced in Sect. 4.2, uses the in-focus condition to make it possible to accurately map the magnetic induction of nanostructured sample with its size as small as tens of nanometers. This advance provides a unique opportunity to test the predicted existence of twisted edge states [55, 56].

At zero magnetic field, the nanostripe shows a helical ground state with complex arrangements (Fig. 5.13b). Under the action of magnetic field, the spin helices follow the common evolution of helical magnets and change into skyrmions. Due to the strong spatial confinement, only a chain of skyrmions appears (Fig. 5.13c). Similar to the numerical results, shrunk, circular and elongated skyrmions appear in turn with the increase of the width of nanostripe. As the size of skyrmions will decrease with the increase of magnetic field, two chains of skyrmions at the wide part of the nanostripe appear at high magnetic field (Fig. 5.13d). Notably, skyrmions in the narrow part of the nanostripe disappear or migrate at high magnetic field. This implies that the stability of individual skyrmions in confined geometries depends on the dimension of the sample. Moreover, a complete chiral edge twist, characterized by a single-twist rotation of magnetization, is directly observed in the induction maps [57]. Physically,

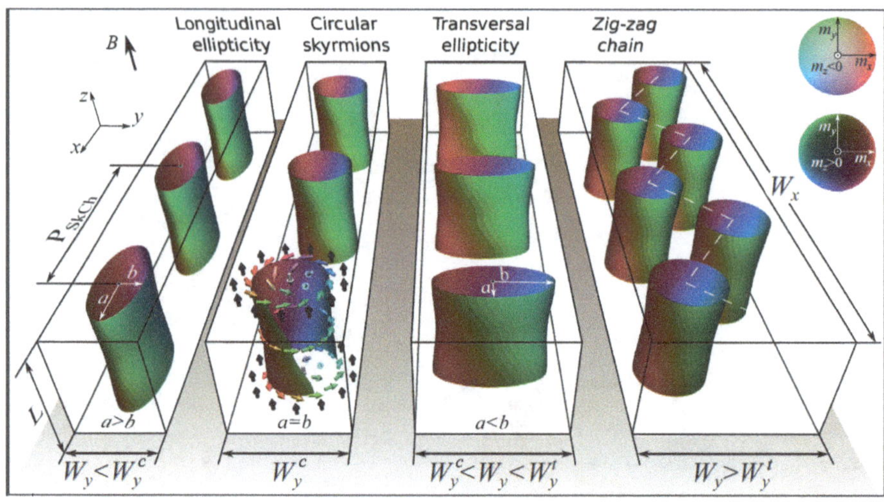

Fig. 5.12 Skyrmion morphology with varied width for nanostripe with width W_y, length W_x and thickness L. The magnetization with its three components m_x, m_y and m_z is presented by two color wheels. For clarity, only isosurfaces for $m_z = 0$ are plotted. Skyrmions in the nanostripe form a nonhomogeneous tube with varied shape from longitudinal to transversal ellipticity with the increase of the sample width. The color wheel stands for the strength and direction of the in-plane magnetization at each point. Images are taken from [54]

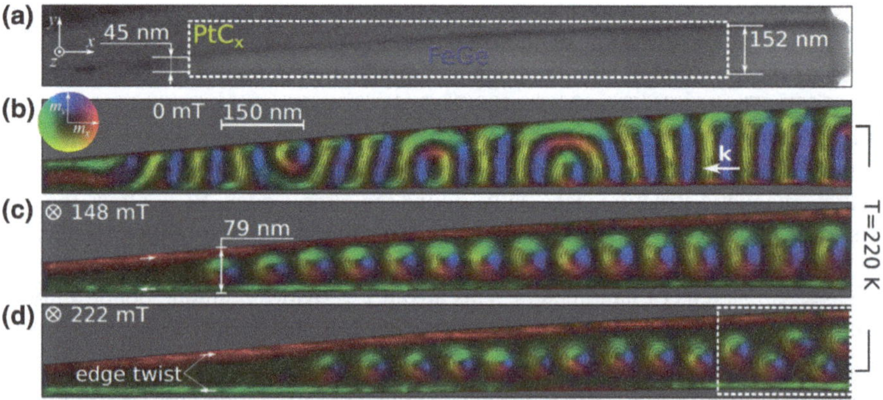

Fig. 5.13 Skyrmion arrangments in a FeGe nanostripe with varied width at T = 220 K. **a** TEM image of the wedge-shaped FeGe nanostripe. The white frames are the regions chosen for the off-axis EH measurements. The magnetic field points downward represented by the symbol "⊗". **b–d** field-driven evolution of spin textures in such a nanostripe. Skyrmions with controllable morphology are clearly observed

such an edge state represents a type of surface state in a chiral magnet to preserve the magnetic chirality of the whole spin texture.

5.5 Conclusions

We have completed a brief introduction through the field of magnetic skyrmions. Starting from a description of limitation of present magnetic memory devices, we have introduced implications of magnetic skyrmions. The demands of skyrmion-based devices require the exploration of new skyrmion materials and understanding of the mechanisms for the formation and stability of geometrically-confined skyrmions.

Acknowledgements This work was supported by the National Key R&D Program of China, Grant No. 2017YFA0303201; the Key Research Program of the Chinese Academy of Science, KJZD-SW-M01; the Natural Science Foundation of China, Grant No. 51622105, 11474290; the Key Research Program of Frontier Sciences, CAS, Grant No. QYZDB-SSW-SLH009; the Youth Innovation Promotion Association CAS No. 2015267; the Major/Innovative Program of Development Foundation of Hefei Center for Physical Science and Technology Grant No. 2016FXCX001.

References

1. D. Weller, A. Moser, IEEE Trans. Mag. **35**, 4423 (1999)
2. A. Brataas, A.D. Kent, H. Ohno, Nat. Mater. **11**, 372–381 (2012) (and reference therein)
3. S. Parkin, S.-H. Yang, Nat. Nanotechnol. **10**, 195 (2015)
4. H.B. Braun, Adv. Phys. **61**(1), 1–116 (2012)
5. O. Tchernyshyov, G.W. Chern, Phys. Rev. Lett. **95**, 197204 (2005)
6. L. Thomas, M. Hayashi, R. Moriya, C. Rettner, S. Parkin, Nat. Commun. **3**, 810 (2012)
7. T.H.R. Skyrme, Nucl. Phys. **31**, 556 (1962)
8. S. Mühlbauer, B. Binz, F. Jonietz, C. Peiderer et al., Science **323**, 915 (2009)
9. X.Z. Yu, Y. Onose, N. Kanazawa, J.H. Park et al., Nature **465**, 901 (2010)
10. N. Nagaosa, Y. Tokura, Nat. Nanotechnol. **8**, 899–911 (2013)
11. S. Rohart, J. Miltat, A. Thiaville, Phys. Rev. B **93**, 214412 (2016)
12. P. Milde, D. Köhler, J. Seidel, L.M. Eng et al., Science **340**, 1076–1080 (2013)
13. M. Lee, W. Kang, Y. Onose, Y. Tokura et al., Phys. Rev. Lett. **102**, 186601 (2009)
14. A. Neubauer, C. Pfleiderer, B. Binz, A. Rosch et al., Phys. Rev. Lett. **102**, 186602 (2009)
15. N. Kanazawa, Y. Onose, T. Arima, D. Okuyama et al., Phys. Rev. Lett. **106**, 156603 (2011)
16. W.J. Jiang, X.C. Zhang, G.Q. Yu, W. Zhang et al., Nat. Phys. **13**, 162–169 (2017)
17. K. Litzius, I. Lemesh, B. Krüger, P. Bassirian et al., Nat. Phys. **13**, 170–175 (2017)
18. A. Fert, V. Cros, J. Sampaio, Nat. Nanotechnol. **8**, 152–156 (2013)
19. W.J. Jiang, P. Upadhyaya, W. Zhang, G.Q. Yu et al., Science **349**, 283–286 (2015)
20. U.K. Rößler, A.A. Leonov, A.N. Bogdanov, J. Phys.: Conf. Ser. **303**, 012105 (2011)
21. T. Okubo, S. Chung, H. Kawamura, Phys. Rev. Lett. **108**, 017206 (2012)
22. S. Heinze, K. von Bergmann, M. Menzel, J. Brede et al., Nat. Phys. **7**, 713 (2011)
23. K. Shibata, X.Z. Yu, T. Hara, D. Morikawa et al., Nat. Nanotechnol. **8**, 723 (2013)
24. H.F. Du, R.C. Che, L.Y. Kong, X.B. Zhao et al., Nat. Commun. **6**, 8504 (2015)
25. T. Moriya, Phys. Rev. **120**, 91 (1960)
26. X.Z. Yu, N. Kanazawa, Y. Onose, K. Kimoto et al., Nat. Mater. **10**, 106 (2011)
27. F.N. Rybakov, A.B. Borisov, A.N. Bogdanov, Phys. Rev. B **87**, 094424 (2013)
28. X.Z. Yu, J.P. DeGrave, Y. Hara, T. Hara et al., Nano Lett. **13**, 3755–3759 (2013)
29. A.B. Butenko, A.A. Leonov, U.K. Roessler, A.N. Bogdanov, Phys. Rev. B **82**, 052403 (2010)
30. M.N. Wilson, E.A. Karhu, A.S. Quigley, U.K. Rößler et al., Phys. Rev. B **86**, 144420 (2012)
31. S. Seki, X.Z. Yu, S. Ishiwata, Y. Tokura, Science **336**, 198 (2012)
32. A.N. Bogdanov, D.A. Yablonskii, Sov. Phys. JETP **68**, 101 (1989)

33. R.R.Birss, *Symmetry and Magnetism*, vol. III (North-Holland Publishing Company, 1964)
34. A.N. Bogdanov, A. Hubert, Phys. Status Solidi B **186**, 527 (1994)
35. W. Li, C.M. Jin, R.C. Che, W.S. Wei et al., Phys. Rev. B **93**, 060409 (2016)
36. A.N. Bogdanov, A. Hubert, J. Magn. Magn. Mater. **195**, 182 (1995)
37. A.N. Bogdanov, A. Hubert, J. Magn. Magn. Mater. **138**, 255 (1994)
38. W.S. Wei, G.J. Zhao, D.R. Kim, C.M. Jin et al., Phys. Rev. B **94**, 104503 (2016)
39. See the details in the internet: http://www.fei.com/products/dualbeam/helios-nanolab
40. D.C. Beaulieu, Electron beam chemical vapor deposition of platinum and carbon. The thesis for degree master (2005)
41. F. Büttner, C. Moutafis, M. Schneider, B. Krüger et al., Nat. Phys. **11**, 225–228 (2015)
42. J.N. Chapman, J. Phys. D: Appl. Phys. **17**, 623–647 (1984)
43. D.B. Williams, C.B. Carter, *Transmission Electron Microscopy: A Textbook for Materials Science* (Springer, New York, 2009)
44. J.N. Chapman, M.R. Scheinfein, J. Magn. Magn. Mater. **200**, 729–740 (1999)
45. D. Cortés-Ortuño, W. Wang, M. Beg, R.A. Pepper, et al. arXiv:1611.07079 (2016)
46. M.N. Wilson, E.A. Karhu, D.P. Lake, A.S. Quigley et al., Phys. Rev. B **88**(21), 214420 (2013)
47. M. Ezawa, Phys. Rev. B **83**(10), 100408(R) (2011)
48. A.O. Leonov, U.K. Rößler, M. Mostovoy, EPJ Web Conf. **75**, 05002 (2014)
49. X.B. Zhao, C.M. Jin, C. Wang, H.F. Du et al., PNAS **113**, 18 (2016)
50. H. Wilhelm, M. Baenitz, M. Schmidt, U.K. Rossler et al., Phys. Rev. Lett. **107**, 127203 (2011)
51. X. Zhang, G.P. Zhao, H. Fangohr, J.P. Liu et al., Sci. Rep. **5**, 7643 (2015)
52. H.F. Du, D. Liang, C.M. Jin, L.Y. Kong et al., Nat. Commun. **6**, 7637 (2015)
53. H.F. Du, W. Ning, M.L. Tian, Y.H. Zhang, Phys. Rev. B **87**(1), 014401 (2013)
54. H.F. Du, Z.A. Li, A. Kovacs, J. Caron et al., Nat. Commun. **8**, 15569 (2017)
55. J. Sampaio, V. Cros, S. Rohart, A. Thiaville et al., Nat. Nanotechnol. **8**, 839 (2013)
56. J. Iwasaki, M. Mochizuki, N. Nagaosa, Nat. Nanotechnol. **8**, 742 (2013)
57. S.A. Meynell, M.N. Wilson, H. Fritzsche, A.N. Bogdanov et al., Phys. Rev. B **90**, 014406 (2014)

Chapter 6
Topological Phases of Quantum Matter

Wei-Feng Tsai, Hsin Lin and Arun Bansil

Abstract Role of topology in generating exotic topological phases of quantum matter is discussed. Illustrative examples of 2D quantum spin-Hall insulators, 3D topological insulators, topological crystalline insulators, and topological Weyl and Dirac semi-metals are presented. We also comment on topological superconductors and on the effects of strong electron correlations in driving topological phases.

6.1 Introduction

It is a recent and quite surprising discovery that quantum matter can harbor electronic states protected by topological considerations not unlike those familiar in the more common geometrical context. The first such crystalline materials, the topological insulators (TIs), support topological states protected by constraints of time-reversal symmetry in the presence of spin-orbit coupling effects. However, it has become clear since that many new classes of protected states can be created in quantum matter through combined effects of time-reversal, crystalline and particle-hole symmetries, and the original field of topological insulators has grown more generally into that of topological materials. Edges of two-dimensional (2D) topological materials and surfaces of three-dimensional (3D) topological materials support novel electronic states. For example, the 1D topological edge states in 2D TIs are forbidden to scatter due to constraints of time-reversal symmetry. Similarly, the surfaces of 3D TIs support metallic topological surface states, which are robust against perturbations from non-magnetic impurities and disorder. The symmetry-protected topological states provide new pathways for addressing fundamental scientific questions. The topolog-

W.-F. Tsai
School of Physics, Sun Yat-Sen University, Guangzhou 510275, China

H. Lin
Institute of Physics, Academia Sinica, Taipei 11529, Taiwan

A. Bansil (✉)
Department of Physics, Northeastern University, Boston, MA 02115, USA
e-mail: ar.bansil@northeastern.edu

S. Gupta and A. Saxena (eds.), *The Role of Topology in Materials*,
Springer Series in Solid-State Sciences 189,
https://doi.org/10.1007/978-3-319-76596-9_6

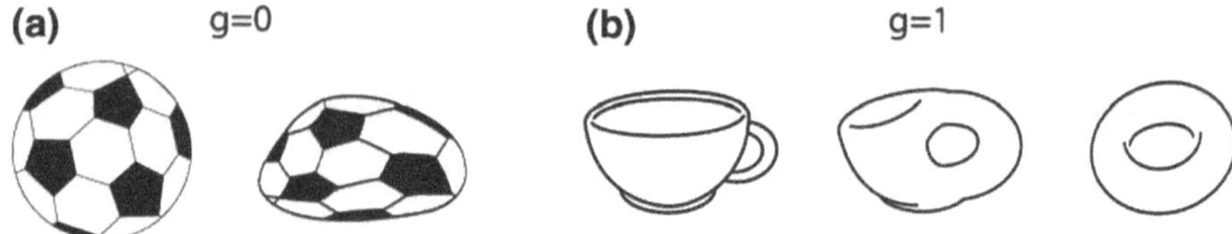

Fig. 6.1 **a** Crumbled and spherical balls have the same genus of $g = 0$, making them topologically equivalent. **b** Similarly, both the doughnut and the coffee cup with one hole are characterized by $g = 1$ (from [5])

ical materials are also expected to provide platforms for a new generation of devices for wide-ranging applications based on exploiting unique properties of protected states.

In providing an overview of the field, this chapter starts with clarifying how topological considerations enter condensed matter physics, followed by a discussion of illustrative examples of 2D quantum spin-Hall insulators, 3D topological insulators, topological crystalline insulators, and topological Weyl and Dirac semi-metals. We then comment on topological superconductivity and strongly correlated topological materials.

The list of references is intended to be minimal. We refer the reader to three comprehensive articles that have appeared on the subject in the Reviews of Modern Physics for extensive discussions and as windows on the large literature on various aspects of topological materials [5, 56, 128].

6.2 Topology in Condensed Matter Physics

Topology in mathematics addresses questions of invariance of global properties of objects under continuous geometrical deformations. For example, surfaces of spherical and crumpled balls in Fig. 6.1a are topologically equivalent because one surface can be deformed into the other without requiring any cutting or gluing of the surfaces. The coffee cup and the doughnut shown in Fig. 6.1b are also topologically equivalent, but the ball and the doughnut are not. The topological distinctiveness here can be coded in terms of the so-called genus number g, which counts the number of holes in the object. The formal connection between geometry and the related topological invariant g can be rigorously expressed via the celebrated Gauss-Bonnet theorem, $\int_S K dS = 2\pi\chi$, where K is the local Gaussian curvature, the integration is over the closed surface, and $\chi = 2(g - 1)$. In particular, the ball has a g number of 0, while the coffee cup or a doughnut a value of 1.

The introduction of topological concepts in condensed matter physics can be traced back to the 1980s when the integer and fractional quantum Hall (QH) effects were discovered [83, 154]. The integer QH state occurs when a strong magnetic field is applied to a two-dimensional (2D) electron gas in a semiconductor. This state hosts completely filled (Landau) bands like an insulator, but it also shows the presence of a non-vanishing Hall conductivity,

$$\sigma_{xy} = N\frac{e^2}{h}, \tag{6.1}$$

which has been found experimentally to be quantized to a precision of 1 part in 10^9 [84] independent of band structure details. These results clearly show that the QH state is fundamentally different from an ordinary insulator. The standard Ginzburg-Landau-Wilson paradigm of symmetry broken phases (emergence of a magnetic phase, for example, through spin SU(2) symmetry breaking) cannot be used to characterize the QH state.

Robustness of the quantization of Hall conductivity in the QH state is reminiscent of the topological invariance seen in the geometric context in Fig. 6.1. Indeed, in 1982 Thouless, Kohmoto, Nightingale, and den Nijs (TKNN) provided a new way to look at the integer QH state by relating a topological invariant, the first Chern number C, to a physical observable, the Hall conductivity [151]. The key is to recognize that the 2D band structure of an insulator can be viewed as a mapping between the crystal momentum $\mathbf{k}$ in the first Brillouin zone (FBZ) and the Bloch Hamiltonian $H(\mathbf{k})$. The gapped band structures can then be classified "topologically" into inequivalent classes in the sense that within a given class the band structures can be deformed continuously without closing the band gap. For QH states, the topological invariant C belongs to the set of integers defined by

$$C = N = \frac{1}{2\pi}\int_{FBZ} F(\mathbf{k})d^2k, \tag{6.2}$$

where $F(\mathbf{k}) = [\nabla_k \times \mathbf{A}(\mathbf{k})]_z$ is the Berry curvature derived from the Berry connection, $\mathbf{A}(\mathbf{k}) = i\sum_n \subset \text{occ.}\langle u_n(\mathbf{k})|\nabla_k u_n(\mathbf{k})\rangle$. Here, $u_n(\mathbf{k})$ are the Bloch states for the nth band and the summation is over all occupied bands. The integer nature of C is easy to understand: The surface integral in (6.2) counts the total vorticity of the U(1) phase of the Bloch wave function in the (magnetic) FBZ [57]. We emphasize that the integrals here are in *momentum space*, not the real space.

If one further constrains the 2D system by time-reversal symmetry (TRS), it can be shown that the Chern number must vanish or that the associated insulating state must be trivial. However, in 2005, Kane and Mele [77] for graphene and Zhang et al. [7] for strained semiconductors showed that even when $C = 0$, the ground state can be a topologically non-trivial 2D *topological insulator* (TI), also referred to as a *quantum spin Hall insulator* (QSHI). This topological state can be intuitively viewed as a superposition of two copies of the same integer QH state but with opposite spin polarizations or Hall conductivities. In other words, assuming S_z^{tot} is conserved and the two spin components are decoupled, the Chern numbers for spin-up ($C\uparrow$) and spin-down ($C\downarrow$) can be defined separately such that $C\uparrow = -C\downarrow$ as required by TRS. Obviously, now the charge Chern number $C\uparrow + C\downarrow = 0$, but $(C\uparrow - C\downarrow)/2$, which we can call as a *spin Chern number*, can be non-vanishing and encodes a non-trivial topological invariant. In general, however, S_z^{tot} is not conserved, and the relevant topological invariant proposed by Fu et al. [48, 50] and others [51, 106, 127], is the Z_2 number ν_0,

$$\nu_0 = \frac{1}{2\pi}\left[\oint_{\partial\tau} \mathbf{A}(\mathbf{k}) \cdot d\mathbf{l} - \int_{\tau} F(\mathbf{k}) d^2k\right], \quad (6.3)$$

where the integration is over half of the FBZ τ and its boundary $\partial\tau$. Non-trivial topological number $\nu_0 = 1$, similar to the Chern number, also links to a quantized physical quantity, namely, the change of the "time-reversal polarization" over τ [49].

When the time-reversal invariant topological insulator is generalized to the three-dimensional (3D) case, the ground state is characterized by four Z_2 topological invariants, see Sect. 6.4 below for further details. One of these four invariants (the strong index) is associated with the axion term of the electrodynamics of the insulator in compact space [127, 128], $H_{\text{axion}} = P^3\mathbf{E} \cdot \mathbf{B}$, which is measurable, in principle, via electromagnetic induction experiments [123] since the TRS quantizes P_3 to be either zero (Z_2 trivial) or one-half (Z_2 non-trivial).

A remarkable property of time-reversal invariant TIs is the existence of gapless edge states in 2D or surface states in 3D, which is not necessarily the case in other types of TIs [43]. This *bulk-edge correspondence* has been confirmed experimentally [61, 62, 64]. In particular, the number of Dirac-like edge or surface modes must be *odd* in the non-trivial Z_2 topological phase, a property that can be used to identify a Z_2 TI. In addition to displaying a linear dispersion around the Dirac points, edge and surface modes show absence of backscattering for *non-magnetic* impurities due to the spin-momentum locking required by constraints of TRS [88, 89, 96, 57] in the presence of spin-orbit coupling (SOC) effects. As a result, edge modes yield nearly quantized spin currents, making these modes highly suited for spintronics applications.

The introduction of crystalline and particle-hole symmetries in the mix greatly expands the menu of symmetry protected states that can be harbored by quantum matter. Schnyder et al. give a list of bulk topological invariants possible for each of the ten classes of Hamiltonians in the presence/absence of three internal symmetries: TRS, particle-hole symmetry, and chiral (sublattice) symmetry for one-, two-, and three spatial dimensions [139]. Other examples of studies along these lines are as follows. Existence of a Z_2 phase is found in 3D spinless TRS TIs with C_4 point group symmetry [45], inversion symmetric (IS) TIs [72, 156], reflection symmetry protected TIs [24, 65, 142], point group symmetric (PGS) TIs [39], and space group symmetric (SGS) TIs [94, 143, 145].

6.3 HgTe/CdTe Quantum Wells and Quantum Spin-Hall Insulators

HgTe/CdTe quantum well is the first materials realization of a QSHI, predicted by Bernevig et al. [7] and then experimentally confirmed by König et al. [88, 89]. In most common semiconductors, the conduction band is derived from *s*-type electrons, while the valence band is composed of *p*-type electrons. However, for a material such as HgTe with strong SOC, this "natural order" is inverted as *p*-like valence states get pushed *above* the *s*-like conduction states. In contrast, this inversion does not occur in CdTe. Therefore, in a HgTe/CdTe quantum well in which HgTe layers of thickness d sandwich CdTe layers, the effective strength of the SOC increases with increasing value of d. Beyond a critical thickness $d > d_c$ of the HgTe layers, the effectively 2D band structure of the quantum well heterostructure becomes inverted at Γ, and leads to a bulk nontrivial Z_2 number. This theoretical proposal was verified by Molenkamp's group by observing a non-vanishing, nearly quantized longitudinal conductivity plateau as a function of the gate voltage for thickness of the HgTe layers greater than 6.5 nm. In addition to HgTe/CdTe, InAs/GaSb/AlSb quantum well systems have also been the subject of much interest [31, 85, 86].

Graphene was one of the earliest proposals for realizing a 2D TI [77], although its band gap is too small to be accessible experimentally [196]. A possible route to overcome this difficulty is to introduce adatoms on a graphene sublattice [67]. In the absence of the SOC, the low-energy spectrum of graphene consists of two gapless (massless) Dirac cones centered at the BZ corners K and K′, which are not time-reversal invariant momentum (TRIM) points. When the SOC is turned on [77], a band gap opens up in the spectrum and massive Dirac fermions are generated. SOC does not induce changes in the parity of eigenstates or band inversions at the TRIM points Γ and M. Since the band structure at Γ is already inverted with respect to M, the ground state can be shown to be a QSHI with $\nu_0 = 1$.

The preceding discussion suggests that graphene-like lattice or structure would provide a natural breeding ground for realizing 2D TI phases. There have been many such 'beyond graphene' theoretical proposals [14, 102], although the experimental realization beyond the quantum-well systems has remained elusive. For example, like graphene a single layer of Si, Ge and Sn atoms can form stable, atomically thin crystals yielding silicene, germanene and stanene with advantage over graphene of a stronger SOC. Moreover, unlike the flat structure of graphene, the atomic bonds in most beyond graphene 2D materials are naturally buckled [93]. Honeycomb III-V thin films are a natural extension of silicene, and low-buckled GaBi, InBi, and TlBi thin films are predicted to be 2D TIs [26]. Similarly, first-principles calculations predict films of Sn compounds SnX (X= H, I, Br, Cl, F or OH) to be in the QSH phase [187], with the hydrogenated version called stanane. The insulating (topological) gap is predicted to be as large as 300 meV in SnI. Functional thin films of Bi and Sb have been predicted to be large gap TIs with band gaps ranging from 0.74–1.08 eV [147].

We consider silicene and the related germanene and stanene films to illustrate advantages of beyond graphene 2D materials for potential spintronics applications.

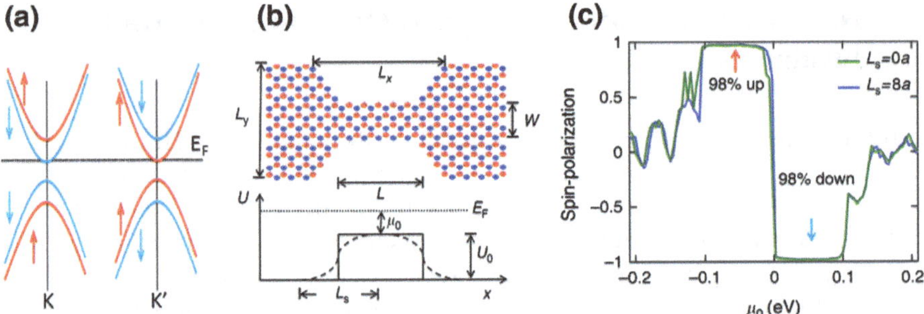

Fig. 6.2 **a** A schematic representation of the band structure of silicene around the K and K′ symmetry points in an applied perpendicular electric field. **b** Design of a spin-filter based on a quantum point contact geometry with a potential profile $U(x)$. **c** Spin-polarization of the filter as a function of the potential barrier height μ_0. (from [152])

Band structures of silicene, germanene and stanene are similar, but the SOC-induced gap predicted in germanene is 23 meV, while that in stanene is 72 meV, which are large enough to withstand room-temperature excitations [93, 152]. Moreover, the inversion symmetry (i.e. symmetry of A and B sublattices) in beyond graphene films can be broken by a perpendicular electric field [30, 152]. As a result, topological phase transitions between the QSHI and ordinary insulator phases can be realized and spin degeneracies can be lifted near the K and K′ points through gating control (see Fig. 6.2a). The spin-split states are nearly 100% spin-polarized and can be tuned by external fields. These states are thus well-suited for the design of devices for filtering spins (see Fig. 6.2b, c), and manipulating spin currents [55, 152].

6.4 Z_2 Topological Insulators in Three Dimensions

As we already noted in Sect. 6.2 above, one needs four Z_2 indices, $(\nu_0; \nu_1\nu_2\nu_3)$, to completely classify 3D topological insulators [50]. When all four indices are zero, the system is an ordinary insulator but if any of the indices is non-zero then we have a non-trivial TI. $(\nu_1\nu_2\nu_3)$ are 'weak' indices and their non-zero values signify a 3D TI, which is obtained by stacking 2D TIs. ν_0 is the "strong" index, and when it assumes a non-zero value, we have a 3D TI without a 2D analogue. Any surface of a strong 3D TI is guaranteed to host gapless (Dirac) surface bands.

The Z_2 indices can be computed by considering the unitary matrix $w_{mn}(\mathbf{k}) = \langle u_m(\mathbf{k})| \Theta |u_n(-\mathbf{k})\rangle$, where $|u_m(\mathbf{k})\rangle$ are occupied Bloch functions, and Θ denotes the antiunitary time-reversal symmetry operator. The strong index ν_0 is then determined by the formula [50].

$$(-1)^{\nu_0} = \prod_{i=1}^{8} \delta_i, \tag{6.4}$$

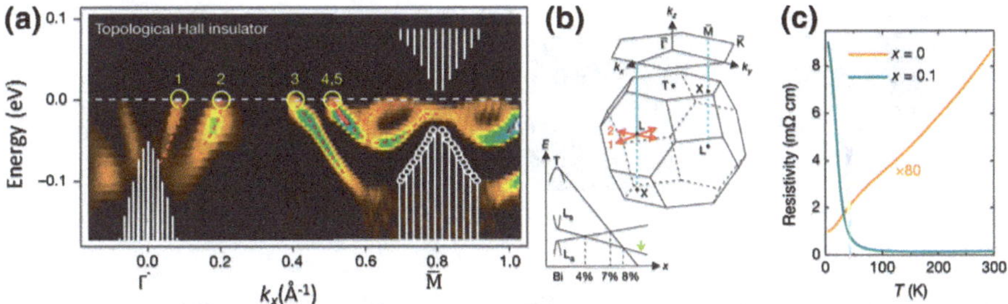

Fig. 6.3 **a** The (111) occupied surface bands of $Bi_{0.9}Sb_{0.1}$ probed by ARPES. The surface bands cross the Fermi energy five times (excluding time-reversed partners), indicating the non-trivial topological nature of the system. **b** 3D Brillouin zone of the alloy and its (111) surface projection. Schematic evolution of bulk band energies as a function of the replacement percentage x of Bi is shown. **c** Resistivity of Bi and the alloy with $x = 0.1$. (from [61])

where $\delta_i = \mathrm{Pf}\,[w(\mathbf{\Lambda}_i)]\,/\sqrt{\mathrm{Det}\,[w(\mathbf{\Lambda}_i)]} = \pm 1$, is evaluated at the TRIM points Λ_i in the FBZ in terms of the Pfaffian of the matrix $w(\mathbf{\Lambda}_i)$. The other three Z_2 indices are obtained from partial products of sets of four δ_i's, similar to those of (6.4), corresponding to TRIM points lying in three independent planes of the FBZ in 3D. Notably, (6.4) can be simplified for an inversion symmetric system [48]:

$$\delta_i = \prod_m \xi_m(\mathbf{\Lambda}_i), \tag{6.5}$$

where the product is over the parities of pairs of occupied Kramer's doublets resulting from TRS at the TRIM points without the corresponding TR partners.

$Bi_{1-x}Sb_x$ alloys were the first material system predicted to host the 3D TI phase [48], which was later shown to be realized experimentally [61]. The development of the non-trivial topological phase in this case can be understood through a band inversion mechanism. The key observation is that although Bi and Sb are both semimetals (see Fig. 6.3c), the orbital nature of conduction and valence bands in Bi and Sb at the three L-points in the rhombohedral FBZ is opposite (see Fig. 6.3b). As a result, with increasing Sb content x in the alloy, the band gaps close and reopen at the three L-points with a critical composition at $x \approx 4\%$, and the system becomes a direct-gap semiconductor at $x \approx 8\%$ (see Fig. 6.3b). The telltale topological surface states have been probed in $Bi_{0.9}Sb_{0.1}$ via angle-resolved photoemission spectroscopy (ARPES) [61]. As seen in Fig. 6.3a, the surface state crosses the Fermi level an odd number of times (excluding time-reversed partners). Interestingly, the quasi-particle interference (QPI) patterns in scanning-tunneling-spectroscopy (STS) experiments show the expected suppression of the backscattering channels due to spin-momentum locking of the topological surface states [53, 130].

Unlike the multiple surface states found in $Bi_{1-x}Sb_x$, the second generation 3D TIs, Bi_2Se_3, Bi_2Te_3, and Sb_2Te_3 host a single Dirac cone (Fig. 6.4a) on their naturally cleaved (111) surfaces [21, 62, 63, 176, 202]. These TIs contain building blocks of quintuple layers (QLs) and their non-trivial nature results from band inversions

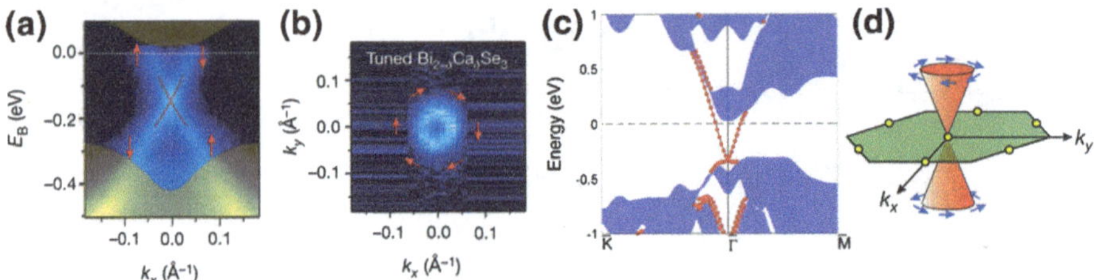

Fig. 6.4 **a** A single spin-polarized, surface Dirac cone is revealed by ARPES in Ca-doped Bi_2Se_3. **b** A chiral left-handed spin texture is observed on the surface Fermi surface. **c** Band structure of Bi_2Se_3 via first-principles calculations is shown with the blue shaded regions representing the bulk states, and the red dots representing the surface states. **d** A schematic plot of the spin-momentum locking surface states in Bi_2X_3 (1;000) topological insulators. (from [62, 176])

driven by SOC in the *p*-orbital manifold at the Γ-point. An important issue for practical applications of TIs concerns the realization of the true bulk insulating state and the manipulation of the position of the Fermi level (E_F) in the presence of intrinsic defects in the material. In this connection, it has been shown, for example, that Ca doping and NO_2 surface deposition can tune E_F to the Dirac point and fully remove the bulk conducting band from E_F [59, 62], while Sb doping in Bi_2Te_3 [87, 204] and Bi_2Se_3 [4] has been shown to control the carrier density and E_F. In $(Bi_{1-x}Sb_x)_2Te_3$ alloy, increasing Sb content shifts E_F down from n- to p-type regime. DFT calculations predict that the position of the topological surface state can be tuned in heterostructures of a TI with various band insulators [105, 174, 205].

Spin-momentum locking of topological surface states in a 3D TI has been observed via spin-ARPES [62, 176] (see Fig. 6.4b, d). Although the Dirac node here is expected to be robust against non-magnetic impurities, doping with magnetic impurities can open a gap, and such a gap opening has been observed in ARPES measurements on Fe/Mn doped Bi_2Se_3 [22, 173, 184]. In particular, spin-ARPES reveals a hedgehog spin texture of the gapped Dirac cones in Mn-doped Bi_2Se_3, which is distinct from that of a gapped Dirac cone due to confinement effects in thin films [114, 184]. Finally, the quantum anomalous Hall effect has been demonstrated in $(Bi,Sb)_2Te_3$ thin films [15, 203], where the surface state tunneling gap in the undoped system can be closed/reopened via Cr doping, and a quantized Hall signal of the expected value is seen.

Among other TI materials, we note the large family of tetradymite-like layered TIs with formulae B_2X_2X', AB_2X_4, $A_2B_2X_5$, and AB_4X_7 (A = Pb, Sn, Ge; B = Bi, Sb; X, X′ = S, Se, Te), which offer substantially greater chemical and materials tunability compared to their binary cousins discussed above. Many of these compounds have been synthesized and ARPES results are available [35, 116, 122, 185, 204]. Another example is Tl-based III-V-VI_2 chalcogenides $MM'X_2$ [M = Tl, M′ = Bi or Sb, and X = Te, Se, or S] [23, 92, 135, 186, 189].

6.5 Topological Crystalline Insulators

Topological crystalline insulators (TCIs) can be viewed as a natural extension of the Z_2 TIs. However, unlike the Z_2 TIs whose topological nature is protected by TRS, the source of topological protection in TCIs mainly comes from the *spatial crystalline symmetries*. Classifications of TCIs based on mirror [24, 65, 142], point-group [39], non-symmorphic [94, 143] and space group symmetries [145] have been delineated. TCIs provide a number of features distinct from the Z_2 TIs. For instance, TCIs can support topological surface states with non-linear energy dispersions, and their band gaps could be opened and controlled by external electric field or strain. Such surface states, therefore, provide a new playground for exploring novel physics with potential applications as field-effect transistors, photodetectors, and nanoelectromechanical devices [95, 200].

A TCI could be realized without SOC. An example is a 3D TCI model in a tetragonal lattice for spinless fermions, where the topological ground state is protected by the combination of TRS and C_4 point-group symmetries [45]. The TRS operator with $\Theta^2 = -1$, which plays a central role in protecting a Z_2 TI, is now replaced by $(\Theta U)^2 = -1$, where U is the unitary operator for C_4 rotation with respect to z-axis and $\Theta^2 = 1$ for spinless fermions. Therefore, $(\Theta U)^2 = -1$ guarantees two-fold degeneracy (per spin) at the four ΘU-invariant momentum points, Γ, M, Z, and A in the FBZ, similar to Kramers theorem. The Z_2 topological invariant is then given by

$$(-1)^{\nu_0} = \delta_{\Gamma M}\delta_{AZ}, \tag{6.6}$$

where

$$\delta_{\mathbf{k}_1\mathbf{k}_2} = e^{i\int_{\mathbf{k}_1}^{\mathbf{k}_2} d\mathbf{k}\cdot\mathbf{A}(\mathbf{k})}\frac{\mathrm{Pf}\,[w(\mathbf{k}_2)]}{\mathrm{Pf}\,[w(\mathbf{k}_1)]} \tag{6.7}$$

in terms of the Berry connection $\mathbf{A}(\mathbf{k})$ (see Sect. 6.2) and Pfaffians of the antisymmetric matrix $w(\mathbf{k}_i)$, where $w_{mn}(\mathbf{k}_i) = \langle u_m(\mathbf{k}_i)|\, U\Theta\, |u_n(-\mathbf{k}_i)\rangle$. The line integrals are between $\mathbf{k}_1$ and $\mathbf{k}_2$ in the corresponding 2D planes.

As to systems with SOC, we focus on SnTe and $Pb_{1-x}Sn_x$(Se,Te) alloys. These are the first TCI materials proposed theoretically [65] and later realized experimentally [34, 149, 181]. SnTe is a narrow-band semiconductor with rocksalt structure (Fig. 6.5a). Its fundamental band gaps are located at the four L points in the face-centered-cubic (fcc) Brillouin zone, and the ordering of the conduction and valence bands at these points in SnTe is inverted relative to PbTe [65]. However, this band inversion occurs at an even number of L points, so that neither SnTe nor PbTe is a TI but due to the presence of the extra mirror symmetry with respect to the (110) plane, SnTe is distinct from PbTe and supports a mirror-symmetry protected TCI phase (Fig. 6.5b). The corresponding topological invariant is the mirror Chern number $n_M = -2$ [65, 150], which is defined as the difference between the usual Chern numbers calculated in subspaces separated by the distinct mirror eigenvalues. The nonzero value of -2 of the mirror Chern number indicates the existence of two pairs

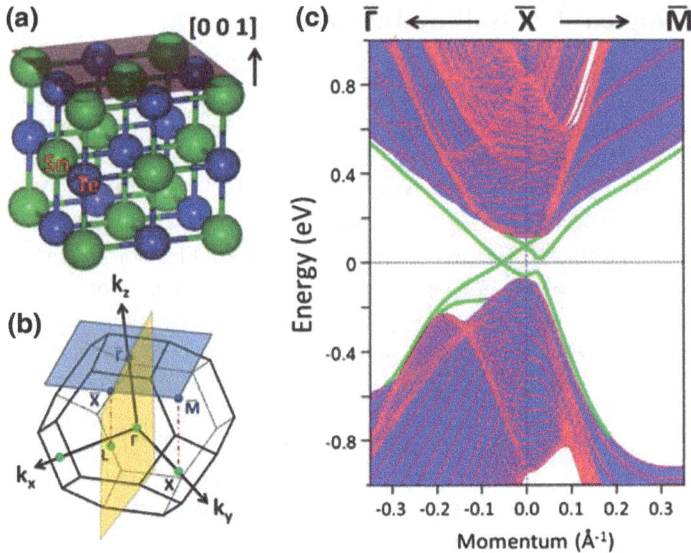

Fig. 6.5 Crystal and electronic structure of SnTe. **a** Rocksalt crystal structure. A (001) surface plane is shown. **b** FCC Brillouin zone (BZ) and the (001) surface BZ. The (110) mirror plane crosses the 2D surface BZ along $\bar{\Gamma}\bar{X}$. **c** Electronic structure of SnTe (001) surface around $\bar{X}$. The surface bands are indicated by thick-green lines and bulk bands by the shaded purple area. A Dirac point of the surface state appears along the $\bar{\Gamma}\bar{X}$ direction, while the surface band is gapped along $\bar{X}\bar{M}$. (From [164])

of counter-propagating gapless modes on any crystal surface symmetric about the (110) mirror plane. On the (001) surface, first-principles results show gapless surface Dirac cones with non-trivial spin-textures, where each Dirac point sits along $\bar{\Gamma}\bar{X}$ on one of the two sides of the $\bar{X}$ point at E_F, while the surface states along $\bar{\Gamma}\bar{M}$ are gapped (see Figs. 6.5c and 6.6d) [164]. Note that there is another $(1\bar{1}0)$ mirror plane so that there is a total of four Dirac points in the FBZ (Fig. 6.6a), as also observed in ARPES experiments [115, 181]. This is in sharp contrast to the case of a Z_2 TI such as Bi_2Se_3, which is protected by TRS, and contains only a single Dirac cone at the center of the (111) surface plane.

It is interesting to consider how the FS evolves as the E_F is lowered below the Dirac point. Initially, the FS consists of two separate hole pockets located away from the $\bar{X}$ point. As we go to lower energies below the E_F, these two disconnected hole pockets reconnect to yield a large $\bar{X}$-centered hole pocket and a small $\bar{X}$-centered electron pocket, see Fig. 6.6b, and the system undergoes a change in FS topology (i.e a Lifshitz transition). Insight into the complex surface electronic structure shown in Figs. 6.5 and 6.6 can be obtained through a 4×4 model Hamiltonian, which involves two coaxial Dirac cones [164]. Here, as seen in Fig. 6.6c, one starts with two non-interacting $\bar{X}$-centered "parent" Dirac cones, which are offset vertically in energy. In SnTe, for example, the lower parent Dirac cone is mainly derived from Sn-p_z orbitals, while the states involved in the higher Dirac cone originate primarily from

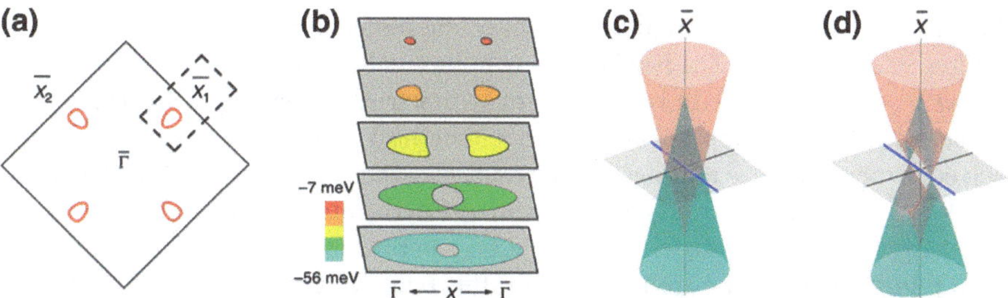

Fig. 6.6 **a** The (001) surface states of SnTe at Fermi level. **b** Surface Fermi surface evolution as the chemical potential decreases (from top to bottom). The change of Fermi surface topology indicates the presence of a Lifshitz transition. **c** Schematic of two non-interacting coaxial Dirac cones centered at $\bar{X}$. **d** As the two Dirac cones interact, gaps open up except at the two "child" Dirac points, protected by the mirror symmetry (bold blue line) on the two sides of the $\bar{X}$ point. [a, b from [65]; c, d from [164]

Te-p_x orbitals. As Fig. 6.6d shows, when the two parent cones hybridize, a gap opens up in the spectrum with the exception of the mirror line and leads to the formation of two "child" Dirac cones away from the $\bar{X}$ symmetry points, which lie at a lower energy. Note that since the TRS protects the parent Dirac points, these states cannot be gapped by breaking the mirror symmetry. The child Dirac points, on the other hand, can be gapped by removing the mirror symmetry.

In view of the differences in orbital characters of the upper and lower Dirac cones noted in the preceding paragraph, we would expect intensity asymmetries in the QPI patterns derived from STS spectra. In particular, the scattering between the states of p_z character in the electron sheet will be strong, but the scattering will be suppressed between states in the hole sheet [199]. Also, removal of one of the two mirror planes in the system will yield massive Dirac cones only along one direction, while we will continue to have massless Dirac cones in the other direction [38, 68, 141]. The simultaneous presence of massive and massless Dirac cones has been adduced through the observation of three non-dispersive features in the STS spectra from TCIs [121].

6.6 Topological Semi-metals

The defining feature of topological semimetals is the existence of band touching points at the Fermi energy, where two or more bands are degenerate at distinct values of the crystal momentum in the FBZ. In 3D, if one expands around one of these touching points (assuming double degeneracy), the effective Hamiltonian takes the following form similar to the Weyl equation in high-energy physics [171]:

$$H(\mathbf{k}) = \sum_{ij} v_{ij} k_i \sigma_j, \tag{6.8}$$

where $i, j = 1, 2, 3$, the parameters v_{ij} have the dimension of velocity, and σ_j are the three Pauli matrices. The band touching points, called *Weyl points* or Weyl *nodes*, are *stable* without the need to invoke extra symmetries as long as each band is non-degenerate, which requires the breaking of either the TRS or the inversion symmetry. Each cone then becomes non-degenerate (except at the Weyl point) and, importantly, each Weyl point becomes associated with a Chern number $\text{sgn}(\text{Det}[v_{ij}]) = \pm 1$ (called *Weyl or chiral charge*), which can be computed by using a formula similar to (6.1). In view of its non-trivial topology, this phase of quantum matter is referred to as a topological Weyl semimetal. The Weyl points must appear in pairs with ± 1 charges in the FBZ [117, 118]. As a consequence, creation or annihilation of Weyl points always involves a pair of points with opposite charges [157]. The total number of Weyl points come in multiples of four for TRS-preserved and two for TRS-broken systems [12, 120].

The band-touching at Weyl nodes discussed in the preceding paragraph involves two non-degenerate bands. In contrast, when these two bands are degenerate, we obtain a fourfold degenerate touching point, which is called a Dirac point or a Dirac node. There are three distinct possibilities here [191]. (1) The presence of an *accidental* degeneracy in the band structure, which will occur normally at the critical phase in transitioning between the non-trivial and trivial insulator phases [108, 109] (2) The TRIM points when an extra non-symmorphic symmetry is added for protection [198], a case that has not been observed experimentally. And, (3) when an extra rotational symmetry is added for protection, where instead of a single Dirac point, we obtain two Dirac points located at a pair of time-reversal invariant crystal momenta (time-reversal partners) lying along the rotational axis. Note that a Dirac node can be viewed as a composite object containing two Weyl nodes of opposite "charges" and thus possesses zero net charge or Chern number.

One remarkable consequence of bulk Weyl nodes carrying "charges" (+1 as a source and −1 as a sink) is that the FS associated with surface bands exhibits open line features or *Fermi arcs*, which connect the projected Weyl nodes with opposite charges on the surface BZ (Fig. 6.7d). Fermi arcs provide a distinctive feature for identifying the Weyl semimetal phase. It is not necessary to have such Fermi arcs in Dirac semimetals.

The first theoretically predicted and experimentally realized family of Weyl semimetal phases in wide current use is the TaAs family of compounds (TaAs, NbAs, TaP, and NbP) [71, 99, 169, 179, 194] These compounds assume a body-centered tetragonal structure that contains two mirror planes, M_x and M_y but not M_z, resulting in the breaking of the inversion symmetry (Fig. 6.7a–c). Although there is no C_4 rotational symmetry, the structure harbors a screw symmetry consisting of a C_4 rotation with a *c/2* translation along the *z*-direction. When SOC is turned off, first-principles calculations show that band crossing occurs on the $k_x = 0$ and $k_y = 0$ mirror planes, forming four nodal rings in the FBZ, see Fig. 6.7e. When SOC is turned on, the resulting spin splittings in the absence of inversion symmetry gap out the nodal rings leaving 12 pairs of gapless band-touching (Weyl) points near the original nodal rings but away from the mirror planes. Each nodal ring evolves into three pairs of nodes that can be classified into two groups, one pair in the $k_z = \frac{2\pi}{c}$ plane (labeled

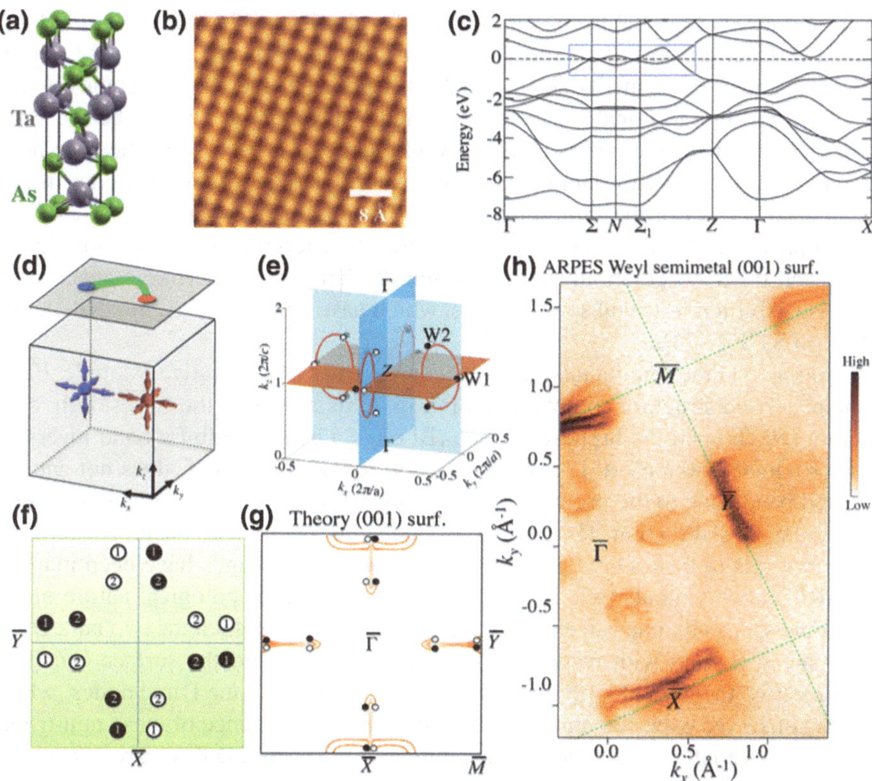

Fig. 6.7 **a** Crystal structure of TaAs. **b** Topographic image of TaAs's (001) surface by STM at the bias voltage −300 mV. **c** First-principles band structure calculations for bulk TaAs without SOC. The band crossings near Fermi level are highlighted using a blue box. **d** Schematic of the presence of a surface Fermi arc in the simplest Weyl semimetal state where there are only two single Weyl nodes with opposite charges in the BZ. **e** Without SOC, there are four nodal rings on the mirror planes; with SOC, nodal rings are gapped out with only 24 Weyl nodes (small black and white circles denoting opposite charges) left. **f** The (schematic) projected Weyl nodes onto (001) surface. **g** The calculated Fermi surface on the (001) surface. **h** ARPES measured Fermi surface of the (001) cleaving plane of TaAs (from [179])

W1) and two pairs away from $k_z = \frac{2\pi}{c}$ plane (labeled W2). There are 24 Weyl nodes in total, all with linear dispersions and each carries a chiral charge ±1. When the bulk band structure is projected onto the (001) surface, see Fig. 6.7f, W1 nodes project in the vicinity of the surface BZ edges, $\bar{X}$ and $\bar{Y}$. The pairs of W2 nodes with the same chiral charge, on the other hand, project onto the same point on the surface BZ. We thus obtain 8 projected W2 nodes, each with a total projected chiral charge of ±2, which are located near the midpoints of the $\bar{\Gamma}\bar{X}$ and the $\bar{\Gamma}\bar{Y}$ lines.

Further confirmation of the characteristics of the Weyl phase is obtained by examining surface states on the (001) surface. Theoretical predictions as well as ARPES experiments in Fig. 6.7g, h show the presence of surface states. The non-trivial nature of these states can be established by checking for consistency based on the bulk-edge correspondence arguments as follows. Two distinct parts in the surface BZ can be observed, one of crescent-like shape and the other bowtie-like. We expect the number of Fermi arcs that terminate on each projected W1/W2 Weyl node to be the same as the magnitude of the corresponding Weyl charge as is seen to be the case here. Moreover, in the ARPES data the two curves terminating at a W2 node are found to move (disperse) in the same direction as the initial state energy is varied. This "co-propagating" behavior indicates that these two energy contours are not closed. These and other observations justify the non-triviality of the surface states. Fermi arcs also exhibit non-trivial spin-textures, which have been observed in ARPES [98, 180] and STS [6, 73].

Turning to Dirac semimetals, we consider Na_3Bi as an illustrative example. First-principles band structure computations in Fig. 6.8b show that the lowest bulk conduction (Na 3s) and the highest valence (Bi $6p_{x,y,z}$) bands exhibit a band inversion of approximately 0.3 eV at Γ in the FBZ (Fig. 6.8c). Strong SOC does not gap the inverted bands due to the protection of an additional threefold rotational symmetry around the [001] direction. Therefore, the bulk bands touch at two locations along the rotational axis to form Dirac nodes; the associated Dirac bands have been mapped via ARPES experiments [97, 182]. In order to reveal the topological nature of the electronic spectrum, one can examine the (100) surface on which the two bulk Dirac nodes are separated with respect to $\widetilde{\Gamma}$ after projection onto the surface BZ. This results in two theoretically predicted Fermi arcs connecting the Dirac nodes, which are also observed experimentally, see Fig. 6.8c, d. The presence of these non-trivial surface states confirms the topological nature of Na_3Bi with 2D Z_2 invariant $\nu_{2D} = 1$ [165, 191].

Two other types of Weyl/Dirac semimetals should be noted. The first are type-II Weyl/Dirac semimetals, which do not respect Lorentz symmetry, while the type-I semimetals discussed in the preceding paragraphs respect this symmetry. This possibility arises when we take $i, j = 0, 1, 2, 3$ in (6.8) with σ_0 as an identity matrix. Unlike the type-I case, where the FS eventually shrinks to the Weyl/Dirac points, in the type-II semimetals the Weyl/Dirac cone exhibits strong tilting such that at the Fermi level the Weyl/Dirac node simply appears as the contact point between an electron and a hole pocket. Type-II Weyl and Dirac phases have been theoretically predicted in (W, Mo) Te_2 [9, 28, 70, 160, 175] and VAl_3 family [16], respectively. The second type are the multi-Weyl semimetals in which the Weyl node can carry a Weyl charge of magnitude greater than one. For instance, the ferromagnetic $HgCr_2Se_4$ spinel with nodes protected by C_4 point-group symmetry are predicted to carry charges of ± 2 [37].

Weyl/Dirac semimetals can exhibit universal transport phenomena such as large negative magneto-resistance [11, 13, 146, 177], anomalous Hall effect [12, 186, 193], and chiral magnetic effect [20, 52, 158]. These defining features of Weyl/Dirac semimetals suggest their unique potential for a new generation of low-power-consuming applications.

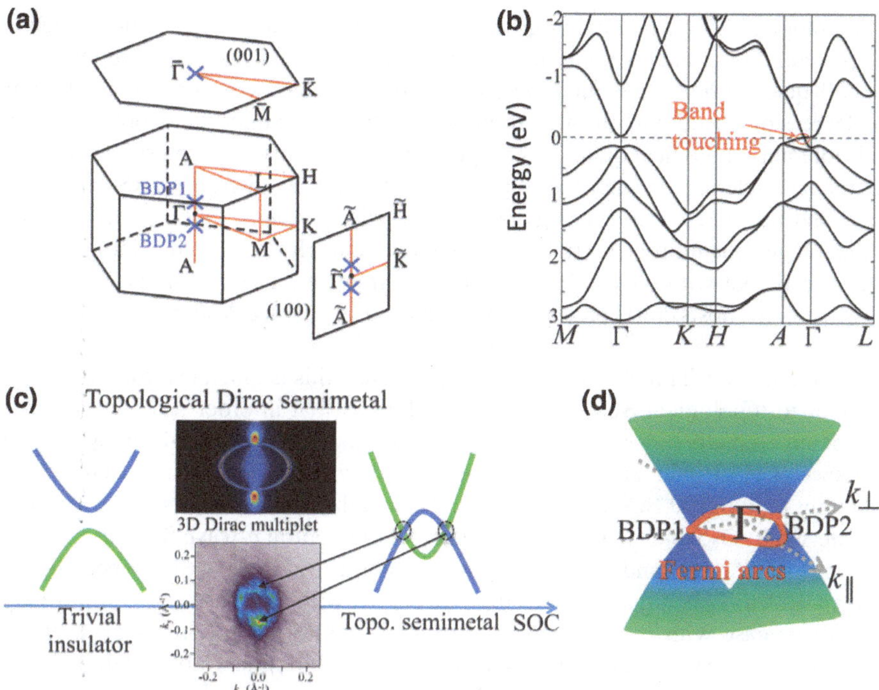

Fig. 6.8 **a** The bulk BZ of Na_3Bi with Dirac nodes marked by blue crosses and the projected surface BZs of the (001) and (100) surfaces. **b** First-principles bulk band calculation with SOC for Na_3Bi. **c** Schematic of the formation of a topological Dirac semimetal phase due to a band inversion. The band crossing points are stable without fine tuning because of additional rotation symmetry along [001]. The inset of **c** shows the ARPES measured and theoretically calculated surface Fermi surface. **d** Schematic band structure from the (100) surface of a topological Dirac semimetal with red curves representing surface Fermi arc states. Note that only surface states at the bulk Dirac point energy are shown. (from [183])

6.7 Topological Superconductivity

As we noted in Sect. 6.2 above, the "non-interacting" gapped electronic systems with a quadratic form of Hamiltonian can belong to ten possible classes, and characterized by either a Z or Z_2 topological invariant [81, 139]. The ten classes can be separated into either insulators or superconductors (SCs). A superconductor in the weak coupling limit can be described by the Bogoliubov-de Gennes (BdG) form with a pairing gap in its spectrum, similar to the insulating gap of an insulator. However, the SC state is distinct from an insulator by having both the additional particle-hole symmetry and a broken charge U(1) symmetry. When the SC state is topologically non-trivial, it supports the existence of topology-protected gapless boundary modes in accord with the bulk-edge correspondence.

Topological SCs are drawing great current interest mainly due to the existence of protected Majorana boundary modes, which result from a non-trivial topology of the bulk quasiparticle wave functions. These boundary fermions contain only half of the degrees of freedom in ordinary fermions and may have important applications in fault-tolerant topological quantum computation [82, 112]. The simplest well-known 2D model systems for such topological SCs are the chiral $p+ip$ (TRS breaking) [129] and $p\pm ip$ (TRS preserving) [126, 131] pairing states, which are analogous to the integer quantum Hall (IQH) and quantum spin Hall (QSH) insulating states, respectively, but with only half the degrees of freedom. The TRS-breaking (TRS-preserving) SC hosts chiral propagating (counter-propagating) Majorana edge modes [126].

Although many TI materials have been predicted, this is much less the case when it comes to topological SCs. The challenge here is that in order to obtain a topological superconductor, we must either have an unconventional pairing symmetry or a special electronic structure in the normal state. One current approach toward obtaining topological SCs looks for natural or synthesized materials [1]. Examples include: Sr_2RuO_4 which has been predicted to harbor chiral p-wave SC [101]; doped graphene [8, 76, 111] and SrPtAs [41] may exhibit chiral d-wave SC; doped Bi_2Se_3 with Cu [60, 172] is argued to be a topological SC with odd-parity [44, 46] as may also be the case with In-doped SnTe [119, 133], Chiral or time-reversal invariant triplet SC has been predicted in doped BC_3 [19], Sb thin films [69], and doped BiH [192]. Another approach toward topological SCs involves engineering composite systems. An example is to consider obtaining the topological SC state via the proximity effect between the surface of a 3D TI and an ordinary s-wave SC [47]. Such a proximity effect has been observed in Bi_2Se_3 on a $NbSe_2$ SC substrate [162] as well as an unconventional d-wave $Bi_2Sr_2CaCu_2O_{8+\delta}$ SC substrate [161]. 1D topological SC could also be realized by putting a 1D chain of magnetic impurities or adatoms or a helical Shiba chain on an SC substrate [58, 110, 125].

One can go beyond the ten-fold classification scheme that underlies our discussion so far by invoking additional symmetries. For instance, a new type of topological mirror SC, which is dominated by the mirror symmetry, has been predicted [201]. More complete classification tables based on various crystalline symmetries have also been developed [24, 36, 142, 143, 163] Another approach is to reexamine the nodal superconductors from the viewpoint of topology [136]. In fact, SCs with unconventional pairing displaying point- or line-nodes are not uncommon. Although the gapless systems lack global topological invariants, they nevertheless could be classified via momentum-dependent topological numbers. Along these lines, nontrivial topology may be found in the high-T_c cuprates [66, 132, 153], heavy fermion systems [2, 74, 79, 107, 206], noncentrosymmetric SCs [10, 134, 137, 138, 188], and Weyl/Dirac superconductors [104, 195].

6.8 Strongly Correlated Topological Materials

The role of strong electronic correlations in generating exotic topological phases of quantum matter is an area of intense current interest, which we only touch upon with brief comments on materials predictions where strong correlation is the key ingredient that drives the non-interacting system into the topologically non-trivial phase. The topological phase then has a non-interacting counterpart within the aforementioned ten-fold classification scheme. Many other possibilities include symmetry-protected topological phases with short-range entanglement [17, 18, 40], long-range entangled topological ordered phases [167, 168], and symmetry-enriched topological phases with fractionalized excitations [25, 90, 100, 148]. We refer the reader to various review articles for further discussion [140, 155, 166].

A number of correlated materials have been suggested as candidate topological insulators. Among these, topological Kondo insulators (TKIs) are a relatively simple type of heavy-fermion compound, where the insulating gap at low temperatures open up via hybridization between the nearly flat (highly correlated) f-bands and the more dispersive d bands. The heavy-fermion compound SmB_6 has drawn considerable theoretical and experimental interest in search of a TKI [32, 33, 80]. Experimentally, optical [54] and transport [42] studies on SmB_6 provide evidence for the existence of a gap. Note, however, that even though the transition to the insulating state starts below 50 K, conductivity remains finite and saturates below 4 K, suggesting the presence of in-gap states [3, 27]. A spin-polarized surface state at Γ lying inside the bulk band gap and X-centered electron-like bands spanning the gap are seen in ARPES experiments [29, 75, 113, 178]. These results support a possible TKI ground state in SmB_6. Although not confirmed experimentally, other predicted materials include SmS [91], as a TKI, and YB_6 and YB_{12} as topological Kondo crystalline insulators [170].

5*d* orbitals of Ir in iridium oxides (iridates) provide another attractive playground for generating correlated SOC effects for realizing topological phases, and iridates with pyrocholore, perovskite, and layered honeycomb structures have been predicted to host topological ground states. In particular, Ir-pyrochlores $R_2Ir_2O_7$ (R = Nd, Sm, Eu, and Y) could undergo metal-insulator transitions [103]. An LDA+U calculation predicts the magnetic phase of $Y_2Ir_2O_7$ to be a Weyl semimetal [159]. Model studies of Ir-pyrochlores reveal a rich topological phase diagram as a function of the strengths of electron-electron interaction and SOC [124]. With crystal field splittings induced by distortion of the IrO_6 octahedra, the material could also be driven into topological insulator [78] and topological Mott insulator [190] phases. In the honeycomb lattice of Na_2IrO_3, a QSH insulator phase and a fractionalized QSH state have been proposed [144, 197]. The predicted topological phases in the iridates have, however, eluded clear experimental confirmation.

6.9 Outlook and Conclusions

An urgent challenge is to find viable new topological materials of various types so that the very limited currently available menu of experimentally realized materials can be expanded, and the many transformational opportunities for fundamental science and applications potential of these remarkable materials can be explored effectively. The field is still in its infancy and we can expect many exciting discoveries to emerge for years to come.

Acknowledgements W.F.T. is supported by the National Thousand-Young-Talents Program, China. The work at Northeastern University was supported by the US Department of Energy (DOE), Office of Science, Basic Energy Sciences grant number DE-FG02-07ER46352 (core research), and benefited from Northeastern University's Advanced Scientific Computation Center (ASCC), the NERSC supercomputing center through DOE grant number DE-AC02-05CH11231, and support (applications to layered materials) from the DOE EFRC: Center for the Computational Design of Functional Layered Materials (CCDM) under DE-SC0012575.

References

1. J. Alicea, New directions in the pursuit of Majorana fermions in solid state systems. Rep. Prog. Phys. **75**, 076501 (2012). https://doi.org/10.1088/0034-4885/75/7/076501
2. M.P. Allan, F. Massee, D.K. Morr, J. Van Dyke, A.W. Rost, A.P. Mackenzie, C. Petrovic, J.C. Davis, Imaging Cooper pairing of heavy fermions in CeCoIn5. Nat. Phys. **9**, 468–473 (2013). https://doi.org/10.1038/nphys2671
3. J.W. Allen, B. Batlogg, P. Wachter, Large low-temperature Hall effect and resistivity in mixed-valent SmB_6. Phys. Rev. B **20**, 4807–4813 (1979). https://doi.org/10.1103/PhysRevB.20.4807
4. J.G. Analytis, R.D. McDonald, S.C. Riggs, J.-H. Chu, G.S. Boebinger, I.R. Fisher, Two-dimensional surface state in the quantum limit of a topological insulator. Nat. Phys. **6**, 960–964 (2010). https://doi.org/10.1038/nphys1861
5. A. Bansil, H. Lin, T. Das, Colloquium: topological band theory. Rev. Mod. Phys. **88**, 021004 (2016). https://doi.org/10.1103/RevModPhys.88.021004
6. R. Batabyal, N. Morali, N. Avraham, Y. Sun, M. Schmidt, C. Felser, A. Stern, B. Yan, H. Beidenkopf, Visualizing weakly bound surface Fermi arcs and their correspondence to bulk Weyl fermions. Sci. Adv. **2**, e1600709 (2016). https://doi.org/10.1126/sciadv.1600709
7. B.A. Bernevig, T.L. Hughes, S.-C. Zhang, Quantum spin hall effect and topological phase transition in HgTe quantum wells. Science **314**, 1757–1761 (2006). https://doi.org/10.1126/science.1133734
8. A.M. Black-Schaffer, S. Doniach, Resonating valence bonds and mean-field d-wave superconductivity in graphite. Phys. Rev. B **75**, 134512 (2007). https://doi.org/10.1103/PhysRevB.75.134512
9. F.Y. Bruno, A. Tamai, Q.S. Wu, I. Cucchi, C. Barreteau, A. de la Torre, S. McKeown Walker, S. Riccò, Z. Wang, T.K. Kim, M. Hoesch, M. Shi, N.C. Plumb, E. Giannini, A.A. Soluyanov, F. Baumberger, Observation of large topologically trivial Fermi arcs in the candidate type-II Weyl semimetal WTe_2. Phys. Rev. B **94**, 121112 (2016). https://doi.org/10.1103/PhysRevB.94.121112
10. P.M.R. Brydon, A.P. Schnyder, C. Timm, Topologically protected flat zero-energy surface bands in noncentrosymmetric superconductors. Phys. Rev. B **84**, 020501 (2011). https://doi.org/10.1103/PhysRevB.84.020501

11. A.A. Burkov, Negative longitudinal magnetoresistance in Dirac and Weyl metals. Phys. Rev. B **91**, 245157 (2015). https://doi.org/10.1103/PhysRevB.91.245157
12. A.A. Burkov, L. Balents, Weyl semimetal in a topological insulator multilayer. Phys. Rev. Lett. **107**, 127205 (2011). https://doi.org/10.1103/PhysRevLett.107.127205
13. A.A. Burkov, Y.B. Kim, Z2 and chiral anomalies in topological dirac semimetals. Phys. Rev. Lett. **117**, 136602 (2016). https://doi.org/10.1103/PhysRevLett.117.136602
14. S.Z. Butler, S.M. Hollen, L. Cao, Y. Cui, J.A. Gupta, H.R. Gutiérrez, T.F. Heinz, S.S. Hong, J. Huang, A.F. Ismach, E. Johnston-Halperin, M. Kuno, V.V. Plashnitsa, R.D. Robinson, R.S. Ruoff, S. Salahuddin, J. Shan, L. Shi, M.G. Spencer, M. Terrones, W. Windl, J.E. Goldberger, Progress, challenges, and opportunities in two-dimensional materials beyond graphene. ACS Nano **7**, 2898–2926 (2013). https://doi.org/10.1021/nn400280c
15. C.-Z. Chang, J. Zhang, X. Feng, J. Shen, Z. Zhang, M. Guo, K. Li, Y. Ou, P. Wei, L.-L. Wang, Z.-Q. Ji, Y. Feng, S. Ji, X. Chen, J. Jia, X. Dai, Z. Fang, S.-C. Zhang, K. He, Y. Wang, L. Lu, X.-C. Ma, Q.-K. Xue, Experimental observation of the quantum anomalous hall effect in a magnetic topological insulator. Science **340**, 167–170 (2013). https://doi.org/10.1126/science.1234414
16. T.-R. Chang, S.-Y. Xu, D.S. Sanchez, S.-M. Huang, G. Chang, C.-H. Hsu, G. Bian, I. Belopolski, Z.-M. Yu, X. Xu, C. Xiang, S.A. Yang, T. Neupert, H.-T. Jeng, H. Lin, M.Z. Hasan, *Type-II Topological Dirac Semimetals: Theory and Materials Prediction (VAl3 family)*. arXiv:160607555 Cond-Mat
17. X. Chen, Z.-C. Gu, X.-G. Wen, Complete classification of one-dimensional gapped quantum phases in interacting spin systems. Phys. Rev. B **84**, 235128 (2011). https://doi.org/10.1103/PhysRevB.84.235128
18. X. Chen, Z.-C. Gu, X.-G. Wen, Classification of gapped symmetric phases in one-dimensional spin systems. Phys. Rev. B **83**, 035107 (2011). https://doi.org/10.1103/PhysRevB.83.035107
19. X. Chen, Y. Yao, H. Yao, F. Yang, J. Ni, Topological p+ip superconductivity in doped graphene-like single-sheet materials BC_3. Phys. Rev. B **92**, 174503 (2015). https://doi.org/10.1103/PhysRevB.92.174503
20. Y. Chen, S. Wu, A.A. Burkov, Axion response in Weyl semimetals. Phys. Rev. B **88**, 125105 (2013). https://doi.org/10.1103/PhysRevB.88.125105
21. Y.L. Chen, J.G. Analytis, J.-H. Chu, Z.K. Liu, S.-K. Mo, X.L. Qi, H.J. Zhang, D.H. Lu, X. Dai, Z. Fang, S.C. Zhang, I.R. Fisher, Z. Hussain, Z.-X. Shen, Experimental realization of a three-dimensional topological insulator, Bi2Te3. Science **325**, 178–181 (2009). https://doi.org/10.1126/science.1173034
22. Y.L. Chen, J.-H. Chu, J.G. Analytis, Z.K. Liu, K. Igarashi, H.-H. Kuo, X.L. Qi, S.K. Mo, R.G. Moore, D.H. Lu, M. Hashimoto, T. Sasagawa, S.C. Zhang, I.R. Fisher, Z. Hussain, Z.X. Shen, Massive dirac fermion on the surface of a magnetically doped topological insulator. Science **329**, 659–662 (2010). https://doi.org/10.1126/science.1189924
23. Y.L. Chen, Z.K. Liu, J.G. Analytis, J.-H. Chu, H.J. Zhang, B.H. Yan, S.-K. Mo, R.G. Moore, D.H. Lu, I.R. Fisher, S.C. Zhang, Z. Hussain, Z.-X. Shen, Single dirac cone topological surface state and unusual thermoelectric property of compounds from a new topological insulator family. Phys. Rev. Lett. **105**, 266401 (2010). https://doi.org/10.1103/PhysRevLett.105.266401
24. C.-K. Chiu, H. Yao, S. Ryu, Classification of topological insulators and superconductors in the presence of reflection symmetry. Phys. Rev. B **88**, 075142 (2013). https://doi.org/10.1103/PhysRevB.88.075142
25. G.Y. Cho, Y.-M. Lu, J.E. Moore, Gapless edge states of background field theory and translation-symmetric $\mathbb{Z}_2$ spin liquids. Phys. Rev. B **86**, 125101 (2012). https://doi.org/10.1103/PhysRevB.86.125101
26. F.-C. Chuang, L.-Z. Yao, Z.-Q. Huang, Y.-T. Liu, C.-H. Hsu, T. Das, H. Lin, A. Bansil, Prediction of large-gap two-dimensional topological insulators consisting of bilayers of group III elements with Bi. Nano Lett. **14**, 2505–2508 (2014). https://doi.org/10.1021/nl500206u
27. J.C. Cooley, M.C. Aronson, Z. Fisk, P.C. Canfield, SmB_6: Kondo insulator or exotic metal? Phys. Rev. Lett. **74**, 1629–1632 (1995). https://doi.org/10.1103/PhysRevLett.74.1629

28. K. Deng, G. Wan, P. Deng, K. Zhang, S. Ding, E. Wang, M. Yan, H. Huang, H. Zhang, Z. Xu, J. Denlinger, A. Fedorov, H. Yang, W. Duan, H. Yao, Y. Wu, S. Fan, H. Zhang, X. Chen, S. Zhou, Experimental observation of topological Fermi arcs in type-II Weyl semimetal MoTe2. Nat. Phys. **12**, 1105–1110 (2016). https://doi.org/10.1038/nphys3871
29. J.D. Denlinger, J.W. Allen, J.-S. Kang, K. Sun, B.-I. Min, D.-J. Kim, Z. Fisk, *SmB6 Photoemission: Past and Present* (2013). arXiv:13126636 Cond-Mat
30. N.D. Drummond, V. Zólyomi, V.I. Fal'ko, Electrically tunable band gap in silicene. Phys. Rev. B **85**, 075423 (2012). https://doi.org/10.1103/PhysRevB.85.075423
31. L. Du, I. Knez, G. Sullivan, R.-R. Du, Robust helical edge transport in gated InAs/GaSb bilayers. Phys. Rev. Lett. **114**, 096802 (2015). https://doi.org/10.1103/PhysRevLett.114.096802
32. M. Dzero, K. Sun, P. Coleman, V. Galitski, Theory of topological Kondo insulators. Phys. Rev. B **85**, 045130 (2012). https://doi.org/10.1103/PhysRevB.85.045130
33. M. Dzero, K. Sun, V. Galitski, P. Coleman, Topological Kondo insulators. Phys. Rev. Lett. **104**, 106408 (2010). https://doi.org/10.1103/PhysRevLett.104.106408
34. P. Dziawa, B.J. Kowalski, K. Dybko, R. Buczko, A. Szczerbakow, M. Szot, E. Łusakowska, T. Balasubramanian, B.M. Wojek, M.H. Berntsen, O. Tjernberg, T. Story, Topological crystalline insulator states in Pb1−xSnxSe. Nat. Mater. **11**, 1023–1027 (2012). https://doi.org/10.1038/nmat3449
35. S.V. Eremeev, G. Landolt, T.V. Menshchikova, B. Slomski, Y.M. Koroteev, Z.S. Aliev, M.B. Babanly, J. Henk, A. Ernst, L. Patthey, A. Eich, A.A. Khajetoorians, J. Hagemeister, O. Pietzsch, J. Wiebe, R. Wiesendanger, P.M. Echenique, S.S. Tsirkin, I.R. Amiraslanov, J.H. Dil, E.V. Chulkov, Atom-specific spin mapping and buried topological states in a homologous series of topological insulators. Nat. Commun. **3**, 635 (2012). https://doi.org/10.1038/ncomms1638
36. C. Fang, B.A. Bernevig, M.J. Gilbert, Topological crystalline superconductors with linearly and projectively represented C_n symmetry (2017). arXiv:170101944 Cond-Mat
37. C. Fang, M. Gilbert, X. Dai, B. Bernevig, Multi-Weyl topological semimetals stabilized by point group symmetry. Phys. Rev. Lett. **108**, 266802 (2012). https://doi.org/10.1103/PhysRevLett.108.266802
38. C. Fang, M.J. Gilbert, B.A. Bernevig, Large-Chern-number quantum anomalous Hall effect in thin-film topological crystalline insulators. Phys. Rev. Lett. **112**, 046801 (2014). https://doi.org/10.1103/PhysRevLett.112.046801
39. C. Fang, M.J. Gilbert, B.A. Bernevig, Bulk topological invariants in noninteracting point group symmetric insulators. Phys. Rev. B **86**, 115112 (2012). https://doi.org/10.1103/PhysRevB.86.115112
40. L. Fidkowski, A. Kitaev, Topological phases of fermions in one dimension. Phys. Rev. B **83**, 075103 (2011). https://doi.org/10.1103/PhysRevB.83.075103
41. M.H. Fischer, T. Neupert, C. Platt, A.P. Schnyder, W. Hanke, J. Goryo, R. Thomale, M. Sigrist, Chiral d-wave superconductivity in SrPtAs. Phys. Rev. B **89**, 020509 (2014). https://doi.org/10.1103/PhysRevB.89.020509
42. K. Flachbart, K. Gloos, E. Konovalova, Y. Paderno, M. Reiffers, P. Samuely, P. Švec, Energy gap of intermediate-valent SmB_6 studied by point-contact spectroscopy. Phys. Rev. B **64**, 085104 (2001). https://doi.org/10.1103/PhysRevB.64.085104
43. M. Freedman, C. Nayak, K. Shtengel, K. Walker, Z. Wang, A class of P, T-invariant topological phases of interacting electrons. Ann. Phys. **310**, 428–492 (2004). https://doi.org/10.1016/j.aop.2004.01.006
44. L. Fu, Odd-parity topological superconductor with nematic order: Application to $Cu_xBi_2Se_3$. Phys. Rev. B **90**, 100509 (2014). https://doi.org/10.1103/PhysRevB.90.100509
45. L. Fu, Topological crystalline insulators. Phys. Rev. Lett. **106**, 106802 (2011). https://doi.org/10.1103/PhysRevLett.106.106802
46. L. Fu, E. Berg, Odd-parity topological superconductors: theory and application to $Cu_xBi_2Se_3$. Phys. Rev. Lett. **105**, 097001 (2010). https://doi.org/10.1103/PhysRevLett.105.097001
47. L. Fu, C.L. Kane, Superconducting proximity effect and majorana fermions at the surface of a topological insulator. Phys. Rev. Lett. **100**, 096407 (2008). https://doi.org/10.1103/PhysRevLett.100.096407

48. L. Fu, C.L. Kane, Topological insulators with inversion symmetry. Phys. Rev. B **76**, 045302 (2007). https://doi.org/10.1103/PhysRevB.76.045302
49. L. Fu, C.L. Kane, Time reversal polarization and a Z_2 adiabatic spin pump. Phys. Rev. B **74**, 195312 (2006). https://doi.org/10.1103/PhysRevB.74.195312
50. L. Fu, C.L. Kane, E.J. Mele, Topological insulators in three dimensions. Phys. Rev. Lett. **98**, 106803 (2007). https://doi.org/10.1103/PhysRevLett.98.106803
51. T. Fukui, Y. Hatsugai, Quantum spin hall effect in three dimensional materials: lattice computation of Z_2 topological invariants and its application to Bi and Sb. J. Phys. Soc. Jpn. **76**, 053702 (2007). https://doi.org/10.1143/JPSJ.76.053702
52. K. Fukushima, D.E. Kharzeev, H.J. Warringa, Chiral magnetic effect. Phys. Rev. D **78**, 074033 (2008). https://doi.org/10.1103/PhysRevD.78.074033
53. K.K. Gomes, W. Ko, W. Mar, Y. Chen, Z.-X. Shen, H.C. Manoharan, Quantum imaging of topologically unpaired spin-polarized dirac fermions (2009) 0909.0921
54. B. Gorshunov, N. Sluchanko, A. Volkov, M. Dressel, G. Knebel, A. Loidl, S. Kunii, Low-energy electrodynamics of SmB_6. Phys. Rev. B **59**, 1808–1814 (1999). https://doi.org/10.1103/PhysRevB.59.1808
55. G. Gupta, H. Lin, A. Bansil, M.B.A. Jalil, C.-Y. Huang, W.-F. Tsai, G. Liang, Y-shape spin-separator for two-dimensional group-IV nanoribbons based on quantum spin hall effect. Appl. Phys. Lett. **104**, 032410 (2014). https://doi.org/10.1063/1.4863088
56. M.Z. Hasan, C.L. Kane, Colloquium: topological insulators. Rev. Mod. Phys. **82**, 3045–3067 (2010). https://doi.org/10.1103/RevModPhys.82.3045
57. Y. Hatsugai, Topological aspects of the quantum hall effect. J. Phys.: Condens. Mat. **9**, 2507–2549 (1997)
58. A. Heimes, P. Kotetes, G. Schön, Majorana fermions from Shiba states in an antiferromagnetic chain on top of a superconductor. Phys. Rev. B **90**, 060507 (2014). https://doi.org/10.1103/PhysRevB.90.060507
59. Y.S. Hor, A. Richardella, P. Roushan, Y. Xia, J.G. Checkelsky, A. Yazdani, M.Z. Hasan, N.P. Ong, R.J. Cava, p-type Bi_2Se_3 for topological insulator and low-temperature thermoelectric applications. Phys. Rev. B **79**, 195208 (2009). https://doi.org/10.1103/PhysRevB.79.195208
60. Y.S. Hor, A.J. Williams, J.G. Checkelsky, P. Roushan, J. Seo, Q. Xu, H.W. Zandbergen, A. Yazdani, N.P. Ong, R.J. Cava, Superconductivity in $Cu_xBi_2Se_3$ and its implications for pairing in the undoped topological insulator. Phys. Rev. Lett. **104**, 057001 (2010). https://doi.org/10.1103/PhysRevLett.104.057001
61. D. Hsieh, D. Qian, L. Wray, Y. Xia, Y.S. Hor, R.J. Cava, M.Z. Hasan, A topological Dirac insulator in a quantum spin Hall phase. Nature **452**, 970 (2008). https://doi.org/10.1038/nature06843
62. D. Hsieh, Y. Xia, D. Qian, L. Wray, J.H. Dil, F. Meier, J. Osterwalder, L. Patthey, J.G. Checkelsky, N.P. Ong, A.V. Fedorov, H. Lin, A. Bansil, D. Grauer, Y.S. Hor, R.J. Cava, M.Z. Hasan, A tunable topological insulator in the spin helical Dirac transport regime. Nature **460**, 1101–1105 (2009). https://doi.org/10.1038/nature08234
63. D. Hsieh, Y. Xia, D. Qian, L. Wray, F. Meier, J. Dil, J. Osterwalder, L. Patthey, A. Fedorov, H. Lin, A. Bansil, D. Grauer, Y. Hor, R. Cava, M. Hasan, Observation of time-reversal-protected single-Dirac-cone topological-insulator states in Bi2Te3 and Sb2Te3. Phys. Rev. Lett. **103** (2009). https://doi.org/10.1103/physrevlett.103.146401
64. D. Hsieh, Y. Xia, L. Wray, D. Qian, A. Pal, J.H. Dil, J. Osterwalder, F. Meier, G. Bihlmayer, C.L. Kane, Y.S. Hor, R.J. Cava, M.Z. Hasan, Observation of unconventional quantum spin textures in topological insulators. Science **323**, 919–922 (2009). https://doi.org/10.1126/science.1167733
65. T.H. Hsieh, H. Lin, J. Liu, W. Duan, A. Bansil, L. Fu, Topological crystalline insulators in the SnTe material class. Nat. Commun. **3**, 982 (2012). https://doi.org/10.1038/ncomms1969
66. C.-R. Hu, Midgap surface states as a novel signature for $d_{x}a^2 - x_b{}^2$-wave superconductivity. Phys. Rev. Lett. **72**, 1526–1529 (1994). https://doi.org/10.1103/PhysRevLett.72.1526
67. J. Hu, J. Alicea, R. Wu, M. Franz, Giant topological insulator gap in graphene with 5d Adatoms. Phys. Rev. Lett. **109**, 266801 (2012). https://doi.org/10.1103/PhysRevLett.109.266801

68. C.-Y. Huang, H. Lin, Y.J. Wang, A. Bansil, W.-F. Tsai, Hedgehog spin texture and competing orders associated with strains on the surface of a topological crystalline insulator. Phys. Rev. B **93**, 205304 (2016). https://doi.org/10.1103/PhysRevB.93.205304
69. J.-Q. Huang, C.-H. Hsu, H. Lin, D.-X. Yao, W.-F. Tsai, Chiral p-wave superconductivity in Sb(111) thin films close to Van Hove singularities. Phys. Rev. B **93**, 155108 (2016). https://doi.org/10.1103/PhysRevB.93.155108
70. L. Huang, T.M. McCormick, M. Ochi, Z. Zhao, M.-T. Suzuki, R. Arita, Y. Wu, D. Mou, H. Cao, J. Yan, N. Trivedi, A. Kaminski, Spectroscopic evidence for a type II Weyl semimetallic state in MoTe2. Nat. Mater. **15**, 1155–1160 (2016). https://doi.org/10.1038/nmat4685
71. S.-M. Huang, S.-Y. Xu, I. Belopolski, C.-C. Lee, G. Chang, B. Wang, N. Alidoust, G. Bian, M. Neupane, C. Zhang, S. Jia, A. Bansil, H. Lin, M.Z. Hasan, A Weyl Fermion semimetal with surface Fermi arcs in the transition metal monopnictide TaAs class. Nat. Commun. **6**, 7373 (2015)
72. T.L. Hughes, E. Prodan, B.A. Bernevig, Inversion-symmetric topological insulators. Phys. Rev. B **83**, 245132 (2011). https://doi.org/10.1103/PhysRevB.83.245132
73. H. Inoue, A. Gyenis, Z. Wang, J. Li, S.W. Oh, S. Jiang, N. Ni, B.A. Bernevig, A. Yazdani, Quasiparticle interference of the Fermi arcs and surface-bulk connectivity of a Weyl semimetal. Science **351**, 1184–1187 (2016). https://doi.org/10.1126/science.aad8766
74. K. Izawa, H. Yamaguchi, Y. Matsuda, H. Shishido, R. Settai, Y. Onuki, Angular position of nodes in the superconducting gap of Quasi-2D heavy-fermion superconductor $CeCoIn_5$. Phys. Rev. Lett. **87**, 057002 (2001). https://doi.org/10.1103/PhysRevLett.87.057002
75. J. Jiang, S. Li, T. Zhang, Z. Sun, F. Chen, Z.R. Ye, M. Xu, Q.Q. Ge, S.Y. Tan, X.H. Niu, M. Xia, B.P. Xie, Y.F. Li, X.H. Chen, H.H. Wen, D.L. Feng, Observation of possible topological in-gap surface states in the Kondo insulator SmB_6 by photoemission. Nat. Commun. **4**, ncomms4010 (2013). https://doi.org/10.1038/ncomms4010
76. Y. Jiang, D.-X. Yao, E.W. Carlson, H.-D. Chen, J. Hu, Andreev conductance in the d+id' wave superconducting states of graphene. Phys. Rev. B **77**, 235420 (2008). https://doi.org/10.1103/PhysRevB.77.235420
77. C.L. Kane, E.J. Mele, Quantum Spin Hall effect in graphene. Phys. Rev. Lett. **95**, 226801 (2005). https://doi.org/10.1103/PhysRevLett.95.226801
78. M. Kargarian, J. Wen, G.A. Fiete, Competing exotic topological insulator phases in transition-metal oxides on the pyrochlore lattice with distortion. Phys. Rev. B **83**, 165112 (2011). https://doi.org/10.1103/PhysRevB.83.165112
79. Y. Kasahara, T. Iwasawa, H. Shishido, T. Shibauchi, K. Behnia, Y. Haga, T.D. Matsuda, Y. Onuki, M. Sigrist, Y. Matsuda, Exotic superconducting properties in the electron-hole-compensated heavy-fermion "Semimetal" URu_2Si_2. Phys. Rev. Lett. **99**, 116402 (2007). https://doi.org/10.1103/PhysRevLett.99.116402
80. D.J. Kim, J. Xia, Z. Fisk, Topological surface state in the Kondo insulator samarium hexaboride. Nat. Mater. **13**, 466–470 (2014). https://doi.org/10.1038/nmat3913
81. A. Kitaev, V. Lebedev, M. Feigel'man, Periodic table for topological insulators and superconductors. AIP Conf. Proc. **1134**, 22–30 (2009). https://doi.org/10.1063/1.3149495
82. A.Y. Kitaev, Fault-tolerant quantum computation by anyons. Ann. Phys. **303**, 2–30 (2003). https://doi.org/10.1016/S0003-4916(02)00018-0
83. K.V. Klitzing, G. Dorda, M. Pepper, New method for high-accuracy determination of the fine-structure constant based on quantized hall resistance. Phys. Rev. Lett. **45**, 494–497 (1980). https://doi.org/10.1103/PhysRevLett.45.494
84. K. von Klitzing, Developments in the quantum Hall effect. Philos. Trans. R. Soc. Lond. Math. Phys. Eng. Sci. **363**, 2203–2219 (2005). https://doi.org/10.1098/rsta.2005.1640
85. I. Knez, R.-R. Du, G. Sullivan, Evidence for helical edge modes in inverted InAs/GaSb quantum wells. Phys. Rev. Lett. **107**, 136603 (2011). https://doi.org/10.1103/PhysRevLett.107.136603
86. I. Knez, C.T. Rettner, S.-H. Yang, S.S.P. Parkin, L. Du, R.-R. Du, G. Sullivan, Observation of edge transport in the disordered regime of topologically insulating InAs/GaSb quantum wells. Phys. Rev. Lett. **112**, 026602 (2014). https://doi.org/10.1103/PhysRevLett.112.026602

87. D. Kong, Y. Chen, J.J. Cha, Q. Zhang, J.G. Analytis, K. Lai, Z. Liu, S.S. Hong, K.J. Koski, S.-K. Mo, Z. Hussain, I.R. Fisher, Z.-X. Shen, Y. Cui, Ambipolar field effect in the ternary topological insulator (BixSb1-x)2Te3 by composition tuning. Nat. Nanotechnol. **6**, 705–709 (2011). https://doi.org/10.1038/nnano.2011.172
88. M. Konig, S. Wiedmann, C. Brune, A. Roth, H. Buhmann, L.W. Molenkamp, X.-L. Qi, S.-C. Zhang, Quantum spin Hall insulator state in HgTe quantum wells. Science **318**, 766–770 (2007). https://doi.org/10.1126/science.1148047
89. W.-C. Lee, C. Wu, D.P. Arovas, S.-C. Zhang, Quasiparticle interference on the surface of the topological insulator Bi_2Te_3. Phys. Rev. B **80**, 245439 (2009). https://doi.org/10.1103/PhysRevB.80.245439
90. M. Levin, A. Stern, Classification and analysis of two-dimensional Abelian fractional topological insulators. Phys. Rev. B **86**, 115131 (2012). https://doi.org/10.1103/PhysRevB.86.115131
91. Z. Li, J. Li, P. Blaha, N. Kioussis, Predicted topological phase transition in the SmS Kondo insulator under pressure. Phys. Rev. B **89**, 121117 (2014). https://doi.org/10.1103/PhysRevB.89.121117
92. H. Lin, R.S. Markiewicz, L.A. Wray, L. Fu, M.Z. Hasan, A. Bansil, Single-Dirac-Cone topological surface states in the $TlBiSe_2$ class of topological semiconductors. Phys. Rev. Lett. **105**, 036404 (2010). https://doi.org/10.1103/PhysRevLett.105.036404
93. C.-C. Liu, W. Feng, Y. Yao, Quantum Spin Hall effect in silicene and two-dimensional Germanium. Phys. Rev. Lett. **107**, 076802 (2011). https://doi.org/10.1103/PhysRevLett.107.076802
94. C.-X. Liu, R.-X. Zhang, B.K. VanLeeuwen, Topological nonsymmorphic crystalline insulators. Phys. Rev. B **90**, 085304 (2014). https://doi.org/10.1103/PhysRevB.90.085304
95. J. Liu, T.H. Hsieh, P. Wei, W. Duan, J. Moodera, L. Fu, Spin-filtered edge states with an electrically tunable gap in a two-dimensional topological crystalline insulator. Nat. Mater. **13**, 178–183 (2014). https://doi.org/10.1038/nmat3828
96. Q. Liu, C.-X. Liu, C. Xu, X.-L. Qi, S.-C. Zhang, Magnetic impurities on the surface of a topological insulator. Phys. Rev. Lett. **102**, 156603 (2009). https://doi.org/10.1103/PhysRevLett.102.156603
97. Z.K. Liu, B. Zhou, Y. Zhang, Z.J. Wang, H.M. Weng, D. Prabhakaran, S.-K. Mo, Z.X. Shen, Z. Fang, X. Dai, Z. Hussain, Y.L. Chen, Discovery of a three-dimensional topological dirac semimetal, Na_3Bi. Science **343**, 864–867 (2014). https://doi.org/10.1126/science.1245085
98. B.Q. Lv, S. Muff, T. Qian, Z.D. Song, S.M. Nie, N. Xu, P. Richard, C.E. Matt, N.C. Plumb, L.X. Zhao, G.F. Chen, Z. Fang, X. Dai, J.H. Dil, J. Mesot, M. Shi, H.M. Weng, H. Ding, Observation of Fermi-Arc spin texture in TaAs. Phys. Rev. Lett. **115**, 217601 (2015). https://doi.org/10.1103/PhysRevLett.115.217601
99. B.Q. Lv, H.M. Weng, B.B. Fu, X.P. Wang, H. Miao, J. Ma, P. Richard, X.C. Huang, L.X. Zhao, G.F. Chen, Z. Fang, X. Dai, T. Qian, H. Ding, Experimental discovery of Weyl semimetal TaAs. Phys. Rev. X **5**, 031013 (2015). https://doi.org/10.1103/PhysRevX.5.031013
100. J. Maciejko, X.-L. Qi, A. Karch, S.-C. Zhang, Fractional topological insulators in three dimensions. Phys. Rev. Lett. **105**, 246809 (2010). https://doi.org/10.1103/PhysRevLett.105.246809
101. A.P. Mackenzie, Y. Maeno, The superconductivity of Sr_2RuO_4 and the physics of spin-triplet pairing. Rev. Mod. Phys. **75**, 657–712 (2003). https://doi.org/10.1103/RevModPhys.75.657
102. R. Mas-Ballesté, C. Gómez-Navarro, J. Gómez-Herrero, F. Zamora, 2D materials: to graphene and beyond. Nanoscale **3**, 20–30 (2011). https://doi.org/10.1039/C0NR00323A
103. K. Matsuhira, M. Wakeshima, R. Nakanishi, T. Yamada, A. Nakamura, W. Kawano, S. Takagi, Y. Hinatsu, Metal-insulator transition in pyrochlore iridates Ln2Ir2O7 (Ln = Nd, Sm, and Eu). J. Phys. Soc. Jpn. **76**, 043706 (2007). https://doi.org/10.1143/JPSJ.76.043706
104. T. Meng, L. Balents, Weyl superconductors. Phys. Rev. B **86**, 054504 (2012). https://doi.org/10.1103/PhysRevB.86.054504
105. T.V. Menshchikova, M.M. Otrokov, S.S. Tsirkin, D.A. Samorokov, V.V. Bebneva, A. Ernst, V.M. Kuznetsov, E.V. Chulkov, Band structure engineering in topological insulator based heterostructures. Nano Lett. **13**, 6064–6069 (2013). https://doi.org/10.1021/nl403312y

106. J.E. Moore, L. Balents, Topological invariants of time-reversal-invariant band structures. Phys. Rev. B **75**, 121306 (2007). https://doi.org/10.1103/PhysRevB.75.121306
107. R. Movshovich, M. Jaime, J.D. Thompson, C. Petrovic, Z. Fisk, P.G. Pagliuso, J.L. Sarrao, Unconventional superconductivity in $CeIrIn_5$ and $CeCoIn_5$: specific heat and thermal conductivity studies. Phys. Rev. Lett. **86**, 5152–5155 (2001). https://doi.org/10.1103/PhysRevLett.86.5152
108. S. Murakami, Phase transition between the quantum spin Hall and insulator phases in 3D: emergence of a topological gapless phase. New J. Phys. **9**, 356 (2007). https://doi.org/10.1088/1367-2630/9/9/356
109. S. Murakami, S. Iso, Y. Avishai, M. Onoda, N. Nagaosa, Tuning phase transition between quantum spin Hall and ordinary insulating phases. Phys. Rev. B **76**, 205304 (2007). https://doi.org/10.1103/PhysRevB.76.205304
110. S. Nadj-Perge, I.K. Drozdov, J. Li, H. Chen, S. Jeon, J. Seo, A.H. MacDonald, B.A. Bernevig, A. Yazdani, Observation of Majorana fermions in ferromagnetic atomic chains on a superconductor. Science **346**, 602–607 (2014). https://doi.org/10.1126/science.1259327
111. R. Nandkishore, L.S. Levitov, A.V. Chubukov, Chiral superconductivity from repulsive interactions in doped graphene. Nat. Phys. **8**, 158–163 (2012). https://doi.org/10.1038/nphys2208
112. C. Nayak, S.H. Simon, A. Stern, M. Freedman, S. Das Sarma, Non-Abelian anyons and topological quantum computation. Rev. Mod. Phys. **80**, 1083–1159 (2008). https://doi.org/10.1103/RevModPhys.80.1083
113. M. Neupane, N. Alidoust, S.-Y. Xu, T. Kondo, Y. Ishida, D.J. Kim, C. Liu, I. Belopolski, Y.J. Jo, T.-R. Chang, H.-T. Jeng, T. Durakiewicz, L. Balicas, H. Lin, A. Bansil, S. Shin, Z. Fisk, M.Z. Hasan, Surface electronic structure of the topological Kondo-insulator candidate correlated electron system SmB6. Nat. Commun. **4**, 3010 (2013). https://doi.org/10.1038/ncomms3991
114. M. Neupane, A. Richardella, J. Sánchez-Barriga, S. Xu, N. Alidoust, I. Belopolski, C. Liu, G. Bian, D. Zhang, D. Marchenko, A. Varykhalov, O. Rader, M. Leandersson, T. Balasubramanian, T.-R. Chang, H.-T. Jeng, S. Basak, H. Lin, A. Bansil, N. Samarth, M.Z. Hasan, Observation of quantum-tunnelling-modulated spin texture in ultrathin topological insulator Bi2Se3 films. Nat. Commun. **5**, 3841 (2014). https://doi.org/10.1038/ncomms4841
115. M. Neupane, S.-Y. Xu, R. Sankar, Q. Gibson, Y.J. Wang, I. Belopolski, N. Alidoust, G. Bian, P.P. Shibayev, D.S. Sanchez, Y. Ohtsubo, A. Taleb-Ibrahimi, S. Basak, W.-F. Tsai, H. Lin, T. Durakiewicz, R.J. Cava, A. Bansil, F.C. Chou, M.Z. Hasan, Topological phase diagram and saddle point singularity in a tunable topological crystalline insulator. Phys. Rev. B **92**, 075131 (2015). https://doi.org/10.1103/PhysRevB.92.075131
116. M. Neupane, S.-Y. Xu, L.A. Wray, A. Petersen, R. Shankar, N. Alidoust, C. Liu, A. Fedorov, H. Ji, J.M. Allred, Y.S. Hor, T.-R. Chang, H.-T. Jeng, H. Lin, A. Bansil, R.J. Cava, M.Z. Hasan, Topological surface states and Dirac point tuning in ternary topological insulators. Phys. Rev. B **85**, 235406 (2012). https://doi.org/10.1103/PhysRevB.85.235406
117. H.B. Nielsen, M. Ninomiya, Absence of neutrinos on a lattice. Nucl. Phys. B **185**, 20–40 (1981). https://doi.org/10.1016/0550-3213(81)90361-8
118. H.B. Nielsen, M. Ninomiya, Absence of neutrinos on a lattice (II). Nucl. Phys. B **193**, 173–194 (1981). https://doi.org/10.1016/0550-3213(81)90524-1
119. M. Novak, S. Sasaki, M. Kriener, K. Segawa, Y. Ando, Unusual nature of fully gapped superconductivity in In-doped SnTe. Phys. Rev. B **88**, 140502 (2013). https://doi.org/10.1103/PhysRevB.88.140502
120. T. Ojanen, Helical Fermi arcs and surface states in time-reversal invariant Weyl semimetals. Phys. Rev. B **87**, 245112 (2013). https://doi.org/10.1103/PhysRevB.87.245112
121. Y. Okada, M. Serbyn, H. Lin, D. Walkup, W. Zhou, C. Dhital, M. Neupane, S. Xu, Y.J. Wang, R. Sankar, F. Chou, A. Bansil, M.Z. Hasan, S.D. Wilson, L. Fu, V. Madhavan, Observation of Dirac node formation and mass acquisition in a topological crystalline insulator. Science **341**, 1496–1499 (2013). https://doi.org/10.1126/science.1239451
122. K. Okamoto, K. Kuroda, H. Miyahara, K. Miyamoto, T. Okuda, Z.S. Aliev, M.B. Babanly, I.R. Amiraslanov, K. Shimada, H. Namatame, M. Taniguchi, D.A. Samorokov, T.V. Menshchikova,

E.V. Chulkov, A. Kimura, Observation of a highly spin-polarized topological surface state in GeBi2Te4. Phys. Rev. B **86**, 195304 (2012). https://doi.org/10.1103/PhysRevB.86.195304
123. H. Ooguri, M. Oshikawa, Instability in magnetic materials with a dynamical axion field. Phys. Rev. Lett. **108**, 161803 (2012). https://doi.org/10.1103/PhysRevLett.108.161803
124. D. Pesin, L. Balents, Mott physics and band topology in materials with strong spin–orbit interaction. Nat. Phys. **6**, 376–381 (2010). https://doi.org/10.1038/nphys1606
125. F. Pientka, L.I. Glazman, F. von Oppen, Topological superconducting phase in helical Shiba chains. Phys. Rev. B **88**, 155420 (2013). https://doi.org/10.1103/PhysRevB.88.155420
126. X.-L. Qi, T.L. Hughes, S. Raghu, S.-C. Zhang, Time-reversal-invariant topological superconductors and superfluids in two and three dimensions. Phys. Rev. Lett. **102**, 187001 (2009). https://doi.org/10.1103/PhysRevLett.102.187001
127. X.-L. Qi, T.L. Hughes, S.-C. Zhang, Topological field theory of time-reversal invariant insulators. Phys. Rev. B **78**, 195424 (2008). https://doi.org/10.1103/PhysRevB.78.195424
128. X.-L. Qi, S.-C. Zhang, Topological insulators and superconductors. Rev. Mod. Phys. **83**, 1057–1110 (2011). https://doi.org/10.1103/RevModPhys.83.1057
129. N. Read, D. Green, Paired states of fermions in two dimensions with breaking of parity and time-reversal symmetries and the fractional quantum Hall effect. Phys. Rev. B **61**, 10267–10297 (2000). https://doi.org/10.1103/PhysRevB.61.10267
130. P. Roushan, J. Seo, C.V. Parker, Y.S. Hor, D. Hsieh, D. Qian, A. Richardella, M.Z. Hasan, R.J. Cava, A. Yazdani, Topological surface states protected from backscattering by chiral spin texture. Nature **460**, 1106 (2009). https://doi.org/10.1038/nature08308
131. R. Roy, Topological superfluids with time reversal symmetry (2008). arXiv:08032868 Cond-Mat
132. S. Ryu, Y. Hatsugai, Topological origin of zero-energy edge states in particle-hole symmetric systems. Phys. Rev. Lett. **89**, 077002 (2002). https://doi.org/10.1103/PhysRevLett.89.077002
133. S. Sasaki, Z. Ren, A.A. Taskin, K. Segawa, L. Fu, Y. Ando, Odd-parity pairing and topological superconductivity in a strongly spin-orbit coupled semiconductor. Phys. Rev. Lett. **109**, 217004 (2012). https://doi.org/10.1103/PhysRevLett.109.217004
134. M. Sato, Nodal structure of superconductors with time-reversal invariance and Z_2 topological number. Phys. Rev. B **73**, 214502 (2006). https://doi.org/10.1103/PhysRevB.73.214502
135. T. Sato, K. Segawa, H. Guo, K. Sugawara, S. Souma, T. Takahashi, Y. Ando, Direct evidence for the Dirac-Cone topological surface states in the ternary chalcogenide $TlBiSe_2$. Phys. Rev. Lett. **105**, 136802 (2010). https://doi.org/10.1103/PhysRevLett.105.136802
136. A.P. Schnyder, P.M.R. Brydon, Topological surface states in nodal superconductors. J. Phys.: Condens. Matter **27**, 243201 (2015). https://doi.org/10.1088/0953-8984/27/24/243201
137. A.P. Schnyder, P.M.R. Brydon, C. Timm, Types of topological surface states in nodal noncentrosymmetric superconductors. Phys. Rev. B **85**, 024522 (2012). https://doi.org/10.1103/PhysRevB.85.024522
138. A.P. Schnyder, S. Ryu, Topological phases and surface flat bands in superconductors without inversion symmetry. Phys. Rev. B **84**, 060504 (2011). https://doi.org/10.1103/PhysRevB.84.060504
139. A.P. Schnyder, S. Ryu, A. Furusaki, A.W.W. Ludwig, Classification of topological insulators and superconductors in three spatial dimensions. Phys. Rev. B **78**, 195125 (2008). https://doi.org/10.1103/PhysRevB.78.195125
140. T. Senthil, Symmetry-protected topological phases of quantum matter. Annu. Rev. Condens. Matter Phys. **6**, 299–324 (2015). https://doi.org/10.1146/annurev-conmatphys-031214-014740
141. M. Serbyn, L. Fu, Symmetry breaking and Landau quantization in topological crystalline insulators. Phys. Rev. B **90**, 035402 (2014). https://doi.org/10.1103/PhysRevB.90.035402
142. K. Shiozaki, M. Sato, Topology of crystalline insulators and superconductors. Phys. Rev. B **90**, 165114 (2014). https://doi.org/10.1103/PhysRevB.90.165114
143. K. Shiozaki, M. Sato, K. Gomi, Topology of nonsymmorphic crystalline insulators and superconductors. Phys. Rev. B **93**, 195413 (2016). https://doi.org/10.1103/PhysRevB.93.195413

144. A. Shitade, H. Katsura, J. Kuneš, X.-L. Qi, S.-C. Zhang, N. Nagaosa, Quantum Spin Hall effect in a transition metal oxide Na_3IrO_3. Phys. Rev. Lett. **102**, 256403 (2009). https://doi.org/10.1103/PhysRevLett.102.256403
145. R.-J. Slager, A. Mesaros, V. Juričić, J. Zaanen, The space group classification of topological band-insulators. Nat. Phys. **9**, 98–102 (2013). https://doi.org/10.1038/nphys2513
146. D.T. Son, B.Z. Spivak, Chiral anomaly and classical negative magnetoresistance of Weyl metals. Phys. Rev. B **88**, 104412 (2013). https://doi.org/10.1103/PhysRevB.88.104412
147. Z. Song, C.-C. Liu, J. Yang, J. Han, M. Ye, B. Fu, Y. Yang, Q. Niu, J. Lu, Y. Yao, Quantum spin Hall insulators and quantum valley Hall insulators of BiX/SbX (X=H, F, Cl and Br) monolayers with a record bulk band gap. NPG Asia Mater. **6**, e147 (2014). https://doi.org/10.1038/am.2014.113
148. B. Swingle, M. Barkeshli, J. McGreevy, T. Senthil, Correlated topological insulators and the fractional magnetoelectric effect. Phys. Rev. B **83**, 195139 (2011). https://doi.org/10.1103/PhysRevB.83.195139
149. Y. Tanaka, Z. Ren, T. Sato, K. Nakayama, S. Souma, T. Takahashi, K. Segawa, Y. Ando, Experimental realization of a topological crystalline insulator in SnTe. Nat. Phys. **8**, 800–803 (2012). https://doi.org/10.1038/nphys2442
150. J.C.Y. Teo, L. Fu, C.L. Kane, Surface states and topological invariants in three-dimensional topological insulators: application to $Bi_{1-x}Sb_x$. Phys. Rev. B **78**, 045426 (2008). https://doi.org/10.1103/PhysRevB.78.045426
151. D.J. Thouless, M. Kohmoto, M.P. Nightingale, M. den Nijs, Quantized Hall conductance in a two-dimensional periodic potential. Phys. Rev. Lett. **49**, 405–408 (1982). https://doi.org/10.1103/PhysRevLett.49.405
152. W.-F. Tsai, C.-Y. Huang, T.-R. Chang, H. Lin, H.-T. Jeng, A. Bansil, Gated silicene as a tunable source of nearly 100% spin-polarized electrons. Nat. Commun. **4**, 1500 (2013). https://doi.org/10.1038/ncomms2525
153. C.C. Tsuei, J.R. Kirtley, Pairing symmetry in cuprate superconductors. Rev. Mod. Phys. **72**, 969–1016 (2000). https://doi.org/10.1103/RevModPhys.72.969
154. D.C. Tsui, H.L. Stormer, A.C. Gossard, Two-dimensional magnetotransport in the extreme quantum limit. Phys. Rev. Lett. **48**, 1559–1562 (1982). https://doi.org/10.1103/PhysRevLett.48.1559
155. A.M. Turner, A. Vishwanath, Beyond band insulators: topology of semimetals and interacting phases. Contemp. Concepts Condens. Matter Sci Topol. Insul. **6**, 293–324 (2013). https://doi.org/10.1016/B978-0-444-63314-9.00011-1
156. A.M. Turner, Y. Zhang, R.S.K. Mong, A. Vishwanath, Quantized response and topology of magnetic insulators with inversion symmetry. Phys. Rev. B **85**, 165120 (2012). https://doi.org/10.1103/PhysRevB.85.165120
157. O. Vafek, A. Vishwanath, Dirac fermions in solids: from high-Tc cuprates and graphene to topological insulators and Weyl semimetals. Annu. Rev. Condens. Matter Phys. **5**, 83–112 (2014). https://doi.org/10.1146/annurev-conmatphys-031113-133841
158. M.M. Vazifeh, M. Franz, Electromagnetic response of weyl semimetals. Phys. Rev. Lett. **111**, 027201 (2013). https://doi.org/10.1103/PhysRevLett.111.027201
159. X. Wan, A.M. Turner, A. Vishwanath, S.Y. Savrasov, Topological semimetal and Fermi-arc surface states in the electronic structure of pyrochlore iridates. Phys. Rev. B **83**, 205101 (2011). https://doi.org/10.1103/PhysRevB.83.205101
160. C. Wang, Y. Zhang, J. Huang, S. Nie, G. Liu, A. Liang, Y. Zhang, B. Shen, J. Liu, C. Hu, Y. Ding, D. Liu, Y. Hu, S. He, L. Zhao, L. Yu, J. Hu, J. Wei, Z. Mao, Y. Shi, X. Jia, F. Zhang, S. Zhang, F. Yang, Z. Wang, Q. Peng, H. Weng, X. Dai, Z. Fang, Z. Xu, C. Chen, X.J. Zhou, Observation of Fermi arc and its connection with bulk states in the candidate type-II Weyl semimetal WTe_2. Phys. Rev. B **94**, 241119 (2016). https://doi.org/10.1103/PhysRevB.94.241119
161. E. Wang, H. Ding, A.V. Fedorov, W. Yao, Z. Li, Y.-F. Lv, K. Zhao, L.-G. Zhang, Z. Xu, J. Schneeloch, R. Zhong, S.-H. Ji, L. Wang, K. He, X. Ma, G. Gu, H. Yao, Q.-K. Xue, X. Chen, S. Zhou, Fully gapped topological surface states in Bi2Se3 films induced by a

d-wave high-temperature superconductor. Nat. Phys. **9**, 621–625 (2013). https://doi.org/10.1038/nphys2744
162. M.-X. Wang, C. Liu, J.-P. Xu, F. Yang, L. Miao, M.-Y. Yao, C.L. Gao, C. Shen, X. Ma, X. Chen, Z.-A. Xu, Y. Liu, S.-C. Zhang, D. Qian, J.-F. Jia, Q.-K. Xue, The coexistence of superconductivity and topological order in the Bi_2Se_3 thin films. Science **336**, 52–55 (2012). https://doi.org/10.1126/science.1216466
163. Q.-Z. Wang, C.-X. Liu, Topological nonsymmorphic crystalline superconductors. Phys. Rev. B **93**, 020505 (2016). https://doi.org/10.1103/PhysRevB.93.020505
164. Y.J. Wang, W.-F. Tsai, H. Lin, S.-Y. Xu, M. Neupane, M.Z. Hasan, A. Bansil, Nontrivial spin texture of the coaxial Dirac cones on the surface of topological crystalline insulator SnTe. Phys. Rev. B **87**, 235317 (2013). https://doi.org/10.1103/PhysRevB.87.235317
165. Z. Wang, Y. Sun, X.-Q. Chen, C. Franchini, G. Xu, H. Weng, X. Dai, Z. Fang, Dirac semimetal and topological phase transitions in A_3Bi (A=Na, K, Rb). Phys. Rev. B **85**, 195320 (2012). https://doi.org/10.1103/PhysRevB.85.195320
166. X.-G. Wen, Zoo of quantum-topological phases of matter (2016). arXiv:161003911 Cond-Mat
167. X.G. Wen, Topological orders in rigid states. Int. J. Mod. Phys. B **04**, 239–271 (1990). https://doi.org/10.1142/S0217979290000139
168. X.G. Wen, Vacuum degeneracy of chiral spin states in compactified space. Phys. Rev. B **40**, 7387–7390 (1989). https://doi.org/10.1103/PhysRevB.40.7387
169. H. Weng, C. Fang, Z. Fang, B.A. Bernevig, X. Dai, Weyl semimetal phase in noncentrosymmetric transition-metal monophosphides. Phys. Rev. X **5**, 011029 (2015). https://doi.org/10.1103/PhysRevX.5.011029
170. H. Weng, J. Zhao, Z. Wang, Z. Fang, X. Dai, Topological crystalline kondo insulator in mixed valence Ytterbium borides. Phys. Rev. Lett. **112**, 016403 (2014). https://doi.org/10.1103/PhysRevLett.112.016403
171. H. Weyl, Electron and gravitation. Z. Phys. **56**, 330–352 (1929). https://doi.org/10.1007/BF01339504
172. L.A. Wray, S.-Y. Xu, Y. Xia, Y.S. Hor, D. Qian, A.V. Fedorov, H. Lin, A. Bansil, R.J. Cava, M.Z. Hasan, Observation of topological order in a superconducting doped topological insulator. Nat. Phys. **6**, 855–859 (2010). https://doi.org/10.1038/NPHYS1762
173. L.A. Wray, S.-Y. Xu, Y. Xia, D. Hsieh, A.V. Fedorov, Y.S. Hor, R.J. Cava, A. Bansil, H. Lin, M.Z. Hasan, A topological insulator surface under strong Coulomb, magnetic and disorder perturbations. Nat. Phys. **7**, 32–37 (2011). https://doi.org/10.1038/NPHYS1838
174. G. Wu, H. Chen, Y. Sun, X. Li, P. Cui, C. Franchini, J. Wang, X.-Q. Chen, Z. Zhang, Tuning the vertical location of helical surface states in topological insulator heterostructures via dual-proximity effects. Sci. Rep. **3**, 1233 (2013). https://doi.org/10.1038/srep01233
175. Y. Wu, D. Mou, N.H. Jo, K. Sun, L. Huang, S.L. Budko, P.C. Canfield, A. Kaminski, Observation of Fermi arcs in the type-II Weyl semimetal candidate WTe_2. Phys. Rev. B **94**, 121113 (2016). https://doi.org/10.1103/PhysRevB.94.121113
176. Y. Xia, D. Qian, D. Hsieh, L. Wray, A. Pal, H. Lin, A. Bansil, D. Grauer, Y.S. Hor, R.J. Cava, M.Z. Hasan, Observation of a large-gap topological-insulator class with a single Dirac cone on the surface. Nat. Phys. **5**, 398–402 (2009). https://doi.org/10.1038/nphys1274
177. J. Xiong, S.K. Kushwaha, T. Liang, J.W. Krizan, M. Hirschberger, W. Wang, R.J. Cava, N.P. Ong, Evidence for the chiral anomaly in the Dirac semimetal Na_3Bi. Science **350**, 413–416 (2015). https://doi.org/10.1126/science.aac6089
178. N. Xu, X. Shi, P.K. Biswas, C.E. Matt, R.S. Dhaka, Y. Huang, N.C. Plumb, M. Radović, J.H. Dil, E. Pomjakushina, K. Conder, A. Amato, Z. Salman, D.M. Paul, J. Mesot, H. Ding, M. Shi, Surface and bulk electronic structure of the strongly correlated system SmB_6 and implications for a topological Kondo insulator. Phys. Rev. B **88**, 121102 (2013). https://doi.org/10.1103/PhysRevB.88.121102
179. S.-Y. Xu, I. Belopolski, N. Alidoust, M. Neupane, G. Bian, C. Zhang, R. Sankar, G. Chang, Z. Yuan, C.-C. Lee, S.-M. Huang, H. Zheng, J. Ma, D.S. Sanchez, B. Wang, A. Bansil, F. Chou, P.P. Shibayev, H. Lin, S. Jia, M.Z. Hasan, Discovery of a Weyl fermion semimetal and topological Fermi arcs. Science **349**, 613–617 (2015). https://doi.org/10.1126/science.aaa9297

180. S.-Y. Xu, I. Belopolski, D.S. Sanchez, M. Neupane, G. Chang, K. Yaji, Z. Yuan, C. Zhang, K. Kuroda, G. Bian, C. Guo, H. Lu, T.-R. Chang, N. Alidoust, H. Zheng, C.-C. Lee, S.-M. Huang, C.-H. Hsu, H.-T. Jeng, A. Bansil, T. Neupert, F. Komori, T. Kondo, S. Shin, H. Lin, S. Jia, M.Z. Hasan, Spin polarization and texture of the fermi arcs in the Weyl fermion semimetal TaAs. Phys. Rev. Lett. **116**, 096801 (2016). https://doi.org/10.1103/PhysRevLett.116.096801
181. S.-Y. Xu, C. Liu, N. Alidoust, M. Neupane, D. Qian, I. Belopolski, J.D. Denlinger, Y.J. Wang, H. Lin, L.A. Wray, G. Landolt, B. Slomski, J.H. Dil, A. Marcinkova, E. Morosan, Q. Gibson, R. Sankar, F.C. Chou, R.J. Cava, A. Bansil, M.Z. Hasan, Observation of a topological crystalline insulator phase and topological phase transition in Pb1−xSnxTe. Nat. Commun. **3**, 1192 (2012). https://doi.org/10.1038/ncomms2191
182. S.-Y. Xu, C. Liu, S.K. Kushwaha, T.-R. Chang, J.W. Krizan, R. Sankar, C.M. Polley, J. Adell, T. Balasubramanian, K. Miyamoto, N. Alidoust, G. Bian, M. Neupane, I. Belopolski, H.-T. Jeng, C.-Y. Huang, W.-F. Tsai, H. Lin, F.C. Chou, T. Okuda, A. Bansil, R.J. Cava, M.Z. Hasan, Observation of a bulk 3D Dirac multiplet, Lifshitz transition, and nestled spin states in Na3Bi (2013). arXiv:13127624 Cond-Mat
183. S.-Y. Xu, C. Liu, S.K. Kushwaha, R. Sankar, J.W. Krizan, I. Belopolski, M. Neupane, G. Bian, N. Alidoust, T.-R. Chang, H.-T. Jeng, C.-Y. Huang, W.-F. Tsai, H. Lin, P.P. Shibayev, F.-C. Chou, R.J. Cava, M.Z. Hasan, Observation of Fermi arc surface states in a topological metal. Science **347**, 294–298 (2015). https://doi.org/10.1126/science.1256742
184. S.-Y. Xu, M. Neupane, C. Liu, D. Zhang, A. Richardella, L. Andrew Wray, N. Alidoust, M. Leandersson, T. Balasubramanian, J. Sánchez-Barriga, O. Rader, G. Landolt, B. Slomski, J. Hugo Dil, J. Osterwalder, T.-R. Chang, H.-T. Jeng, H. Lin, A. Bansil, N. Samarth, M. Zahid Hasan, Hedgehog spin texture and Berry/'s phase tuning in a magnetic topological insulator. Nat. Phys. **8**, 616–622 (2012). https://doi.org/10.1038/nphys2351
185. S.-Y. Xu, L.A. Wray, Y. Xia, R. Shankar, A. Petersen, A. Fedorov, H. Lin, A. Bansil, Y.S. Hor, D. Grauer, R.J. Cava, M.Z. Hasan, Discovery of several large families of Topological Insulator classes with backscattering-suppressed spin-polarized single-Dirac-cone on the surface (2010). arXiv:1007.5111
186. S.-Y. Xu, Y. Xia, L.A. Wray, S. Jia, F. Meier, J.H. Dil, J. Osterwalder, B. Slomski, A. Bansil, H. Lin, R.J. Cava, M.Z. Hasan, Topological phase transition and texture inversion in a tunable topological insulator. Science **332**, 560–564 (2011). https://doi.org/10.1126/science.1201607
187. Y. Xu, B. Yan, H.-J. Zhang, J. Wang, G. Xu, P. Tang, W. Duan, S.-C. Zhang, Large-gap quantum spin hall insulators in tin films. Phys. Rev. Lett. **111**, 136804 (2013). https://doi.org/10.1103/PhysRevLett.111.136804
188. K. Yada, M. Sato, Y. Tanaka, T. Yokoyama, Surface density of states and topological edge states in noncentrosymmetric superconductors. Phys. Rev. B **83**, 064505 (2011). https://doi.org/10.1103/PhysRevB.83.064505
189. B. Yan, C.-X. Liu, H.-J. Zhang, C.-Y. Yam, X.-L. Qi, T. Frauenheim, S.-C. Zhang, Theoretical prediction of topological insulators in thallium-based III-V-VI2 ternary chalcogenides. EPL Europhys. Lett. **90**, 37002 (2010). https://doi.org/10.1209/0295-5075/90/37002
190. B.-J. Yang, Y.B. Kim, Topological insulators and metal-insulator transition in the pyrochlore iridates. Phys. Rev. B **82**, 085111 (2010). https://doi.org/10.1103/PhysRevB.82.085111
191. B.-J. Yang, N. Nagaosa, Classification of stable three-dimensional Dirac semimetals with nontrivial topology. Nat. Commun. **5**, ncomms5898 (2014). https://doi.org/10.1038/ncomms5898
192. F. Yang, C.-C. Liu, Y.-Z. Zhang, Y. Yao, D.-H. Lee, Time-reversal-invariant topological superconductivity in n-doped BiH. Phys. Rev. B **91**, 134514 (2015). https://doi.org/10.1103/PhysRevB.91.134514
193. K.-Y. Yang, Y.-M. Lu, Y. Ran, Quantum Hall effects in a Weyl semimetal: possible application in pyrochlore iridates. Phys. Rev. B **84**, 075129 (2011). https://doi.org/10.1103/PhysRevB.84.075129
194. L.X. Yang, Z.K. Liu, Y. Sun, H. Peng, H.F. Yang, T. Zhang, B. Zhou, Y. Zhang, Y.F. Guo, M. Rahn, D. Prabhakaran, Z. Hussain, S.-K. Mo, C. Felser, B. Yan, Y.L. Chen, Weyl semimetal phase in the non-centrosymmetric compound TaAs. Nat. Phys. **11**, 728–732 (2015). https://doi.org/10.1038/nphys3425

195. S.A. Yang, H. Pan, F. Zhang, Dirac and Weyl superconductors in three dimensions. Phys. Rev. Lett. **113**, 046401 (2014). https://doi.org/10.1103/PhysRevLett.113.046401
196. Y. Yao, F. Ye, X.-L. Qi, S.-C. Zhang, Z. Fang, Spin-orbit gap of graphene: first-principles calculations. Phys. Rev. B **75**, 041401 (2007). https://doi.org/10.1103/PhysRevB.75.041401
197. M.W. Young, S.-S. Lee, C. Kallin, Fractionalized quantum spin Hall effect. Phys. Rev. B **78**, 125316 (2008). https://doi.org/10.1103/PhysRevB.78.125316
198. S.M. Young, S. Zaheer, J.C.Y. Teo, C.L. Kane, E.J. Mele, A.M. Rappe, Dirac semimetal in three dimensions. Phys. Rev. Lett. **108**, 140405 (2012). https://doi.org/10.1103/PhysRevLett.108.140405
199. I. Zeljkovic, Y. Okada, C.-Y. Huang, R. Sankar, D. Walkup, W. Zhou, M. Serbyn, F. Chou, W.-F. Tsai, H. Lin, A. Bansil, L. Fu, M.Z. Hasan, V. Madhavan, Mapping the unconventional orbital texture in topological crystalline insulators. Nat. Phys. **10**, 572–577 (2014). https://doi.org/10.1038/nphys3012
200. I. Zeljkovic, D. Walkup, B.A. Assaf, K.L. Scipioni, R. Sankar, F. Chou, V. Madhavan, Strain engineering Dirac surface states in heteroepitaxial topological crystalline insulator thin films. Nat. Nanotechnol. **10**, 849–853 (2015). https://doi.org/10.1038/nnano.2015.177
201. F. Zhang, C.L. Kane, E.J. Mele, Topological mirror superconductivity. Phys. Rev. Lett. **111**, 056403 (2013). https://doi.org/10.1103/PhysRevLett.111.056403
202. H. Zhang, C.-X. Liu, X.-L. Qi, X. Dai, Z. Fang, S.-C. Zhang, Topological insulators in Bi2Se3, Bi2Te3 and Sb2Te3 with a single Dirac cone on the surface. Nat. Phys. **5**, 438 (2009). https://doi.org/10.1038/nphys1270
203. J. Zhang, C.-Z. Chang, P. Tang, Z. Zhang, X. Feng, K. Li, L. Wang, X. Chen, C. Liu, W. Duan, K. He, Q.-K. Xue, X. Ma, Y. Wang, Topology-driven magnetic quantum phase transition in topological insulators. Science **339**, 1582–1586 (2013). https://doi.org/10.1126/science.1230905
204. J. Zhang, C.-Z. Chang, Z. Zhang, J. Wen, X. Feng, K. Li, M. Liu, K. He, L. Wang, X. Chen, Q.-K. Xue, X. Ma, Y. Wang, Band structure engineering in (Bi1−xSbx)2Te3 ternary topological insulators. Nat. Commun. **2**, 574 (2011). https://doi.org/10.1038/ncomms1588
205. Q. Zhang, Z. Zhang, Z. Zhu, U. Schwingenschlögl, Y. Cui, Exotic topological insulator states and topological phase transitions in Sb2Se3–Bi2Se3 heterostructures. ACS Nano **6**, 2345–2352 (2012). https://doi.org/10.1021/nn2045328
206. B.B. Zhou, S. Misra, E.H. da Silva Neto, P. Aynajian, R.E. Baumbach, J.D. Thompson, E.D. Bauer, A. Yazdani, Visualizing nodal heavy fermion superconductivity in CeCoIn5. Nat. Phys. **9**, 474–479 (2013). https://doi.org/10.1038/nphys2672
207. X. Zhou, C. Fang, W.-F. Tsai, J. Hu, Theory of quasiparticle scattering in a two-dimensional system of helical Dirac fermions: Surface band structure of a three-dimensional topological insulator. Phys. Rev. B **80**, 245317 (2009). https://doi.org/10.1103/PhysRevB.80.245317

Part III
Biology and Mathematics

Chapter 7
Theoretical Properties of Materials Formed as Wire Network Graphs from Triply Periodic CMC Surfaces, Especially the Gyroid

Ralph M. Kaufmann and Birgit Wehefritz-Kaufmann

Abstract We report on our recent results from a mathematical study of wire network graphs that are complements to triply periodic CMC surfaces and can be synthesized in the lab on the nanoscale. Here, we studied all three cases in which the graphs corresponding to the networks are symmetric and self-dual. These are the cubic, diamond and gyroid surfaces. The gyroid is the most interesting case in its geometry and properties as it exhibits Dirac points (in 3d). It can be seen as a generalization of the honeycomb lattice in 2d that models graphene. Indeed, our theory works in more general cases, such as periodic networks in any dimension and even more abstract settings. After presenting our theoretical results, we aim to invite an experimental study of these Dirac points and a possible quantum Hall effect. The general theory also allows to find local symmetry groups which force degeneracies aka level crossings from a finite graph encoding the elementary cell structure. Vice-versa one could hope to start with graphs and then construct matching materials that will then exhibit the properties dictated by such graphs.

7.1 Introduction

We will first start to review the main motivating examples for our analysis and our methods. These are the triply periodic constant mean curvature (CMC) surfaces which are very intriguing objects due to their highly symmetric nature. By a classification result, the only triply periodic minimal surfaces whose complements are given by symmetric and self-dual graphs are the P (cubic), D (diamond) and G (gyroid)

R. M. Kaufmann
Department of Mathematics, Purdue University, 150 N. University Street,
West Lafayette, IN 47907, USA
e-mail: rkaufman@math.purdue.edu

B. Wehefritz-Kaufmann (✉)
Department of Mathematics and Department of Physics and Astronomy,
Purdue University, 150 N. University Street, West Lafayette, IN 47907, USA
e-mail: ebkaufma@math.purdue.edu

S. Gupta and A. Saxena (eds.), *The Role of Topology in Materials*,
Springer Series in Solid-State Sciences 189,
https://doi.org/10.1007/978-3-319-76596-9_7

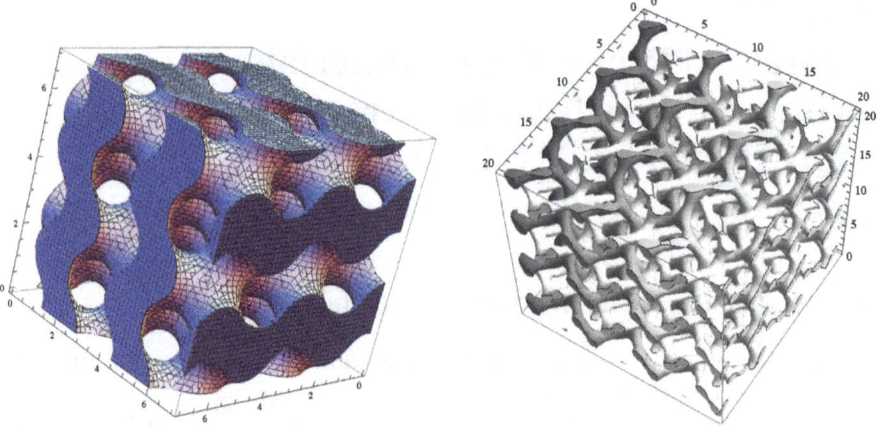

Fig. 7.1 The double gyroid surface (left) and its complement, the two non–intersecting channel systems C_+ and C_- (right)

surfaces (see e.g. [1]). While the cubic and diamond surfaces have been known for almost two centuries (they were already discovered by Schwarz in 1830 [2]), the gyroid was an omission in Schwarz's classification and was only discovered about 50 years ago in 1970 by Alan Schoen [3]. As any naturally occurring surface, a physical version will not be a true 2d object, but will have some, albeit small thickness, which makes it 3d. Such a thick surface has two sides, each a true 2d surface. Thus, the interface actually consists of *two* disconnected surfaces, where *each* of them is a surface of the given type. The double gyroid (DG) for instance is such a configuration of two mutually non–intersecting embedded gyroids which form the boundaries of the thick gyroid surface.

A single gyroid has symmetry group $I4_132$ while the double gyroid has the symmetry group $Ia\bar{3}d$ where the extra symmetry comes from interchanging the two gyroids.[1]

The actual equations of the gyroid are very complicated and initially only locally known by a differential equation since it is a CMC surface. To get the true shape mathematically one can use a computer program, the Brakke Surface Evolver [5]. However, in a good approximation, the surfaces can be visualized by using the level surfaces [6]. An example of a level surface approximation for the double gyroid is given by the following formula:

$$L_t : \sin(x)\cos(y) + \sin(y)\cos(z) + \sin(z)\cos(x) = t$$

which we use in the visualizations. The double gyroid surface is then modeled by L_t and L_{-t} for $0 < t < \sqrt{2}$. It is pictured in Fig. 7.1.

[1]Here $I4_132$ and $Ia\bar{3}d$ are given in the international or Hermann–Mauguin notation for symmetry groups, see e.g. [4].

Recently, it was demonstrated that the gyroid can be synthesized in the lab [7]. Both the surface and its complement, which forms a two-channel network (see Fig. 7.1), can be realized by using nano-porous silica film. It is interesting to note that the gyroid structure appears spontaneously in nature on the wings of certain butterflies or beetles to give them their brilliant color [8].

Several variations of these structures have been synthesized, i.e. semiconductor quantum-wire arrays of PbSe, PbS, and CdSe. The synthesis process involves several steps. First, the actual surface nanostructure is formed by self-assembly in some carefully prepared surfactant or block copolymer systems. The nanopores (channels) are then filled with a semiconductor and the original silica surface is dissolved to yield the nanowire network. A second semiconductor material may potentially be grown in the void space to yield a bulk heterojunction semiconductor. The typical lattice constant of these structures is of the order of 18 nm. This means that they are "supercrystals", with lattice constants far exceeding atomic length scales. However, a quantum mechanical treatment is still applicable and the typical length scale is comparable with graphene (where the length scale is of the order of 10 nm). We will focus on the wire structure in this article.

We performed a theoretical study to predict properties of these materials. Our approach is comprised of an analysis of the symmetries, the singularities and a non-commutative model all of which we will briefly explain. This treatment is not restricted to the particular example of the gyroid and can be used to study any periodic wire network and even further generalizations. We have applied it to the cases of the wire networks derived from the cubic, the diamond and the gyroid triply periodic CMC surfaces as well as other periodic structures such as Bravais lattices and the honeycomb lattice underlying graphene.

7.1.1 Classical Geometry of the Gyroid and Graph Approximation for the Channels

Let us first describe in more detail the classical geometry of the gyroid. Details can be found in [9]. The complement $\mathbb{R}^3 \backslash G$ of a single gyroid G has two components. These components will be called the gyroid wire systems or channels.

There are two distinct channels, one left and one right handed. Each of these 3d channels can be contracted onto an embedded graph, called skeletal graph [3, 10]. We will call these graphs Γ_+ and Γ_-. Each graph is periodic and trivalent. We fix Γ_+ to be the graph which has the node $v_0 = (\frac{5}{8}, \frac{5}{8}, \frac{5}{8})$ in the above approximation. We will give more details on the graph Γ_+ below.

The channel containing Γ_+ is shown in Fig. 7.2a. A (crystal) unit cell of the channel together with the embedded graph Γ_+ is shown in Fig. 7.2b and just the skeletal graph is contained in Fig. 7.2c. The unit cell is obtained by using the simple lattice translations along the x, y and z axes. Such a cell contains 8 vertices which are trivalent. The actual symmetry group is bcc and hence higher. If one mods out

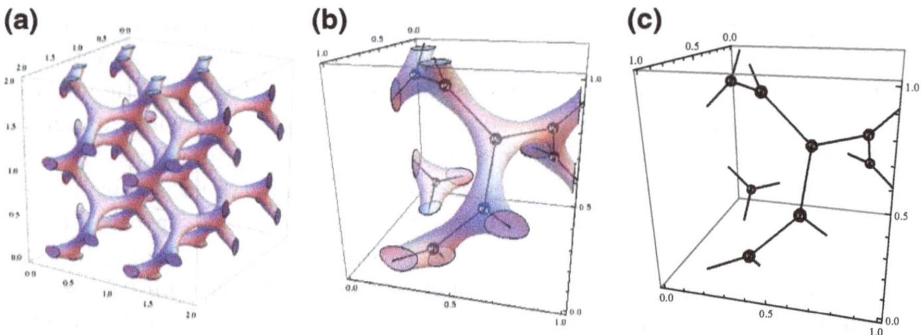

Fig. 7.2 **a** 3d periodic structure of the gyroid wire network **b** skeletal graph inside the channel **c** skeletal graph with labeled vertices

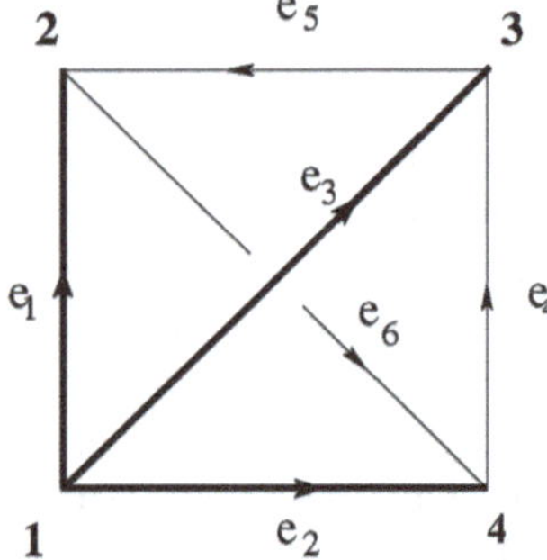

Fig. 7.3 Abstract quotient graph used in our calculation. The bold edges are a spanning tree and the vertex 1 is a root of this spanning tree

by the full symmetry group then an elementary cell will only have 4 points which are trivalent. This is captured by a graph with 4 vertices where each vertex is linked to all other vertices. This graph, which is called the full square or the tetrahedral graph, is abstract. This means that it is not embedded in any real space, but just a combinatorial object, see Fig. 7.3.

Stating things in a more precise fashion: The graph Γ_+ is the graph made up of the following vertices in the unit cell

$$
\begin{aligned}
v_0 &= \left(\frac{5}{8}, \frac{5}{8}, \frac{5}{8}\right) & v_4 &= \left(\frac{7}{8}, \frac{5}{8}, \frac{3}{8}\right) \\
v_1 &= \left(\frac{3}{8}, \frac{7}{8}, \frac{5}{8}\right) & v_5 &= \left(\frac{1}{8}, \frac{7}{8}, \frac{3}{8}\right) \\
v_2 &= \left(\frac{3}{8}, \frac{1}{8}, \frac{7}{8}\right) & v_6 &= \left(\frac{1}{8}, \frac{1}{8}, \frac{1}{8}\right) \\
v_3 &= \left(\frac{5}{8}, \frac{3}{8}, \frac{7}{8}\right) & v_7 &= \left(\frac{7}{8}, \frac{3}{8}, \frac{1}{8}\right),
\end{aligned} \tag{7.1}
$$

and all their translations along the lattice directions and the edges between them according to the incidences that can be read off from Fig. 7.2.

By combinatorial arguments, we have obtained results about the classical geometry of the infinite graph Γ_+ [9]. Namely, there are closed loops on the graph Γ_+. Each minimal loop goes through 10 sites and at each point there are 30 oriented minimal loops or 15 such undirected loops.

As mentioned, the translational symmetry group for both the gyroid and the double gyroid is actually the body-centered cubic (bcc) lattice. In our theoretical calculations, we will deal with a finite graph that is obtained as a quotient graph from Γ_+. We can use the bcc symmetry of the lattice to construct this abstract quotient graph. A set of generators of the bcc symmetry is

$$g_1 = \frac{1}{2}(1, -1, 1), \quad g_2 = \frac{1}{2}(-1, 1, 1), \quad g_3 = \frac{1}{2}(1, 1, -1) \tag{7.2}$$

The passage from the graph of the unit cell with 8 vertices is given by identifying the vertices. $v_0 \leftrightarrow v_6, v_1 \leftrightarrow v_7, v_2 \leftrightarrow v_4$ and $v_3 \leftrightarrow v_5$. The 6 edge vectors are then represented by the 6 vectors

$$e_1 = \frac{1}{4}\begin{pmatrix}-1\\1\\0\end{pmatrix}, e_2 = \tfrac{1}{4}\begin{pmatrix}0\\-1\\1\end{pmatrix}, e_3 = \frac{1}{4}\begin{pmatrix}1\\0\\-1\end{pmatrix} \tag{7.3}$$

$$e_4 = \frac{1}{4}\begin{pmatrix}1\\1\\0\end{pmatrix}, e_5 = \tfrac{1}{4}\begin{pmatrix}0\\-1\\-1\end{pmatrix}, e_6 = \frac{1}{4}\begin{pmatrix}-1\\0\\-1\end{pmatrix} \tag{7.4}$$

Now, the translates of the points $v_i, i = 1, \ldots, 4$ along integer linear combinations of the $g_j, j = 1, 2, 3$ and the translates of the edge vectors $e_k, k = 1, \ldots, 6$ form the graph Γ_+. Taking the quotient by the free Abelian subgroup L that is generated by the vectors g_i, we arrive at the abstract quotient graph $\bar{\Gamma} = \Gamma_+/L$. It is the graph with 4 vertices and 6 edges, where all pairs of distinct vertices are connected by exactly one edge shown in Fig. 7.3. It turns out that this graph basically suffices to capture the essential information about the geometry.

The passage from the channel systems to the graphs retains all homotopical information as does the passage from the thick surface to just one copy of the gyroid, by shrinking the thickness to zero. Any topological information which is homotopy invariant (that is basically the information that is invariant under continuous deformations) is encoded in the gyroid surface and the two skeletal graphs. Furthermore, since the level surface approximation is a deformation of the original gyroid, one can use this simplification for the study. Notice that not all geometric information is retained by such a deformation, for instance being a CMC surface is not. Also, as we have seen, dimensions are not preserved either, what is preserved, however, are topological charges, singularities, homology, K-theory, etc. These deformation independent quantities are of course very desirable as a physically realized version

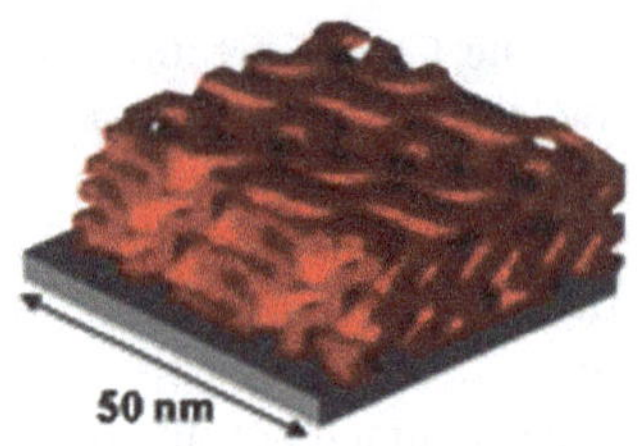

Fig. 7.4 Fabricated gyroid structure squished by gravity [11]

of the gyroid will not be perfect. Indeed the result of the gyroid, the self-assembly is actually a gyroid, that is a bit squished, see Fig. 7.4.

The fact that the skeletal graph approximation is valid for the electronic properties has been shown by a different physical argument: namely, numerical simulations of a simple wave equation [12] have shown that the lowest-energy wavefunctions are supported primarily on the junctions. Thus, one may expect to reproduce the low-energy end of the spectrum by using the tight-binding approximation, in which the junctions are replaced by the vertices, and the segments connecting them by the edges of a graph.

We can and will thus continue the study of the wire system by using the graph approximation.

7.1.2 P and D Surfaces

There are two other triply periodic self–dual and symmetric CMC surfaces- the cubic (P) and the diamond (D) network. They are shown in Fig. 7.5 together with their wire networks obtained in the same way as for the gyroid. Here we summarize the results from [13].

The P surface has a complement which has two connected components each of which can be retracted to the simple cubical graph whose vertices are the integer lattice $\mathbb{Z}^3 \subset \mathbb{R}^3$. The translational group is $\mathbb{Z}^3$ in this embedding, so it reduces to the case of a Bravais lattice. Its abstract quotient graph is shown in Fig. 7.6 on the left.

The D surface has a complement consisting of two channels each of which can be retracted to the diamond lattice $\Gamma_\diamond$. The diamond lattice is given by two copies of the fcc lattice, where the second fcc is the shift by $\frac{1}{4}(1, 1, 1)$ of the standard fcc lattice, see Fig. 7.5. The edges are nearest neighbor edges. The symmetry group is $Fd\bar{3}m$. The quotient graph for the D surface is shown in Fig. 7.6 on the right.

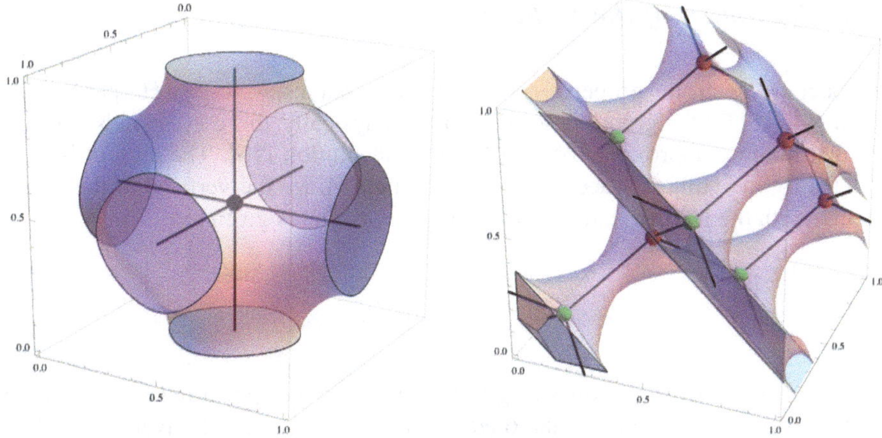

Fig. 7.5 The cubic (P) (left) and the diamond (D) wire network (right)

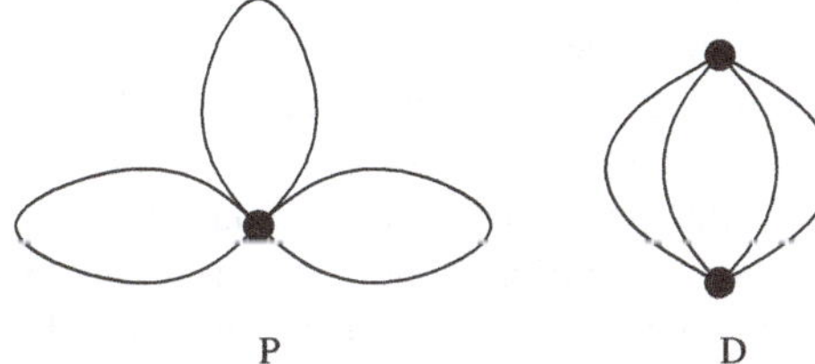

Fig. 7.6 Abstract quotient graphs for the P and D surfaces

7.2 Theory

7.2.1 Overview

Our method is two–pronged depending on whether or not a background magnetic field is present. In the absence of a magnetic field, we use singularity theory and classical geometry to classify topological features such as Dirac points and topological charges. In the presence of a magnetic field, the classical geometry becomes non-commutative. Some of the topological invariants carry over, such as a gap classification, and topological charges aka Chern classes, which could give rise to a quantum Hall effect. Some other new phenomena appear, which could potentially lead to new properties as described below.

7.2.2 Summary of the Methods

To model the electronic properties of the physical systems, we used a Harper Hamiltonian [14, 15] for the graphs described above. Physically, this corresponds to using the tight-binding approximation and Peierls substitution [16]. The system is thus modeled by the Hilbert space $\mathscr{H} = l^2(\Lambda)$, where Λ are the vertices of the graph together with the Harper Hamiltonian acting on $\mathscr{H}$ that is given by

$$H = \sum_{i=1}^{6} U_{e_i} + U^{\star}_{e_i}$$

where the sum is over all edges e_i given above. For later generalization, we note that these edges correspond to the 6 edges of the quotient graph given in Fig. 7.2. We recall that $l^2(\Lambda)$ are the square summable series on Λ. A typical element/state is given by $\phi = (\phi)_{\lambda\in\Lambda}$ with $\phi_\lambda \in \mathbb{C}$ as complex number that can be imagined to sit at the site λ and $\sum_{\lambda\in\Lambda} |\phi_\lambda|^2 < \infty$. In the case without magnetic field U_e is the translation operator with $U_e(\phi)_\lambda = \phi_{\lambda-e}$. Its conjugate U^*_e is the translation along $-e$. In the case with magnetic field the U_e are replaced with so–called magnetic translation operators as explained in Sect. 7.3.

At this point, we wish to remark that the traditional translation operators used in absence of a magnetic field (a) commute with each other and (b) are symmetries and hence commute with the Hamiltonian, whence we call it the commutative case. In contrast, if there is a magnetic field, translations cease to commute with each other as their commutator introduces a phase. Likewise they cease to commute with the Hamiltonian. This is the origin of the non–commutative geometry.

Going back to the commutative case, using Fourier transform the operators can alternatively be considered as depending on quasi-momenta of the Brillouin zone. It is this Brillouin zone geometry that the non–commutative version also captures.

In order to present the Harper Hamiltonian as a matrix, we rewrite the Hilbert space as a direct sum $\mathscr{H} = \bigoplus_v \mathscr{H}_v$ where the sum is over the vertices v in an elementary cell, that is the vertices of the abstract quotient graphs. For instance for the gyroid $\mathscr{H}_{v_0}$ has as elements square summable sequences $(\phi)_\lambda$ where now λ is in the sub–lattice generated by v_0 which are all the vertices v of Γ_+ that are translates of v_0 by integer linear combinations of the $g_i, i = 1, 2, 3$. Now translation along a directed edge between starting at v and ending at w will map $\mathscr{H}_w$ to $\mathscr{H}_v$, by our convention.

After Fourier transform the Harper Hamiltonian becomes dependent on the quasi–momenta of the Brillouin zone. This means that we have a family of Hamiltonians depending on parameters which parameterize the Brillouin zone. For the 3d skeletal graphs this is $T^3 = S^1 \times S^1 \times S^1$, which we can visualize as a cube $[0, 2\pi]^3$ where opposite sides are identified.

We will describe the results separately for the three surfaces P, D and G.

7.2.3 The Gyroid Without Magnetic Field

There are 4 vertices in the elementary cell and accordingly in quotient graph. Hence the Hamiltonian will be a 4×4 matrix with translation operators as entries.

$$H_{\Gamma_+} = \begin{pmatrix} 0 & U_1^* & U_2^* & U_3^* \\ U_1 & 0 & U_6^* & U_5 \\ U_2 & U_6 & 0 & U_4 \\ U_3 & U_5^* & U_4^* & 0 \end{pmatrix} \tag{7.5}$$

Here we use the short hand $U_i = U_{e_i}$ and $U_i^* = U_{-e_i}$.

Using a normalization, essentially a change of basis for the lattice generators and singling out v_0 as a special vertex, by conjugation, the Harper Hamiltonian can be taken to the form [9]

$$H = \begin{pmatrix} 0 & 1 & 1 & 1 \\ 1 & 0 & A & B^* \\ 1 & A^* & 0 & C \\ 1 & B & C^* & 0 \end{pmatrix} \tag{7.6}$$

where A, B and C are combinations of translational operators of the form U_e. We refer to [9] for the details. Using Fourier transform, that is looking at states with fixed quasi–momenta, we can rewrite them as $A = \exp(ia)$, $B = \exp(ib)$, $C = \exp(ic)$, with $a, b, c \in [0, 2\pi]$ taken periodically. Alternatively, we can think of $\exp(ia)$, $\exp(ib)$ and $\exp(ic)$ as a set of generators for the functions on the 3-torus T^3. This point of view will be generalized later.

The eigenvalues of $H(a, b, c)$, that is the energies E_i of the different bands depending on the quasi–momentum $k = (a, b, c)$ are given by the roots of the characteristic polynomial:

$$P(a, b, c, z) = z^4 - 6z^2 + a_1(a, b, c)z + a_0(a, b, c) \tag{7.7}$$

where

$$\begin{aligned} a_1 &= -2\cos(a) - 2\cos(b) - 2\cos(c) - 2\cos(a + b + c) \\ a_0 &= 3 - 2\cos(a + b) - 2\cos(b + c) - 2\cos(a + c) \end{aligned}$$

The geometry of the dispersion relation, i.e. the set of the E_i as a function of the quasi-momenta $k = (a, b, c)$, is that one can view the energy spectrum as a cover of the Brillouin zone. Here over each point k of the Brillouin zone T^3, we have the Eigenvalues of $H(k)$. Moving around k, we get a cover of T^3 which generically, i.e. when there is no degenerate Eigenvalue, has 4 sheets. There are degeneracies, however, when there are less than 4 distinct Eigenvalues and the sheets are glued together giving ramifications.

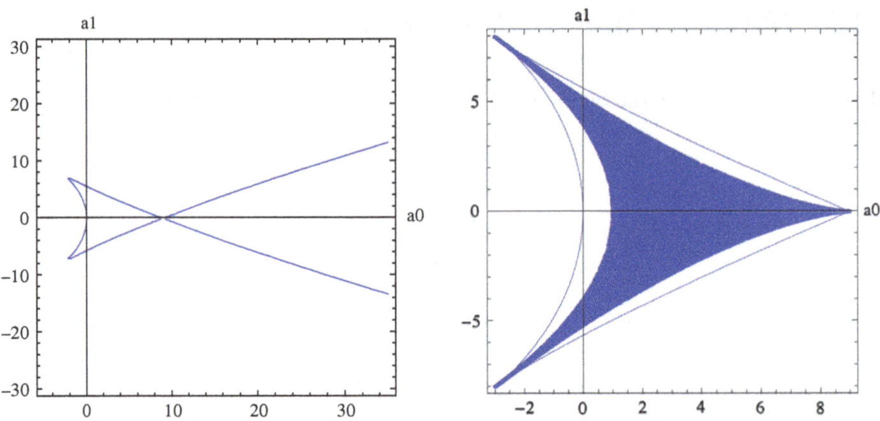

Fig. 7.7 The −6 slice of the swallowtail of A_3 and the region occupied by the gyroid

This geometry is amenable to study via singularity theory [17]. Indeed mathematically the geometry is a pull-back of a miniversal unfolding of an A-type singularity. This allowed us to analytically calculate the singular points in the gyroid spectrum and to classify them using results of Grothendieck [18].

Let us briefly highlight some of the construction. The first step is the realization that in the characteristic polynomial, the coefficients depend on the parameters (a, b, c). Since H is traceless, there is no term of z^3 in the characteristic polynomial and we are left with the coefficients $a_i(a, b, c)$ of z^i for $i = 0, 1, 2$ with $a_2(a, b, c) = -6$ being constant. The latter fact is no coincidence, but is a consequence of the type of the graph (it is simply laced) and the fact that it has 6 edges. The miniversal unfolding of the A_3 singularity is given by the cover of the roots of the polynomial $z^4 + a_2z^2 + a_1z^1 + a_0z^0$, where now the a_i can take any complex value and are regarded as parameters. The base space of the cover is then just the parameter space of the a_i, that is $\mathbb{C}^3$. Again the cover generically has 4 sheets and is ramified over the locus where some roots coincide. This locus is called the discriminant locus or the swallowtail. The second crucial observation is then that the a_i as depending on (a, b, c) define a map from the Brillouin zone to the base of the miniversal unfolding, i.e. $\Xi : T^3 \to \mathbb{C}^3$ which we called the characteristic map. Since all the values $a_i(a, b, c)$ are real, we can restrict to the real locus $\mathbb{R}^3$ of $\mathbb{C}^3$. The spectrum of H is singular at (a, b, c) if and only if $\Xi(a, b, c)$ lies in the discriminant locus. We call these points the singular points, although, of course they are not singular in T^3, but the cover is singular over them.

The characteristic region is the image of Ξ, that is the region that can be reached when a, b, c are varied between 0 and 2π. Since $a_2(a, b, c) = -6$, the information is captured in the a_0, a_1 plane, which should be considered as the slice at $a_2 = -6$, that is the plane parallel to the a_0, a_1 plane in $\mathbb{R}^3$ through $(0, 0, -6)$. This is depicted in Fig. 7.7.

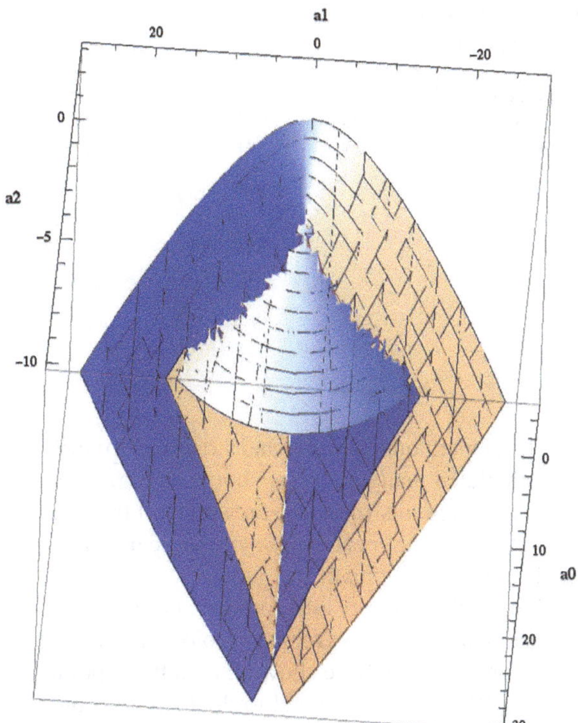

Fig. 7.8 The A_3 singularity (the swallowtail)

The curve shown in the figure is the discriminant locus which is explicitly given by

$$20736\, a_0 - 4608\, a_0^2 + 256\, a_0^3 + 864\, a_1^2 - 864\, a_0\, a_1^2 - 27\, a_1^4 = 0 \tag{7.8}$$

As we showed, the boundaries of the characteristic region are obtained as the collection of points (a_0, a_1) for $a = b = c$ and $a = b = -c$.

The characteristic region is contained in the slice $a_2 = -6$ of the A_3 singularity (the swallowtail, shown in Fig. 7.8) and intersects the discriminant in exactly three isolated points, the two cusps and the double point of that slice of the swallowtail. This result was derived analytically.

The two cusps are in the stratum of type A_2 (where three roots coincide) and the double point is in the stratum of type (A_1, A_1) (where two pairs of roots coincide). As can be shown the fibers of Ξ over all these point are discrete giving rise to finitely many points in T^3 at which there are level crossings for the energies. For the A_2 singularities, this is just one point each at each cusp, giving rise to two triple crossings, while the fiber over (A_1, A_1) consists of two points. Over each of these points there are two double crossings and as each single crossing is of type A_1, these are Dirac points, see below.

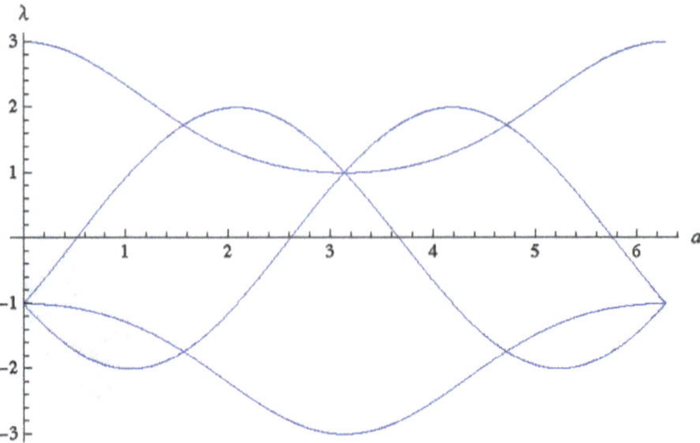

Fig. 7.9 Dispersion relation for the gyroid Harper Hamiltonian along the diagonal in the Brillouin zone. Note the time-reversal symmetry (TRS)

The dispersion relation (band structure) for the gyroid Harper Hamiltonian, which we calculated analytically, is shown in Fig. 7.9 along the diagonal in the Brillouin zone, that is points (a, a, a), on which all the singular points lie.

Notice that the spectrum is symmetric under $k \to -k$, which can be read off from the Harper Hamiltonian.

Summarizing our results for the band structure of the gyroid Harper Hamiltonian, we find two Dirac points for the gyroid at the points $(\frac{\pi}{2}, \frac{\pi}{2}, \frac{\pi}{2})$, $(\frac{3\pi}{2}, \frac{3\pi}{2}, \frac{3\pi}{2})$ in the first Brillouin zone. We have shown that at these points, the dispersion relation becomes linear [19]. They correspond to double crossings in the spectrum. These points are actually double Dirac points, i.e. there are two double crossings at each of these points. They can be seen as the 3d- analogue of Dirac points in graphene. There are further degenerate points in the spectrum at $(0, 0, 0)$ and (π, π, π), which are triple crossings. See Fig. 7.9. All these are forced by level sticking which we found using an enhanced symmetry group of a quantum graph [20].

For a general periodic graph with n vertices in an elementary cell one again obtains a characteristic map, but now to the miniversal unfolding of A_{n-1} singularity. By a theorem of Grothendieck [18] one knows that there is a stratification of the discriminant locus obtained by deleting vertices from the A_{n-1} graphs and by also deleting any edge incident to the vertex.

For instance, in the A_3 case (see Fig. 7.8), deleting the left or right vertex leaves one with the A_2 graph and deleting the middle vertex leaves one with two copies of the A_1 graph. Hence the singularity type (A_1, A_1), the smooth part of the swallowtail, corresponds to deleting two vertices, which gives the 3 parts of the swallowtail as strata over which there are A_1 singularities. The smallest stratum is obtained by not deleting any vertex and this is the A_3 singularity at the origin, see Fig. 7.10.

In general the singularities are then determined by the strata and the fibers of the characteristic map Ξ over the swallowtail.

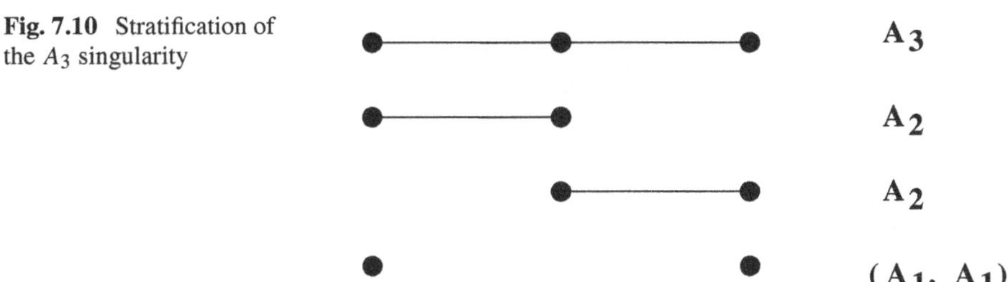

Fig. 7.10 Stratification of the A_3 singularity

7.2.4 Enhanced Symmetries from a Re-gauging Groupoid

Another reason that singular points have to be present is given by symmetries, which lead to level sticking. As we will argue, the momentum space geometry is basically encoded in the quotient graph with certain decorations. In this setting local symmetries for the geometry can be induced by so–called quantum enhanced symmetries of the underlying graph. The procedure for this is not straightforward, though, and proceeds via re–gauging groupoid and a "lift" of its action to the momentum space [17]. The result of this rather elaborate process is the existence of *projective* representations of subgroups of the symmetry group of the graph that appear as stabilizers in the geometric action on the momentum space [21].

We will give a cursory overview of the calculus and refer to [17, 21] for the details. A sample graphical calculation which we now discuss is given in Fig. 7.11. This is a generalization of the methods of [22].

The main new tool we introduced refers to enhanced graph symmetries, which arise from certain re-gaugings of the Harper Hamiltonian [20]. The gauge expresses a choice of basis in which the Hamiltonian becomes a matrix whose entries lie in the same space of operators. The choice in terms of combinatorial data is given by a spanning tree and a root r for this spanning tree. The operators are then all operators on $\mathscr{H}_r$. We used such a gauge with $r = v_0$ to obtain the form (7.6) for the Hamiltonian. The Hamiltonian is then a decoration on the abstract graph. Basically the edge is decorated by the corresponding operator. The coefficients h_{vw} in the matrix of the Hamiltonian are then the sum over all edge decorations of the edges connecting the two vertices v and w. The gauge is reflected in the fact that the edges of the spanning tree are decorated by the identity operator 1. The first entry in Fig. 7.11 illustrates this nicely. Accordingly the entries in the first row and column of (7.6) say that the diagonal entry is 1. Acting by a symmetry of the graph moves the spanning tree and the root. This consequently breaks the gauge. For instance exchanging the vertices v_0 and v_1 of the graph yields the second entry in Fig. 7.11. In order to reestablish the gauge condition of decoration by units on the spanning tree, that is to re–gauge, one employs a decoration on the vertices, this is the third part of the figure. The vertex decorations are again by the same type of operators, which in the commutative case can be seen as phases. This decoration gives the quantum enhancement. Up to the initial condition that the root vertex is decorated by the identity, the decoration is

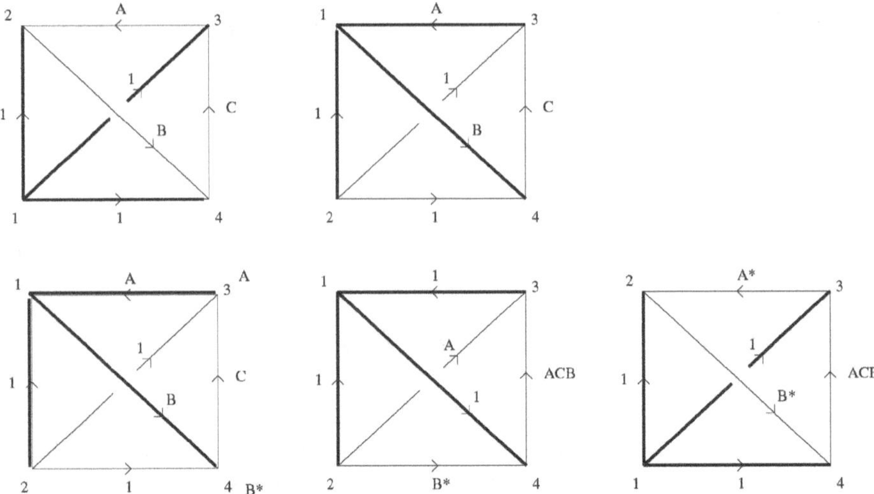

Fig. 7.11 Sample calculation of re-gauging with edge and vertex factors, see [20]

entirely determined by the two different gaugings and satisfies the requirement that when the decorations on the vertices act on the edge decoration, by multiplication from the left by the decoration of the target vertex and on the right by the conjugate of the source decoration, the decoration of the spanning tree vertices is again just by identity operators, see part four of the figure.

This type of calculation produces several bits of data. The permutation matrix, a diagonal matrix which contains the vertex factors and finally a transformation of the decorations of the edges, which in the example above is $(A, B, C) \mapsto (A^*, B^*, ACB)$. This reading off is facilitated by rewriting the abstract graph with the decorations in parallel to the original graph as shown in the last part of the figure. Regarding the A, B, C as providing coordinates (a, b, c) of T^3 the latter information gives an automorphism of T^3. This means that to each element of the symmetry group of the finite graph, we obtain a transformation of T^3 and we can look at the stabilizer subgroups of points on T^3. The product of the two matrices, permutation and re-gauging, then gives a *projective* representation of the stabilizer groups, which are subgroups of the symmetry group of the graph. The fact that these representations are projective is essential and this is due to the fact that we used the quantum enhanced version, which allows for re–gauging by vertex decorations.

Table 7.1 provides a nice summary of our results [20]. It shows the enhanced symmetry groups for each of the level crossings. This explains the degeneracy of each eigenvalue which is also listed in the table. It would not be possible to explain these degeneracies by the group $\mathbb{S}_4$ acting on the graph alone. Indeed the double Dirac point corresponds to 2-dim irreducible *projective* representations. A classical action of $\mathbb{S}_4$ does not have any 2-dim irreps. The projective representations can be further

Table 7.1 Possible choices of parameters (a, b, c) leading to non-Abelian enhanced symmetry groups and degenerate eigenvalues of H. Here A_4 is the alternating group and $2A_4$ its double cover which is also known as the binary tetrahedral group

a, b, c	Group	Iso class of type of extension		Dim of eigenvalues λ irreps	
$(0, 0, 0)$	$\mathbb{S}_4$	$\mathbb{S}_4$	Trivial	1,3	$\lambda = -1$ three times
					$\lambda = 3$ once
(π, π, π)	$\mathbb{S}_4$	$\mathbb{S}_4$	Trivializable Cocycle	1,3	$\lambda = 1$ three times
					$\lambda = -3$ once
$(\frac{\pi}{2}, \frac{\pi}{2}, \frac{\pi}{2})$ $(\frac{3\pi}{2}, \frac{3\pi}{2}, \frac{3\pi}{2})$	A_4	$2A_4$	Isomorphic extension	2,2	$\lambda = \pm\sqrt{3}$ Twice each

characterized by the group extension for which they form a genuine representation, see, e.g. [23].

7.2.5 *Slicing, Chern Classes and Stability Under Perturbations*

In order to find topological charges in 3d, we used Chern classes and a slicing method. We furthermore started to study deformations of the effective geometry and could show that the Dirac points remain effectively stable as they carry a minimal topological charge, which can be detected as jumps in the first Chern classes or the Berry phase on 2-torus slices of the 3-torus at a given height. This is shown in Fig. 7.12. The double points give jumps of 1 unit while the triple points give jumps of 2 units. Furthermore, the triple points do carry a charge, but they can split, and do so, under deformations as a numerical study shows [24]. The type of splitting can be explained if the time reversal symmetry is preserved.

The idea behind this is the following. Going back to [25, 26], we know that the integers in the integer quantum Hall effect (QHE) can be viewed as Chern numbers. Chern classes can be written as differential forms using Chern–Weil theory [27]. In particular the ith Chern class has a degree 2i differential form representative. Integrating the form over a $2i$–dimensional manifold gives a number. The relevant classes here are just the first Chern classes. Here the 2–form can be thought of as the Berry curvature [26, 28]. Since the QHE is on a 2d torus, one can simply integrate over the torus. Now if we are on the 3-torus, we cannot just integrate over it. The first Chern class is a 2–form and the second Chern class would be a 4–form. On the other hand, we can look at a slice 2-torus at fixed height c and integrate the Berry curvature over just this slice. We will get one Chern number for each band. That is all the points of the form (a, b, c), say fixed c and $a, b \in S^1 \times S^1$. In order to have the correct notion of a Berry phase, we should however have a non–degenerate

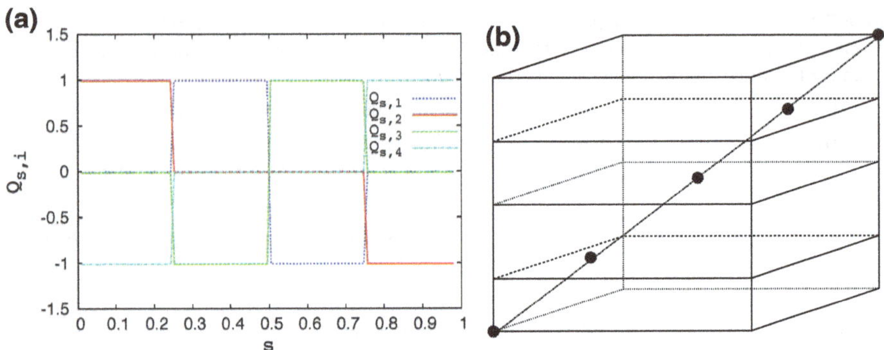

Fig. 7.12 **a** Topological charges as functions of the height of a 2-torus slice of the Brillouin zone. The jumps are step functions and the sloped transition is merely a guide. **b** The Brillouin torus as a cube with periodic boundaries, the position of the Dirac points and triple crossings along the diagonal and two 2-torus slices

Eigenvalue. So this will only work for values of c where no (a, b, c) is singular. In the gyroid example this means that $c \neq 0, \frac{\pi}{4}, \frac{\pi}{2}, \frac{3\pi}{4}$. In this way, we get the Chern number of the ith band as a function of c as a well defined integer outside these special values of c. Since the Chern number is a topological invariant, it is locally constant. It does, however, jump when c crosses one of the bad points above. This jump depends on the local structure of the singularity. It is by ± 1 for Dirac points, but can be more complicated for more complicated singularities. We determined the local structure over the A_2 singular triple crossing points, which told us that the jumps are by $-2, 0, 2$ for the three crossing bands. The numerical check is depicted in the first part of Fig. 7.12, while the second part of the figure shows the slices in four regions of constant Chern classes.

What is special in 3-d is that these charges are now protected in the sense that under perturbations, more singular points may appear, but the net jump over these points is conserved. Numerically we found that the double Dirac points drift apart, while the triple crossings decay into four double crossings under deformations that preserve the time reversal symmetry that is present in the original context [24].

We showed [17] that slicing in all three directions actually totally fixes the homological information obtainable from Berry phases and curvature.

7.2.6 Possible Experimental Verification

It would be intriguing to measure the Dirac cones in this particularly rich 3d geometry experimentally. There are two hurdles that have to be overcome. First, one has to find a material with the right Fermi surface, but this should be possible. The second is that these Dirac points are buried in the continuum, since cutting at this particular energy will form actual Femi surfaces away from the Dirac points. This type of investigation has been done, however, in topological semimetals [29, 30] using angle-resolved photoemission spectroscopy (ARPES). The Dirac points for the gyroid lie along the diagonal slice of the Brillouin zone and it should be possible to find a curve through them that has a pseudo-gap.

A similar study has been performed in photonics [31] using angle resolved transmission measurement.

Moreover, the Dirac points are stable under symmetry breaking deformations [17, 24] and hence give affirmation of the applicability of the theory to real materials. Perhaps these effects, including the decay of the higher order singularity, can be used advantageously. Mathematically there are associated topological charges stemming from first Chern classes. These manifest themselves as jumps in the first Chern classes or the integral over the Berry curvature on 2-torus slices. It would also be very interesting to measure the decay of the triple points that is predicted by global constraints. Finding the triple points in the 3d material experimentally would be a great step in realizing properties forced by interesting topology.

Let us now briefly mention the results for the other geometries:

7.2.7 The P Wire Network Without Magnetic Field

It can be shown that the characteristic polynomial is simply a polynomial of degree 1 [19], $P(t, z) = z - \sum_i t_i$, so that after shifting z we are left with just $z = 0$, which is not critical. Not surprisingly, there are no singularities. That means that the cover itself is just the trivial cover of T^3 by itself. Nevertheless the slice geometry is that of T^2 which supports the quantum Hall effect. Accordingly this example becomes interesting in the presence of a magnetic field.

In terms of representations, there is only the root of the spanning tree which is unique. The $\mathbb{S}_3$ action permutes the edges and their weights. This yields the permutation action on the T^3. There is no nontrivial cover and the eigenvalues remain invariant.

7.2.8 The D Wire Network and the Honeycomb Lattice Without Magnetic Field

We treat the diamond and the honeycomb case in parallel since they have similar quotient graphs. The graphs $\bar{\Gamma}$ are given in Fig. 7.13. The honeycomb lattice is used to model graphene, and indeed we reproduce the known results about Dirac points in graphene with our theory. This is a nice cross–check.

After gauging, the Hamiltonians are

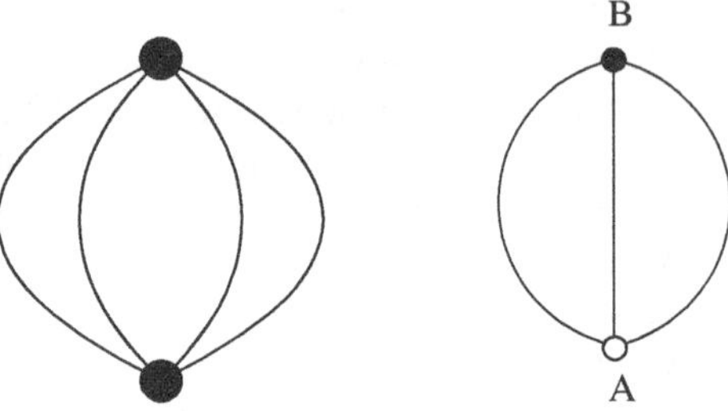

Fig. 7.13 The graphs $\bar{\Gamma}$ for the diamond (left) and the honeycomb case (right)

$$H_{hon} = \begin{pmatrix} 0 & 1+U+V \\ 1+U^*+V^* & \end{pmatrix} \tag{7.9}$$

and

$$H_D = \begin{pmatrix} 0 & 1+U+V+W \\ 1+U^*+V^*+W^* & \end{pmatrix} \tag{7.10}$$

We use $U = exp(iu)$, $V = exp(iv)$, $W = exp(iw)$ with u, v, w real. The fact that there are sums in the entries is due to the fact that there are multiple edges connecting the two vertices.

The polynomials are $P(u, v, z) = z^2 - 3 - 2cos(u) - 2cos(v) - 2\cos(u - v)$ and $P(u, v, w, z) = z^2 - 4 - 2\cos(u) - 2\cos(v) - 2\cos(w) - 2\cos(u - v) - 2\cos(u - w) - 2\cos(v - w)$. The characteristic regions in $\mathbb{R}$ are just the intervals $[-9, 0]$ and $[-16, 0]$. The discriminant is the point 0. From this we see that in both cases we have to have $a_0 = 0$ and the singular locus is simply this fiber.

7.2.8.1 The Honeycomb Case

For the honeycomb, the standard calculation shows that if $a_0 = 0$ then $U = V^*$ and $U \in \{\rho_3 := \exp(2\pi i/3), \bar{\rho}_3\}$, which means that the fiber consists of 2 points. These are the well known Dirac points $(\rho_3, \bar{\rho}_3)$, $(\bar{\rho}_3, \rho_3)$.

In terms of representations, we can look at the graph symmetries which are combinations of the interchange of the two vertices or the three edges (see Fig. 7.13). The vertex interchange renders the fixed points $u = \pm 1, v = \pm 1$ which have eigenvectors $v_1 = (1, 1)$ and $v_2 = (-1, 1)$ and eigenvalues $1 + u + v$ and $-(1 + u + v)$, respectively. The irreps of the C_3 action are $triv \oplus \omega$ where $\omega = \exp(\frac{2i\pi}{3})$.

As far as the edge permutations are concerned the interesting one is the cyclic permutation (123) which yields the equations

$$u = \bar{v}, v = \bar{v}u$$

for fixed points. Hence $u^3 = 1$. We get non–trivial matrices at the two points $(\omega, \bar{\omega})$ and $(\bar{\omega}, \omega)$. At these points e_1, e_2 are eigenvectors with eigenvalue 0 and $H = 0$, since $1 + u + v = 1 + \omega + \bar{\omega} = 0$.

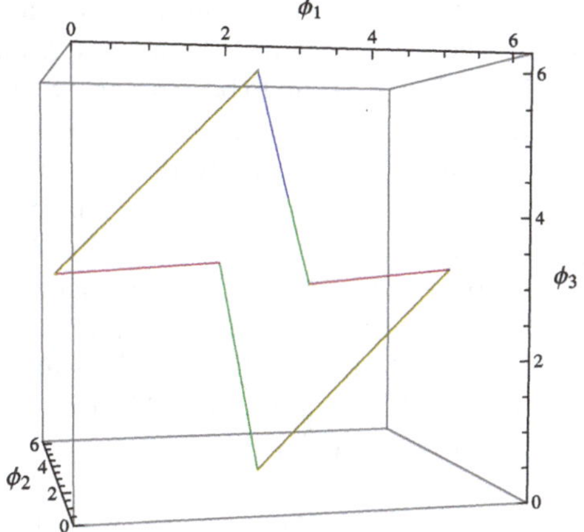

Fig. 7.14 The 3-torus as a cube with opposite sides identified and the singular locus of the diamond case consisting of three circles pairwise intersecting in a point

Denoting the elements of $\mathbb{Z}/2\mathbb{Z}$ again by $+, -$, there is an embedding of $\mathbb{S}_3 \to \mathbb{Z}/2\mathbb{Z} \times \mathbb{S}_3$ given by $(12) \mapsto (-, (12)), (23) \mapsto (-, (23))$. Notice that $(123) \mapsto (+, (123))$. It is then an easy check that the equations for the fixed points are satisfied exactly by $(\omega, \bar{\omega})$ and $(\bar{\omega}, \omega)$. The representation is a projective representation of $\mathbb{S}_3$ cohomologous to the 2–dim irreducible representation of $\mathbb{S}_3$.

The fixed points are exactly the Dirac points of graphene and the symmetry above forces the degeneracies.

7.2.8.2 The Diamond Case

The equation for the fiber over 0,

$$-4 - 2cos(u) - 2\cos(v) - 2\cos(w) - 2\cos(u - v) - 2\cos(u - w) - 2\cos(v - w) = 0$$

has been solved in [19] and the solutions are given by $(u, v, w) = (\phi_i, \phi_j, \phi_k)$ with $\phi_i = \pi, \phi_j \equiv \phi_k + \pi \bmod 2\pi$ with $\{i, j, k\} = \{1, 2, 3\}$. So in this case the fiber of the characteristic map is 1–dimensional and the pull–back has singularities along a locus of dimension 1, which also implies that there are no Dirac points. Geometrically the singular locus are three circles pairwise intersecting in a point. This is shown in Fig. 7.14.

Looking at representations, things again become interesting. Permuting the two vertices, we obtain eight fixed points if $u, v, w \in \{1, -1\}$. The matrix for this transposition is $\begin{pmatrix} 0 & 1 \\ 1 & 0 \end{pmatrix}$. This gives super–selection rules and we know that $v_1 = (1, 1)$ and $v_2 = (-1, 1)$ are eigenvectors. The eigenvalues being $1 + u + v + w$ and $-(1 + u + v + w)$ at these eight points.

We can also permute the edges with the $\mathbb{S}_4$ action. In this case the $\mathbb{S}_3$ action leaving the spanning tree edge invariant acts as a permutation on (u, v, w). The relevant matrices however are just the identity matrices and the representation is trivial. The transposition (12), however, results in the action $(u, v, w) \mapsto (\bar{u}, \bar{u}v, \bar{u}w)$ on T^3, see Fig. 7.11. So to be invariant we have $u = 1$, but this implies that ρ_{12} is the identity matrix. Invariance for (13) and (14) and the three cycles containing 1 are similar. But, if we look at invariance under the element (12)(34) we are lead to the equations $u = \bar{u}, v = \bar{u}w, w = \bar{u}v$. This has solutions $u = 1, v = w$, for these fixed points again we find only a trivial action. But for $u = -1, v = -w$ these give rise to the diagonal matrix $diag(1, -1)$ and hence eigenvectors $e_1 = (1, 0)$ and $e_2 = (0, 1)$, but looking at the Hamiltonian, these are only eigenvectors if it is the zero matrix $H(u, v, w) = 0$. Indeed the conditions above imply $1 + u + v + w = 0$. Similarly, we find a $\mathbb{Z}/2\mathbb{Z}$ group for (13)(24) and (14)(23) yielding the symmetric equations $v = -1, u = -w$ and $w = -1, u = -v$. These are exactly the three circles found in [13].

Going to bigger subgroups of $\mathbb{S}_4$ we only get something interesting if the stabilizer group G_t contains precisely two of the double transpositions above. That is the Klein four group $\mathbb{Z}/2\mathbb{Z} \times \mathbb{Z}/2\mathbb{Z}$. The invariants are precisely the intersection points of the three circles given by $u = v = -1$ and $w = 1$ and its cyclic permutations.

To find the 2–dimensional irreducible representations, we look at different Klein four groups embedded into $\mathbb{Z}/2\mathbb{Z} \times \mathbb{S}_4$. If we denote the elements of $\mathbb{Z}/2\mathbb{Z}$ by $+, -$, then we first look at $(+, id)$, $(+, (12)(34))$, $(-, (13)(24))$, $(-, (14)(23))$. The element $(-, (13)(24))$ is the composition of edge permutation (13)(24) together with the switching of the vertices. It gives the equation $u = v\bar{w}$ for fixed points, while the fixed points of $(-, (14)(23))$ satisfy $u = \bar{v}w$. Combining these equations with the ones for (12)(23) above, we find again the solutions $u = 1, v = w$ and $u = -1, v = -w$. The difference however is that the representation in the case $u = -1$ is the irreducible projective representation of the Klein group corresponding to the irreducible 2–dim representation of its 2–fold cover given by the quaternion group $\pm 1, \pm i, \pm j, \pm k$. For $u = 1$ the irreps are one–dimensional and give no new information. Using the different embeddings of the Klein group we find the 2–dim irreps on the three circles above responsible for the degeneration of the eigenvalues. These are three lines of doubly degenerate eigenvalue 0. They are not Dirac points since there is one free parameter accordingly the fibers of the characteristic map of [19] are one dimensional which implies that the singular point is not isolated.

Additional cases of planar graphs that do not correspond directly to CMC surfaces are discussed in [19]. We do not find any Dirac points, either. So the gyroid remains the only object in this family of 3-d wire networks from CMC surfaces that exhibits 3-d Dirac points.

7.3 Noncommutative Approach in the Presence of a Magnetic Field

In the case with magnetic field, we used an approach from non-commutative geometry [32–34]. Here the geometry is modeled by a C^*–algebra $\mathscr{B}_\Theta$, which is the smallest algebra containing the Hamiltonian and the translational symmetries. The standard choice of the Hamiltonian is again the Harper Hamiltonian [9] above, but where now the U_e are replaced by so called magnetic translation operators, which do not commute, but satisfy

$$U_{e_1}U_{e_2} = \exp(\int_R BdS)U_{e_2}U_{e_1},$$

where R is the rectangle spanned by the two vectors e_1 and e_2. The aim is then to classify the algebras $\mathscr{B}_\Theta$ that appear when the magnetic field is varied. There Θ is the magnetic field parameter. The magnetic fields we considered were constant magnetic fields, so that there are three real parameters for the coefficients of the field. We often use the quadratic form $\hat{\Theta}$ corresponding to the B–field $B = 2\pi\hat{\Theta}$.

The basic ingredients are like in the commutative case. We can gauge by picking a vertex and a spanning tree, so that the decorations along the spanning tree and hence the entries in the Hamiltonian are 1. The other decorations are now some combinations of magnetic translation operators. It turns out, as we showed, that these operators satisfy the relations of a non-commutative torus, so that the Hamiltonian becomes a matrix with coefficients in the non–commutative torus.

In particular for the gyroid, we are dealing with the non–commutative 3-torus. By definition the non-commutative 3-torus $\mathbb{T}^3_\Theta$ is the C^* algebra spanned by 3 linearly independent unitary operators O_i that satisfy the commutation relations $O_iO_j = \exp(2\pi i\theta_{ij})O_jO_i$ for $i, j, k = 1, 2, 3$, where θ_{ij} is a skew-symmetric matrix determined by the magnetic field. The operators in question are the A, B, C of (7.6), which now just do not commute anymore. The relation to the classic commutative geometry is given by the fact that the commutative unital C^* algebra of continuous complex valued functions on $\mathbb{T}^3_\Theta$, that is $C^*(\mathbb{T}^3_\Theta) = C(\mathbb{T}^3_\Theta, \mathbb{C})$ is the C^* algebra generated by the three functions $\exp(ia), \exp(ib), \exp(ic)$. The lattice translations can also be written as matrices with entries in the non–commutative 3-torus. Thus the algebra $\mathscr{B}_\Theta$ is isomorphic to a subalgebra of the C^* algebra of 4×4 matrices with coefficients in the non–commutative (nc) 3-torus.

The inclusion of the nc 3-torus into the algebra $\mathscr{B}_\Theta$ as the lattice translations is by definition the nc geometry based on the gyroid in a magnetic field. It captures the non–commutative deformation of the Eigenvalue cover of the Brillouin zone in the previous discussion. To make the connection, we recall that in general a covering map is a special type of surjection $\pi : E \to B$ which when we regard functions yields a pull–back map going the other way around: $\pi^* : C^*(B) \to C^*(E)$ which sends a function f on B to the function $f \circ \pi$ on E. In our case $B = \mathbb{T}^3_\Theta$ the Brillouin zone and E is the cover by the Eigenvalues aka energies.

Fig. 7.15 Hofstadter's butterfly; the image shows the distribution of energy levels E of Bloch electrons in a magnetic field in 2 dimensions [36]

An upshot of this theory is that some of the methods, like Chern–classes and K-theory, that is the theory of bundles with addition and formal difference, carry over to this setting.

7.3.1 Gyroid in the Presence of a Magnetic Field

There are two series of results. Based on a rather abstract K-theoretic argument following [33], our study predicts a distribution of energy gaps in the gyroid spectrum in terms of the magnetic field that is a 3-d analogue of Hofstadter's butterfly [35], which results in the 2d setting for the Bravais lattice $\mathbb{Z}^2$, see Fig. 7.15. To be more precise, if the magnetic field is a fixed 3-dim vector, for rational values of the magnetic field, there are finitely many gaps, for irrational values there are possibly infinitely many gaps. This is a very interesting result that has not yet been checked experimentally to our knowledge.

The second series of results characterizes the abstract algebra $\mathscr{B}_\Theta$. Our calculations lead to the following conclusions: The algebra $\mathscr{B}_\Theta$ is generically the full matrix algebra and hence Morita equivalent to the non–commutative 3-torus. It is true sub–algebra only at finitely many rational points in the parameter space given by the three parameters of the magnetic field, which we listed explicitly [9]. We will express the results here in terms of the bcc lattice vectors given in (7.2). B represents the 3–dimensional vector of a constant magnetic field. We define the following parameters:

$$\theta_{12} = \frac{1}{2\pi} B \cdot (g_1 \times g_2), \quad \theta_{13} = \frac{1}{2\pi} B \cdot (g_1 \times g_3), \quad \theta_{23} = \frac{1}{2\pi} B \cdot (g_2 \times g_3)$$

$$\alpha_1 := e^{2\pi i \theta_{12}} \bar{\alpha}_2 := e^{2\pi i \theta_{13}} \alpha_3 := e^{2\pi i \theta_{23}}$$

$$\phi_1 = e^{\frac{\pi}{2} i \theta_{12}}, \quad \phi_2 = e^{\frac{\pi}{2} i \theta_{31}}, \quad \phi_3 = e^{\frac{\pi}{2} i \theta_{23}}, \quad \Phi = \phi_1 \phi_2 \phi_3$$

Our results can then be summarized as follows:

1. If $\Phi \neq 1$ or not all α_i are real then $\mathscr{B}_\Theta = M_4(\mathbb{T}^3_\Theta)$.
2. If $\Phi = 1$, all $\alpha_i = \pm 1$, at least one $\alpha_i \neq 1$ and all ϕ_i are different then $\mathscr{B}_\Theta = M_4(\mathbb{T}^3_\Theta)$.
3. If $\phi_i = 1$ for all i then the algebra is the same as in the commutative case.
4. In all other cases (this is a finite list) $\mathscr{B}$ is non–commutative and $\mathscr{B}_\Theta \subsetneq M_4(\mathbb{T}^3_\Theta)$.

It would be very interesting to find and measure special properties of a corresponding material at these values of the magnetic field. There should be a hidden symmetry or charge associated to the reduction of the algebra to a proper sub–algebra.

7.3.2 P Wire Network in a Magnetic Field

For the simple cubic lattice (and also any other Bravais lattice of rank k, the simple cubic lattice has $k = 3$): if $\Theta \neq 0$ then $\mathscr{B}_\Theta$ is simply the noncommutative torus $\mathbb{T}^k_\Theta$ and if $\Theta = 0$ then this $\mathscr{B}_0$ is the C^* algebra of T^k. There are no degenerate points.

The analysis of [14] of the quantum Hall effect, however, suggests that there is a non–trivial noncommutative line bundle in the case of $k = 2$ for *non-zero B-field*. Furthermore, in this case there is a non–trivial bundle, not using the noncommutative geometry, but rather the Eigenfunctions constructed in [25] for the full Hilbert space $\mathscr{H}$. This is what is also considered in [26].

7.3.3 D Wire Network in a Magnetic Field

We express our results in terms of parameters q_i and χ_i defined as follows: Set $e_1 = \frac{1}{4}(1, 1, 1), e_2 = \frac{1}{4}(-1, -1, 1), e_3 = \frac{1}{4}(-1, 1, -1)$. For $B = 2\pi\Theta$ let

$$\Theta(-e_1, e_2) = \varphi_1 \quad \Theta(-e_1, e_3) = \varphi_2 \quad \Theta(e_2, e_3) = \varphi_3 \text{ and } \chi_i = e^{i\varphi_i} \text{ for } i = 1, 2, 3 \tag{7.11}$$

There are three operators U, V, W, given explicitly in [13], which span $\mathbb{T}^3_\Theta$ and have commutation relations

$$UV = q_1 VU \quad UW = q_2 WU \quad VW = q_3 WV \tag{7.12}$$

where the q_i expressed in terms of the χ_i are:

$$q_1 = \bar{\chi}_1^2 \chi_2^2 \chi_3^2 \quad q_2 = \bar{\chi}_1^6 \bar{\chi}_2^2 \bar{\chi}_3^2 \quad q_3 = \bar{\chi}_1^2 \bar{\chi}_2^6 \chi_3^2 \tag{7.13}$$

Vice versa, fixing the values of the q_i fixes the χ_i up to eighth roots of unity:

$$\chi_1^8 = \bar{q}_1 \bar{q}_2 \quad \chi_2^8 = q_1 \bar{q}_3 \quad \chi_3^8 = q_1^2 \bar{q}_2 q_3 \tag{7.14}$$

Other useful relations are $q_2 \bar{q}_3 = \bar{\chi}_1^4 \chi_2^4 \bar{\chi}_3^4$ and $q_2 q_3 = \bar{\chi}_1^8 \bar{\chi}_2^8$. The algebra $\mathscr{B}_\Theta$ is the *full* matrix algebra *except* in the following cases in which it is a proper subalgebra.

1. $q_1 = q_2 = q_3 = 1$ (the special bosonic cases) and one of the following is true:
   a. All $\chi_i^2 = 1$ then $\mathscr{B}_\Theta$ is isomorphic to the commutative algebra in the case of no magnetic field above.
   b. Two of the $\chi_i^4 = -1$, the third one necessarily being equal to 1.
2. If $q_i = -1$ (special fermionic cases) and $\chi_i^4 = 1$. This means that either
   a. all $\chi_i^2 = -1$ or
   b. only one of the $\chi_i^2 = -1$ the other two being 1.
3. $\bar{q}_1 = q_2 = q_3 = \bar{\chi}_2^4$ and $\chi_1^2 = 1$ it follows that $\chi_2^4 = \chi_3^4$. This is a one parameter family.
4. $q_1 = q_2 = q_3 = \bar{\chi}_1^4$ and $\chi_2^2 = 1$ it follows that $\chi_1^4 = \bar{\chi}_3^4$. This is a one parameter family.
5. $q_1 = q_2 = \bar{q}_3 = \bar{\chi}_1^4$ and $\chi_1^2 = \bar{\chi}_2^2$. It follows that $\chi_3^4 = 1$. This is a one parameter family.

The same remark as made above for the gyroid geometry applies. In this case there is an even more interesting structure that appears in the phase diagram of algebras.

7.3.4 Honeycomb in a Magentic Field

Generically $\mathscr{B}_\Theta = \mathbb{T}_\Theta^2$. In order to give the degenerate points, let $-e_1 := (1, 0)$, $e_2 = \frac{1}{2}(1, \sqrt{3})$, $e_3 := \frac{1}{2}(1, -\sqrt{3})$ be the lattice vectors and $f_2 := e_2 - e_1 = \frac{1}{2}(-3, \sqrt{3})$, $f_3 := e_3 - e_1 = \frac{1}{2}(3, \sqrt{3})$ the period vectors of the honeycomb lattice. The parameters we need are

$$\theta := \hat{\Theta}(f_2, f_3), \quad q := e^{2\pi i\theta} \;\; \textit{and} \;\; \phi = \hat{\Theta}(-e_1, e_2), \quad \chi := e^{i\pi\phi}, \textit{thus} \quad q = \bar{\chi}^6 \tag{7.15}$$

where $\hat{\Theta}$ is the quadratic form corresponding to the B–field $B = 2\pi\hat{\Theta}$.

Our results can be summarized as follows:
The algebra $\mathscr{B}_\Theta$ is the full matrix algebra of $M_2(\mathbb{T}_\theta^2)$ except in the following finite list of cases [9]

1. $q = 1$.
2. $q = -1$ and $\chi^4 = 1$.

The precise algebras are given in [9]. We wish to point out that $q = \chi = 1$ is the commutative case and $q = -\chi = 1$ is isomorphic to the commutative case, while the other cases give non–commutative proper subalgebras of $M_2(\mathbb{T}^2_\theta)$.

7.3.5 *Possible 3d Quantum Hall Effect*

Another consequence is that with a suitable material the gyroid channels should exhibit a 3-d quantum Hall effect. The 3-d quantum Hall effect has been predicted in different materials from isotropic 3-d crystals to layered structures like graphite. The gyroid system will be particularly well suited to experimental access given that it is a super-lattice and has a large lattice constant. With the lattice constant being this large, the magnetic field required to observe the quantum Hall effect will be in a normal range (see the discussion in [37]). Indeed the quantum Hall effect for the graphene type lattice was realized at 13–14 nm [38].

Another interesting research direction is provided by the nc geometry of the gyroid in a constant magnetic field which manifests itself in a type of phase diagram. For generic values of the magnetic field, the effective non-commutative geometry (nc) is that of the nc 3-torus. However, there are certain families parameterized by special values of the magnetic field such that the unit fluxes are particularly tuned, which correspond to different geometries. It would be exciting to find special behavior of the materials at these values. The different geometries could entail further topological quantum numbers, such as those corresponding to the QHE.

7.4 General Theory and Possible Material Design

So far, we have discussed how to start from a real world structure such as a periodic wire network and obtain properties from its topology via the use of graphs. In the end, everything devolved to having a finite graph with decorations. We can reverse this question. Can one construct materials from the finite graph and what properties are then forced by this? First, one will have to find a periodic graph Γ in real space $\mathbb{R}^d$, which is a cover of the finite graph $\bar{\Gamma}$. This means that $\bar{\Gamma} = \Gamma/\mathbb{Z}^d$ where $\mathbb{Z}^d$ is the translational symmetry group. By covering theory [39], we can construct such a cover for each normal subgroup of the fundamental group of Γ. Since Γ is connected, the fundamental group is the free group $\mathbb{F}_b$ where b is the Betti number of the graph $\bar{\Gamma}$. This is the number of independent loops, which has the following two interpretations. If we take a spanning tree of $\bar{\Gamma}$ and contract it, we are left with a graph with one vertex and b loops. More technically b is the rank of the first homology group $H_1(\bar{\Gamma})$. By general covering theory [39] there is a cover Γ_N of $\bar{\Gamma}$ for each normal subgroup

$N \subset \mathbb{F}_b$, such that $\bar{\Gamma} = \Gamma_N/(G/N)$. In our case, we are thus looking for normal subgroups N of $\mathbb{F}_b$ such that $\mathbb{F}_b/N \simeq \mathbb{Z}^b$. One choice for N, which yields the so called maximal Abelian cover is the commutator subgroup $[\mathbb{F}_b, \mathbb{F}_b] \subset \mathbb{F}_b$. Modding out by the commutators makes the free group $\mathbb{F}_b$ into the free Abelian group $\mathbb{Z}^b$. There are other smaller covers with the desired property as well.

Following [40, 41] the abstract maximal Abelian cove Γ can actually be embedded as a graph $\Gamma \subset \mathbb{R}^{b_1} \simeq H_1(\bar{\Gamma}, \mathbb{R})$ with $\mathbb{Z}^{b_1}$ acting on the ambient $\mathbb{R}^{b_1}$ inducing the covering action on Γ. This graph is called canonical placement and has an energy minimizing property.

With certain assumptions that are called fully commutative toric non–degenerate case in [20], we can reconstruct all the data from the finite graph using the maximal Abelian cover. This is the case for our examples. For the smaller covers one also needs an embedding $N \subset \mathbb{F}_b$ which has to satisfy certain conditions. One can go one step further to the nc case as well by adding abstract operators along the edges, and then generating a C^* algebra with them by forcing certain commutation relations. For example, use a spanning tree, then each edge in the graph not in the spanning tree is decorated with an operator O_e and these operators are chosen to generate the b–dim non–commutative torus (this is essentially the toric non–degenerate case). There are more possibilities here as well.

A systematic study of such decorated graphs and their quantum enhanced symmetry groups could be used for material design. The cover will tell one which structure to construct in real space and the quantum symmetry groups will then force degeneracies such as Dirac cones.

7.5 Discussion and Conclusion

We have treated three cases of wire networks stemming from minimal CMC surfaces as well as the honeycomb lattice underlying graphene in the context of a Harper Hamiltonian on an abstract quotient graph. We were able to derive level crossings in the band structure analytically and characterize them completely using methods from singularity theory and representation theory. This leads to a complete theoretical understanding of all degeneracies in the spectra. Among other results, we showed that the gyroid wire network is the only geometry exhibiting isolated Dirac points in 3d. In the presence of a magnetic field, we proposed a new approach from non–commutative geometry that characterizes the spectrum for the gyroid wire network as 3d analogue of Hofstadter's butterfly and leads to a classification of the abstract algebra $\mathscr{B}_\Theta$. Combined with specific material dependent properties and knowledge about the Fermi level, we expect that this leads to new results and predictions about a possible phase diagram. Let us briefly discuss the series of results that may be relevant for an experimentalist. The gyroid wire network is a new, complex system that could exhibit a 3-d quantum Hall effect and provide a fractal Hofstadter butterfly, much like a 3d analog of graphene. Furthermore it has a modified behavior at special parameters of the magnetic field. Without magnetic field, again parallel to graphene, the gyroid has a linear dispersion relation at its Dirac points, which are additionally protected by topological charges. The feasibility of these speculations comes from the synthesis of this material at the 18 nm scale, which constitutes a super-lattice that affords the necessary fluxes.

We furthermore propose to use the present study as the starting point to design specific materials that have desired properties. We argue that it is sufficient to impose constraints on the abstract finite quotient graph to obtain them in the full lattice structure. This could be a very powerful method.

More generally our methods apply to a wide range of periodic systems and can be expanded to other lattices and geometries.

Acknowledgements BK thankfully acknowledges support from the NSF under the grants PHY-0969689 and PHY-1255409. RK thanks the Simons Foundation for support under the collaboration grant #317149.

References

1. D.M. Anderson, H.T. Davis, J.C.C. Nitsche, L.E. Scriven, Adv. Chem. Phys. **77**, 337–396 (1990)
2. H.A. Schwarz, *Gesammelte Mathematische Abhandlungen*, vol. I (Springer, Berlin, 1890)
3. A.H. Schoen, NASA TN-D5541 (1970)
4. T. Hahn (ed.), *International Tables for Crystallography*, vol. A (Springer, Berlin, 2005)
5. K.A. Brakke, Exp. Math. **1**, 141–165 (1992)
6. C.A. Lambert, L.H. Radzilowski, E.L. Thomas, Curved surfaces in chemical structure. Philos. Trans. Math. Phys. Eng. Sci. **354**(1715), 2009–2023 (1996)
7. V.N. Urade, T.C. Wei, M.P. Tate, H.W. Hillhouse, Chem. Mater. **19**(4), 768–777 (2007)
8. K. Michielsen, D.G. Stavenga, J. R. Soc. Interface **5**, 85–94 (2008)
9. R.M. Kaufmann, S. Khlebnikov, B. Wehefritz-Kaufmann, J. Noncommutative Geom. **6**, 623–664 (2012)
10. K. Große-Brauckmann, M. Wohlgemuth, Calc. Var. **4**, 499–523 (1996)
11. E. Stach, Private communication
12. S. Khlebnikov, H.W. Hillhouse, Phys. Rev. B **80**(11), 115316 (2009)
13. R.M. Kaufmann, S. Khlebnikov, B. Wehefritz-Kaufmann, J. Phys. Conf. Ser. **343**, 012054 (2012)
14. J. Bellissard, A. van Elst, H. Schulz-Baldes, J. Math. Phys. **35**, 5373–5451 (1994)
15. M. Marcolli, V. Mathai, Towards the fractional quantum Hall effect: a noncommutative geometry perspective, in *Noncommutative Geometry and Number Theory: Where Arithmetic Meets Geometry*, ed. by Caterina Consani, Matilde Marcolli (Vieweg, Wiesbaden, 2006), pp. 235–263
16. G. Panati, H. Spohn, S. Teufel, Commun. Math. Phys. **242**, 547–578 (2003)
17. R.M. Kaufmann, S. Khlebnikov, B. Wehefritz-Kaufmann, J. Singul. **15**, 53–80 (2016)
18. M. Demazure, Classification des germs à point critique isolé et à nombres de modules 0 ou 1 (d'après Arnol'd). Séminaire Bourbaki, 26e année, vol. 1973/74 Exp. No 443, pp. 124–142, Lecture Notes in Mathematics, vol. 431 (Springer, Berlin, 1975)
19. R.M. Kaufmann, S. Khlebnikov, B. Wehefritz-Kaufmann, Ann. Phys. **327**, 2865–2884 (2012)
20. R.M. Kaufmann, S. Khlebnikov, B. Wehefritz-Kaufmann, Ann. Henri Poincaré **17**, 1383–1414 (2016)
21. R.M. Kaufmann, S. Khlebnikov, B. Wehefritz-Kaufmann, J. Phys. Conf. Ser. **597**, 012048 (2015)
22. J.E. Avron, A. Raveh, B. Zur, Rev. Mod. Phys. **60**, 873 (1988)

23. G. Karpilovsky, *Projective Representations of Finite Groups* (Dekker, New York, 1985)
24. R.M. Kaufmann, S. Khlebnikov, B. Wehefritz–Kaufmann, Topologically stable Dirac points in a three-dimensional supercrystal, in preparation
25. D.J. Thouless, M. Kohmoto, M.P. Nightingale, M. den Nijs, Phys. Rev. Lett. **49**, 405–408 (1982)
26. B. Simon, Phys. Rev. Lett. **51**, 2167–2170 (1983)
27. D. Husemöller, *Fibre Bundles*, vol. 20, Graduate Texts in Mathematics (Springer, Berlin, 1993)
28. M.V. Berry, Proc. R. Soc. Lond. A **392**, 45–57 (1984)
29. S.-Y. Xu et al., Science **347**, 294 (2015)
30. B.-J. Yang, N. Nagaosa, Nat. Commun. **5**, 4898 (2014)
31. L. Lu, L. Fu, J.D. Joannopoulo, M. Soljacic, Nat. Photonics **7**, 294–299 (2013)
32. A. Connes, *Noncommutative Geometry* (Academic Press Inc., San Diego, 1994)
33. J. Bellissard, *From Number Theory to Physics* (Springer, Berlin, 1992), pp. 538–630
34. P.G. Harper, Proc. Phys. Soc. Lond. A **68**, 874–878 (1955)
35. D.R. Hofstadter, Phys. Rev. B **14**(1976), 2239–2249 (1976)
36. D. Weiss, Nat. Phys. **9**, 395–396 (2013)
37. M. Koshino, H. Aoki, Phys. Rev. B **67**, 195336 (2003)
38. C.R. Dean, L. Wang, P. Maher, C. Forsythe, F. Ghahari, Y. Gao, J. Katoch, M. Ishigami, P. Moon, M. Koshino, T. Taniguchi, K. Watanabe, K.L. Shepard, J. Hone, P. Kim, Nature **497**, 598 (2013)
39. J.R. Munkres, *Topology: A First Course* (Prentice-Hall Inc., Englewood Cliffs, 1975)
40. M. Kotani, T. Sunada, Trans. AMS **353**, 1–20 (2000)
41. T. Sunada, Jpn. J. Math. **7**, 1–39 (2012)

Chapter 8
Entangled Proteins: Knots, Slipknots, Links, and Lassos

Joanna I. Sulkowska and Piotr Sułkowski

Abstract In recent years the studies of entangled proteins have grown into the whole new, interdisciplinary and rapidly developing field of research. Here we present various types of entangled proteins studied within this field, which form knots, slipknots, links, and lassos. We discuss their geometric features and indicate what biological and physical role the entanglement plays. We also discuss mathematical tools necessary to analyze such structures and present databases and servers assembling information about entangled proteins: KnotProt, LinkProt, and LassoProt.

8.1 Introduction

Entanglement of geometric objects is an important and fascinating phenomenon. On one hand it leads to deep mathematical problems, which are studied within branches of mathematics such as topology; several Fields medals have been awarded for such work, in particular in knot theory. On the other hand entanglement is common in Nature and plays a role in physical, chemical, and biological systems. Nobel prizes related to topology have been awarded in 2016 in physics and chemistry. Topology, and entanglement in particular, are important, because they take into account not just local, but global properties of systems under consideration. This sheds new light on such systems and often requires developing new tools to analyze them.

J. I. Sulkowska (✉)
Centre of New Technologies, University of Warsaw, Banacha 2c,
02-097 Warsaw, Poland
e-mail: jsulkows@cent.uw.edu.pl

P. Sułkowski
Faculty of Physics, University of Warsaw, Pasteura 5, 02-093 Warsaw, Poland

P. Sułkowski
Walter Burke Institute for Theoretical Physics, California Institute
of Technology, Pasadena, CA 91125, USA

S. Gupta and A. Saxena (eds.), *The Role of Topology in Materials*,
Springer Series in Solid-State Sciences 189,
https://doi.org/10.1007/978-3-319-76596-9_8

The study of topology and entanglement is particularly interesting when it concerns real physical systems, and at the same time poses theoretical and mathematical challenges. This is the case of entangled proteins, whose studies we summarize in this work. On one hand, it is unbelievable that such complicated structures would evolve accidentally, so there must be some physical and biological role of entanglement. On the other hand, the study of entangled proteins poses new mathematical challenges, for example a description of knots on open chains. Entangled proteins have been very actively studied in recent years, along with the development of new mathematical tools that enable their characterization.

In this work we summarize studies of entangled proteins, which in recent years have gained the status of a new branch of interdisciplinary research, involving biophysics, biochemistry, mathematics, and computer science. Historically knotted proteins attracted attention first, so – after a brief introduction on what we mean by entanglement in general – we discuss first their properties. Subsequently we discuss properties of other entangled structures identified in proteins: slipknots, links, and lassos. We present both theoretical and mathematical tools to describe such structures, as well as their biological role and physical features.

8.2 Entanglement in Proteins

Proteins are long chains made of amino acids. They are often described in terms of primary, secondary, and tertiary structure. However such a description does not take into account an important feature of a protein chain that we refer to as entanglement. To describe it, it is sufficient to represent a protein as a one-dimensional chain, or a polygonal shape made of a series of segments spanned between consecutive C_α atoms. From a mathematical perspective, when such a one-dimensional chain is embedded in a three-dimensional space, it can non-trivially wind around itself. By the entanglement we understand a pattern of such winding.

There is a branch of mathematics called knot theory, which studies properties of entangled chains. However the most important feature of chains studied in knot theory is that they are closed, i.e. they don't have loose ends. Such entangled closed chains are called knots, and – apart from closing the chain – they resemble knots that we know from our daily life. Similarly as in daily life, one can also consider knots in open chains, especially if the ends of such chains are long enough – however such knots are not uniquely defined, and they can be untied by a sequence of smooth manipulations (which do not involve cutting of the chain). On the other hand, knots on closed chains cannot be untied without cutting a chain – therefore they can be uniquely defined, and a given type of a knot can be assigned to a given entangled configuration. In this sense knots, as understood mathematically, are referred to as topological objects.

It turns out that protein chains can also be knotted [5, 29, 32, 42, 43, 71, 76] – at least in the imprecise sense of knotting that we use in daily life. As proteins do have loose ends, it is not possible to assign types of knots uniquely to their configurations.

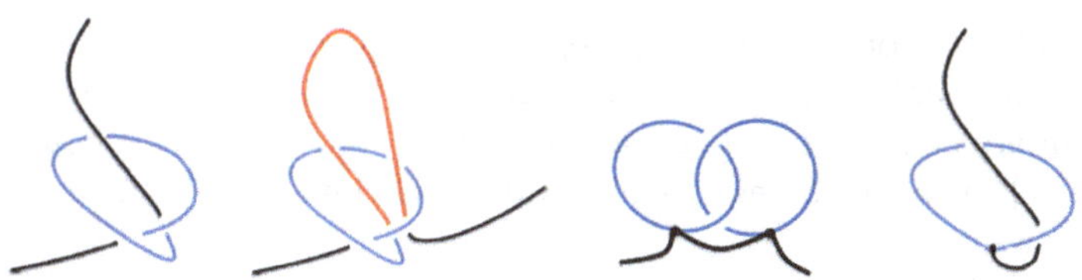

Fig. 8.1 Several classes of entangled structures identified in proteins (from left to right): knots, slipknots, links, lassos. Here we present only one basic example from each of those classes – more complicated examples are discussed in what follows

Nonetheless, it often happens that termini of a protein are long enough, so that a type of a knot can be assigned to a protein, at least approximately. For this reason one can take advantage of various tools and techniques from knot theory when studying entangled proteins. On the other hand, these tools should be used carefully and they are often insufficient to characterize configurations of open chains – some additional information should also be provided in this case, and some new techniques are necessary to study knotted proteins. As within the last decade it turned out that knotted proteins are much more common than originally believed, such techniques have been developed, as we will discuss in what follows. Knotted proteins are the first class of entangled structures that we discuss in this work – a simple example of a knot formed in an open chain (which could be a protein backbone) is shown in Fig. 8.1.

One interesting feature of entangled proteins is that they can form configurations which are trivial from the topological and knot theory viewpoint, however – in appropriate sense – they are still entangled. One example of such a configuration is called a slipknot [32, 63]. It consists of two loops, one threaded through the other, as shown in Fig. 8.1. If two termini of a slipknot are pulled, the two loops disentangle; equivalently, after connecting two termini of a slipknot in the simplest possible way (e.g. by extending them far away), the configuration represents mathematically trivial knot. As in knot theory a slipknot configuration would be regarded as trivial, some novel tools are necessary to describe its geometry – and this is an important task, because it turns out that the entanglement of the two loops of a slipknot has interesting consequences and affects proteins' properties. Proteins with slipknots are the second class of entangled structures that we discuss in this work.

So far we have briefly explained what we mean by entanglement of open chains. However the pattern of entanglement can be more involved once additional linkages, such as disulfide (or other) bridges are taken into account. When such bonds are present, then proper loops in a protein chain can be also considered (i.e. closed loops formed by only a part of the whole chain). In principle such proper loops might themselves be knotted, which however has not been observed to date. Nonetheless, when two or more loops are present in a given chain, they can be entangled and form structures called links in knot theory [11], see Fig. 8.1. Such links are non-trivial topological objects and can be classified using knot theory tools. Links that are formed on chains with disulfide linkages we call as deterministic, because they are made of closed loops and their topology is uniquely determined. In addition, one can also consider links made of several chains that form a dimer, trimer, etc., after their termini are connected (e.g. closed on a large sphere) – we call such links as probabilistic. Proteins with links are the third class of entangled structures discussed in what follows.

Once disulfide bridges are taken into account, one can also consider configurations where one or two termini of a protein chain pierce through a loop closed by such a bridge. Such configurations are called lassos [13, 45], see Fig. 8.1. Even though they can be regarded as topologically trivial and – similarly to knots in open chains – cannot be uniquely defined, they also have interesting biological and physical consequences. Lassos are the last class of entangled structures that we discuss in this work.

One important conclusion of the work summarized here is that, first of all, entangled structures exist in proteins – even though for a long time it was believed that they are too complicated to form [42]. Another very important result is that the entanglement has certain biological and physical role. Moreover, studies of entangled proteins pose new challenges in mathematics, computer science, and other related disciplines. For all these reasons the new, interdisciplinary field of entangled proteins is rapidly growing, and despite many fascinating results found in last years, we have no doubts that a lot more is still to be discovered.

8.3 Proteins with Knots and Slipknots

The first class of entangled proteins that we discuss are proteins with knots. As mentioned above, because proteins have loose ends, it is not possible to identify uniquely knots formed by their chains. It is however possible to identify such knots using probabilistic methods. In this section we present basic properties of knots in mathematical sense, discuss how to describe knots and slipknots in proteins using this language, and summarize which (families of) knotted proteins have been identified to date. We also briefly discuss folding mechanisms and how the presence of knots affects the function of proteins.

8.3.1 Classification and Description of Knots

To present mathematical knots on closed chains it is useful to project them on a two-dimensional plane. A type of a knot is then encoded in a pattern of crossings formed by the projection of the chain. It is always possible to minimize the number of such crossings by smooth transformations of the chain, which do not involve cutting it. On the level of the two-dimensional diagram such transformations can be reduced to three elementary operations, referred to as Reidemeister moves, see Fig. 8.2. Various topological types of knots can be classified in terms of equivalence classes (with

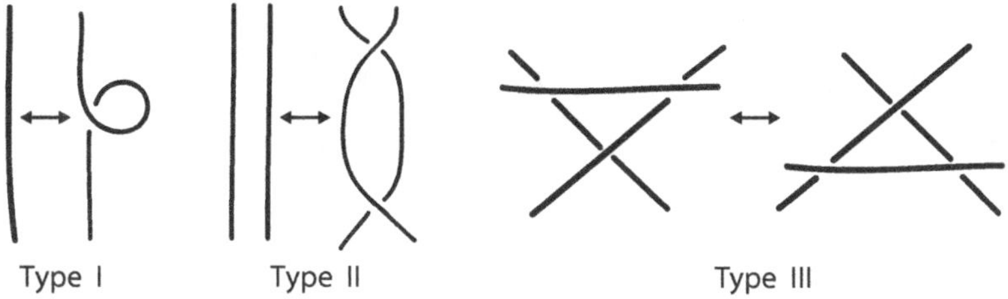

Fig. 8.2 Reidemeister moves

respect to Reidemeister moves) of their diagrams. It is also very useful to introduce so called knot invariants – various mathematical objects, such as numbers or some functions, which can be uniquely assigned to a given knot type (so that they do not change under Reidemester moves). Knot invariants are used to distinguish and classify knots: they can be computed for two given knots, and if the results are different, it means that these knots have different topological type. One ultimate goal of research in knot theory is to construct an invariant which would distinguish all knots, i.e. if it would take the same value for two knots, it would mean that these two knots are of the same type. Such an invariant is still not known, however invariants that we know these days are sufficient to distinguish relatively simple knots found in proteins.

The simplest knot invariant is the minimal number of crossings in a two-dimensional knot diagram obtained from the projection of a knot; this is simply called the number of crossings. For a trivial knot (i.e. unknotted loop), which is denoted 0_1, this number is zero. There is no knot with one or two crossings, i.e. any such configuration can be smoothly reduced to, and represents the unknot. There is one knot with three crossings, called the trefoil and denoted 3_1, and one knot with four crossings, called the figure-eight knot and denoted 4_1. There are two different topological types of knots with five crossings, denoted 5_1 and 5_2, and three types of knots with six crossings 6_1, 6_2 and 6_3. Some of the knots mentioned here (in fact those identified in proteins to date) are shown in Fig. 8.3. The number of different knots with fixed number of crossings grows very rapidly; for example there are 21 knots with eight crossings, and 165 knots with ten crossings. In the notation we just mentioned the main number denotes the number of crossings, and the subscript labels different knots (with a given crossing number) – so that, for example, 3_1 denotes a unique (the only one) knot with three crossings.

In addition, each knot may arise in two forms, which differ only by the mirror image. Sometimes we distinguish the mirror image by adding plus or minus sign in the notation of the knot; for example two versions (mirror images) of the trefoil knot are denoted 3_1^+ and 3_1^-. For some knots their mirror image is identical to the original knot – for example 4_1 knot has this property.

The number of crossings is not a strong invariant – it identifies uniquely only the unknot, trefoil, and figure-eight knot, and for more crossings there are many different, inequivalent topological types of knots with the same number of crossings.

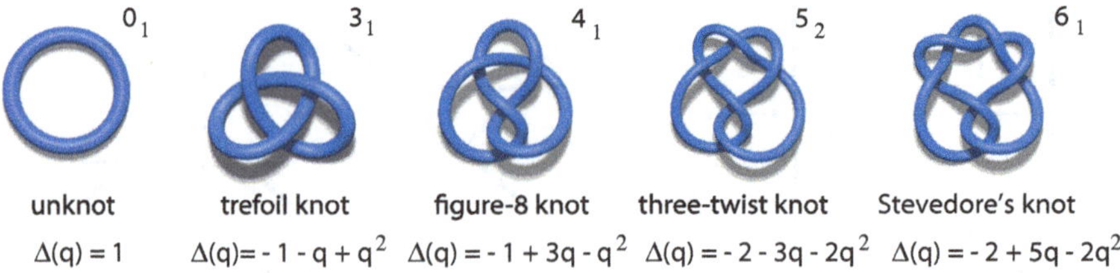

Fig. 8.3 Unknot (denoted 0_1, left) and examples of non-trivial knots (in closed chains), which have been identified in proteins to date: trefoil (denoted 3_1), figure-8 knot (denoted 4_1), 5_2 knot, and Stevedore's knot (denoted 6_1). Below each knot its Alexander polynomial is given

For this reason more powerful knot invariants are introduced. There is a large family of knot invariants known as knot polynomials, which are much stronger, i.e. they have different values for more knots, and thus they enable to distinguish more knots. Such polynomials may depend on one or more variables. For example, the so called Alexander polynomial $\Delta(q)$ and Jones polynomial $J(q)$ depend on one variable q. More intricate HOMFLY-PT polynomial $P(a, q)$ [21, 50] depends on two variables, and reduces to Alexander and Jones polynomials respectively for $a = 1$ and $a = q^2$. These polynomials can be determined from a two-dimensional diagram of a knot, and – as we explain in what follows – we also use them to identify knots in proteins. Alexander polynomials for knots identified in proteins are given in Fig. 8.3.

One challenge in analyzing knots in proteins is the fact that protein chains are open. To assign a type of knot to an open chain formed by a protein, one may transform it into a closed chain by joining its termini. The problem is that such an operation is not unique, and various knot types may be created depending on how two termini are connected. One simple way to form a closed loop is to connect two termini by a straight segment. However such an operation may lead to other knot than intuitively seen, especially in case when the termini are located far from each other and the entangled region. It is therefore more reasonable to connect the termini to two points – or the same point – on a large sphere surrounding a protein [18, 43], see Fig. 8.4. This gives more reasonable results, however the resulting type of knot may still depend on the details of the method and it is not unique (e.g. it may depend on the choice of such a point or points on a sphere). The best way to cope with this problem is to choose many equally distributed points on a sphere, and – connecting the termini through all those points – compute probability of forming various knots. This is the method that we often use in order to determine a type of a protein knot – for example, after connecting the termini as mentioned above, we calculate the knot polynomial which determines the knot type [29]. We also introduce the minimal probability, typically of the order of 40%, necessary to regard a given structure as knotted (in other words, if a probability of detecting each type of a knot by connecting the termini in various ways is smaller than this 40%, we regard the structure as unknotted). Unless otherwise stated, in case one particular knot type is assigned to a protein, this means that this knot is formed with the highest probability.

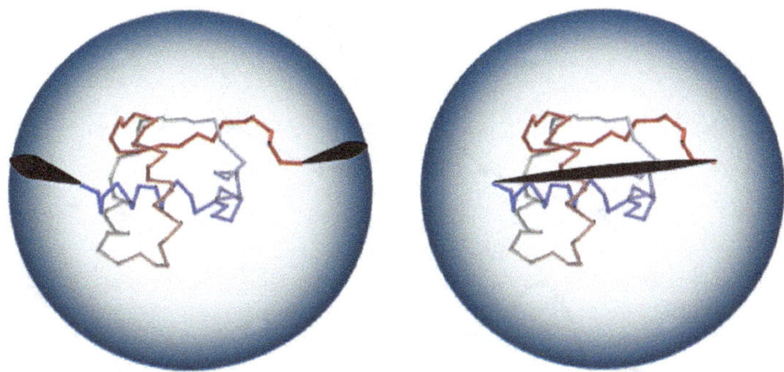

Fig. 8.4 To assign a knot type to an open chain one can either connect two termini to two points on a large sphere surrounding this chain (left), or to connect these two termini directly by a straight segment (right)

Influence of the knot detection method on the type and probability of identified knots was analyzed among others in [14, 73].

Once we can assign knot types to open chains, we can also define geometric quantities that characterize such knots, in particular the knot core and knot tails. The knot core is the shortest subchain of a given chain, for which a given knot type is detected; it can be easily determined by cutting consecutive residues from both termini, and checking whether the remaining chain is still knotted (in the probabilistic sense described above). Knot tails, by definition, consist of all residues which do not form the knot core, i.e. these are parts of the backbone chain between either of its termini and the knot core. We also characterize protein knots as being deep or shallow; the latter ones are those which can be untied by thermal fluctuations, and typically their tails consist of not more than a few residues. Deep knots cannot be untied by thermal fluctuations and they have longer tails.

In the analysis of protein knots we also use the so called KMT algorithm [34], which simplifies a given chain without changing its topology. This algorithm works as follows. We analyze triangles determined by all triples of consecutive residues (represented by C_α atoms). In case such a triangle is not pierced by any other segment of the protein chain, we remove the middle residue from a given triangle (and so reduce the triangle to the segment connecting remaining two residues), see Fig. 8.5. This operation simplifies the (closed) chain without changing its topological type. Such operations are performed as long as possible, leading ultimately to a simplified chain. In particular, if the chain is reduced to only three residues, this means that the original configuration represented the unknot – therefore the KMT algorithm is a simple method to check whether a given chain is not knotted (one should however be careful when interpreting the results of the KMT algorithm – there are certain peculiar unknotted configurations that cannot be reduced simply to three residues). It is also useful to use the KMT algorithm before calculation of knot polynomials – this may significantly reduce their computation time.

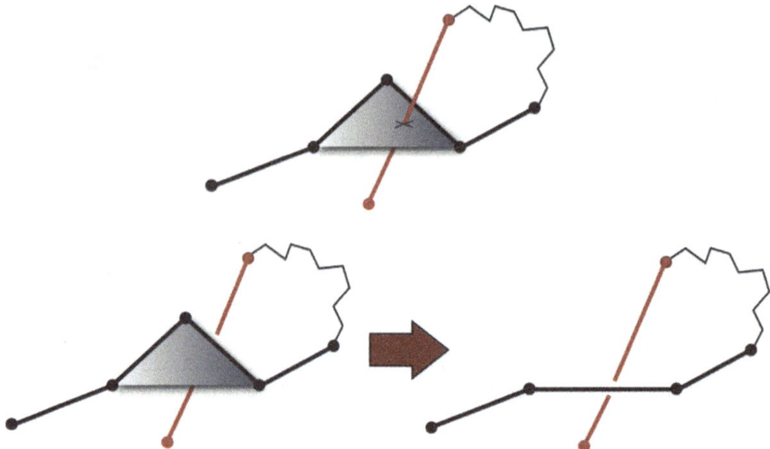

Fig. 8.5 KMT algorithm. If a triangle defined by three consecutive residues (represented by C_α atoms) is not pierced by any other segment of the backbone chain, then the middle atom in this triple is removed. This operation is repeated for all consecutive triples of residues as long as possible, resulting in a simplified chain with the same topology

8.3.2 Proteins with Knots and Slipknots – KnotProt Server and Database

For a long time it was believed that knots cannot be formed in proteins, due to their complexity. A possibility of finding knots in proteins was discussed for the first time by Mansfield, who also found a shallow knot in human carbonic anhydrase [42]. The first example of a deeply knotted protein was found by Taylor in 2000 [71], and subsequently other examples of knotted proteins were identified [32, 39, 43, 76]. The most complicated protein knot is the Stevedore's knot 6_1, identified in [5]. Proteins with slipknots were identified and analyzed e.g. in [32, 63].

These days the main source of data that facilitates detection of knots is the Protein Data Bank (PDB, or RCSB database), which stores geometric configurations of more than hundred thousands proteins. It is impossible to identify knots in such a large set of proteins without the use of mathematical and computer tools. In fact, detection of a knot even in a single protein is usually impossible just by a naked eye, due to the complicated entangled structure of the backbone chain. This is why various algorithms and mathematical techniques that we summarized above, involving in particular computation of knot polynomials, are indispensable in the search of knots in proteins. Several databases that store information about knotted proteins were constructed some years ago, for example [33, 35]. However, a modern database and server that stores up to date (and regularly updated), and much more extensive information about proteins with knots and slipknots, is the KnotProt [29], whose features we summarize in what follows.

The KnotProt database, http://knotprot.cent.uw.edu.pl, contains information about all proteins with knots or slipknots. This database contains not only the information about a type of a knot in a given protein and some of its geometric features (e.g. the length of loose ends and the size of the knotted core), but it also presents an internal geometric structure of a given protein in terms of a matrix diagram, which is called the knotting fingerprint. The form and properties of such diagrams will be introduced in the next section. Moreover, the database presents extensive information about the biological function of proteins with knots and slipknots, structural and homological similarity to other knotted or unknotted proteins, and various statistics. In addition, the KnotProt database enables users to upload protein or polymer structures, analyze whether they form knots or slipknots, and generate their knotting fingerprints.

As of today (spring 2017), there are 993 knotted chains identified in KnotProt, and 473 chains with slipknots. This is around 2% of all proteins deposited in the Protein Data Bank. Knotted proteins are found in all kingdoms of life [30]. Knots are found in globular proteins, including those from mitochondria or a ribosome, and in membrane proteins. Around 90% of knotted proteins act as enzymes, whose active site – responsible e.g. for binding ligands – is located in the knotted core. Some knotted proteins are responsible for DNA binding, and some still have an unknown function. The biggest family of proteins with a deep knot is the SPOUT family [72]. The smallest and rather deeply knotted protein, from methanocaldococcus jannaschii, consists of 92 amino acids and is known as MJ0366. The deepest knot is found in protein with unknown function from T. pallidum (PDB code: 5ijr chain A) – its tails have length of at least 100 amino acids. This protein consists of three domains and the middle one is knotted. A deep knot exists also in the family of membrane proteins, calcium exchanger protein (e.g. protein with PDB code 4k1c). A protein structure with an artificial (designed) knot is also known [31].

We present a list of representative proteins with knots and slipknots, based on the KnotProt database, in Tables 8.1 and 8.2. In the first column of these tables representative entangled proteins are listed, grouped according to their function, and PDB codes of such representative structures are given in the second column. In Table 8.1 in the third column a type of a knot for the whole chain is provided. In the last column of both tables knotting fingerprints (whose meaning is explained below) of respective structures are given.

8.3.3 *Knotting Fingerprint for Knots and Slipknots*

As we discussed earlier, knots in proteins are formed in open chains, which is subtle from mathematical perspective and enables their identification only in a probabilistic manner. However, once a relevant definition of knots in open chains is provided, it can be used not only to determine a type of the knot for the whole chain under consideration, but also for all subchains of such a knot. Such an analysis provides much more accurate and detailed information about entangled proteins. It also enables rigorous detection of slipknots, which can be defined as chains which are

Table 8.1 A list of families of proteins with knots (following [30]). In the second column a representative structure of each family is given (PDB codes include the chain identificator as the last letter, in the subscript). In the third column the knot type is given, and in the last column the knot fingerprint is provided (whose meaning is explained in the text). A sign in the superscript of the knot type denotes its chirality

Protein family	PDB code	Knot	Fingerprint
Enzymes:			
α-haloacid dehalogenase I	$3bjx_A$	$\mathbf{6_1^+}$	$K6_1^+6_1^+4_13_1^+$
Ubiquitin C-terminal hydrolase	$2etl_A$	$\mathbf{5_2^-}$	$K5_2^-3_1^-3_1^-$
Chromophore binding domain	$2o9c_A$	$\mathbf{4_1}$	$K4_14_1$
CII Ketol acid reductoisomerase	$1yve_L$	$\mathbf{4_1}$	$K4_13_1^+$
Mitochondrial apoptosis-inducing factor 1*	$5fmh_A$	$\mathbf{3_1^-}$	$K3_1^-3_1^-$
Methyltransferase (α/β knot)			
tRNA methyltransferase	$1uak_A$	$\mathbf{3_1^+}$	$K3_1^+$
rRNA methyltransferase	$2egv_A$	$\mathbf{3_1^+}$	$K3_1^+$
protein methyltransferase	$5h5f_A$	$\mathbf{3_1^+}$	$K3_1^+$
Carbonic anhydrase II	$1lug_A$	$\mathbf{3_1^+}$	$K3_1^+$
SAM synthetase	$1fug_A$	$\mathbf{3_1^+}$	$K3_1^+$
Transcarbamylase fold	$1js1_X$	$\mathbf{3_1^+}$	$K3_1^+$
N-acetylglucosamine deacetylase	$5bu6_A$	$\mathbf{3_1^+}$	$K3_1^+$
DNA binding:			
Zinc-finger fold	$2k0a_A$	$\mathbf{3_1^-}$	$K3_1^-$
Ribbon-helix-helix superfamily:			
MJO366	$2efv_A$	$\mathbf{3_1^-}$	$K3_1^-$
VirC2	$2rh3_A$	$\mathbf{3_1^-}$	$K3_1^-$
DndE	$4lrv_A$	$\mathbf{3_1^-}$	$K3_1^-$
Unknown function:			
Protein from *T. pallidum*	$5jir_A$	$\mathbf{3_1^+}$	$K3_1^+$
Artificial proteins:			
Artificially (designed) knotted protein	$3mlg_A$	$\mathbf{3_1^-}$	$K3_1^-$
Ribosome subunits:			
Mitochondrial ribosomal protein	$4v1a_w$	$\mathbf{3_1^-}$	$K3_1^-3_1^-3_1^-3_1^-$
Membrane proteins:			
Calcium exchanger protein:			
NCX	$5hwy_A$	$\mathbf{3_1^+}$	$K3_1^+$
2jlo	$4kpp_A$	$\mathbf{3_1^+}$	$K3_1^+$

Table 8.2 A list of families of proteins with slipknots. In the second column a representative structure of each family if given (PDB codes include the chain identificator as the last letter, in the subscript). In the last column the slipknot fingerprint is provided (whose meaning is explained in the text). A sign in the superscript of the knot type denotes its chirality

Protein family	PDB code	Slipknot fingerprint
Membrane proteins:		
Cation symporter – 2	3qe7	$S3_1^+4_13_1^+3_1^+3_1^+$
Neurotransmitter symporter	2a65	$S3_1^+4_13_1^+$
Hydantoin transporter	2jlo	$S3_1^+4_13_1^+$
AA-permease	3gia	$S3_1^+4_13_1^+$
Sodium transporter, SSF	3dh4	$S3_1^+4_13_1^+$
Glycine betaine transporter BetP	2wit	$S3_1^+4_13_1^+$
Glutamate symport protein	4p19	$S3_1^+3_1^+$
Enzymes:		
Colicin-E9	$5ew5_A$	$S3_1^+3_1^+3_1^+3_1^+$
Colicin-E7	1yve	$S3_1^+3_1^+3_1^+3_1^+$
Colicin-E3	1jch	$S3_1^+3_1^+3_1^+3_1^+$
Ectonucleotide pyrophosphatase	4zg6	$S3_1^-$
Cytochrome	$5udy_A$	$S3_1^+$
Arginine decarboxylase	$2qqd_C$	$S3_1^+$
Nucleotide diphosphatase	$3szz_A$	$S3_1^+$
D-ribose pyranase	$1ogf_A$	$S3_1^+$

unknotted as a whole, but which possess a subchain which is knotted. It is convenient to present an information about knotting of all subchains of a given chain in terms of a matrix diagram, called knotting fingerprint, see Fig. 8.6, which was introduced in [67] following [32].

More precisely, the knotting fingerprint presents data about knotting of each subchain of a given chain in a lower-triangular part of matrix of the size $N \times N$, where N is the length of the whole chain. The position (i, j) of this matrix represents a subchain spanned between ith and jth residue of the backbone chain, and it is colored according to the type of the (most probable) knot formed by this subchain. The intensity of the color corresponds to the probability of forming such a knot. Typically such a matrix diagram consists of several approximately rectangular regions, which represent various types of knots spanned by various subchains. Such a diagram is a source of various interesting information. First, it enables identification of the minimal length of the knotted region (i.e. the size of the knotted core), the depth of a knot (i.e. the number of amino acids that can be removed from either end of the protein chain before converting it from a knot to a different type of knot or

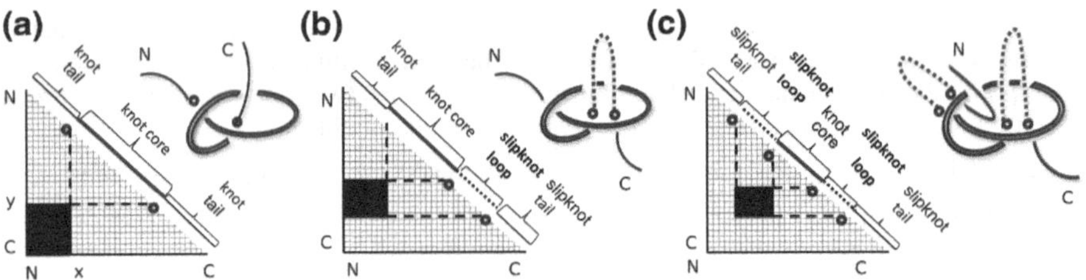

Fig. 8.6 An illustration of knotting fingerprints which shows how to interpret them (following [67]). Horizontal and vertical edges of each triangle represent the backbone chain (from N to C terminus). For each point (i, j) in the matrix diagram we verify whether a subchain stretched between ith and jth residue is knotted – if so, we color this point (in black in this example). If the whole protein chain forms a knot, then the neighborhood of the left bottom corner of the matrix diagram is black (left panel). If the whole chain is not knotted but contains a slipknot, then the left bottom corner is not black (middle and right panels). The location of the black rectangles enables to determine the knotted core (the largest subchain of the whole chain for which a knot is still detected), and remaining parts of the chain are referred to as knot tails (left panel). In case of slipknots, we can analogously determine the location of various parts of the chain, referred to as the slipknot loop, knotted core, and slipknot tails (middle and right panels)

an unknot), and other geometric data characterizing a knot. Second, the location of various knots along the chain indicates their biological and physical role, especially if it is correlated with the location of active sites.

It is also useful to encode the pattern of knots on subchains of a given chain in a simplified notation, which lists all those types of knots identified in the fingerprint matrix. For example by $K4_13_1$ we denote a protein which forms 4_1 knot, and which possesses a subchain forming 3_1 knot; the initial letter K means that the whole chains forms 4_1 knot. In case the whole chain forms a slipknot the letter S is used; for example, $S4_13_1$ means that the whole chain forms a slipknot, and some of its subchains form knots of type 4_1 and 3_1. Knotting patterns (including the orientation (or mirror image) of identified knots, denoted by $\pm$ in the superscript) of some representative proteins with knots and slipknots are provided in the last column in Tables 8.1 and 8.2.

In Fig. 8.7 we present examples of the most complicated fingerprints of a knot and a slipknot found to date. The knotted chain (left) with the fingerprint $K6_16_14_13_1$ is the structure of DehI (PDB code 3bjx chain A). The chain with the slipknot (right) with the fingerprint $S3_14_13_13_13_1$ is the crystal structure of uracil transporter–uraa (PDB code 3qe7A). Fingerprints for all protein structures with knots and slipknots identified to date can be found in the KnotProt database.

8.3.4 Folding of Knotted Proteins

An important challenge in the study of entangled proteins is to understand and describe not only their geometric configurations which determine biological

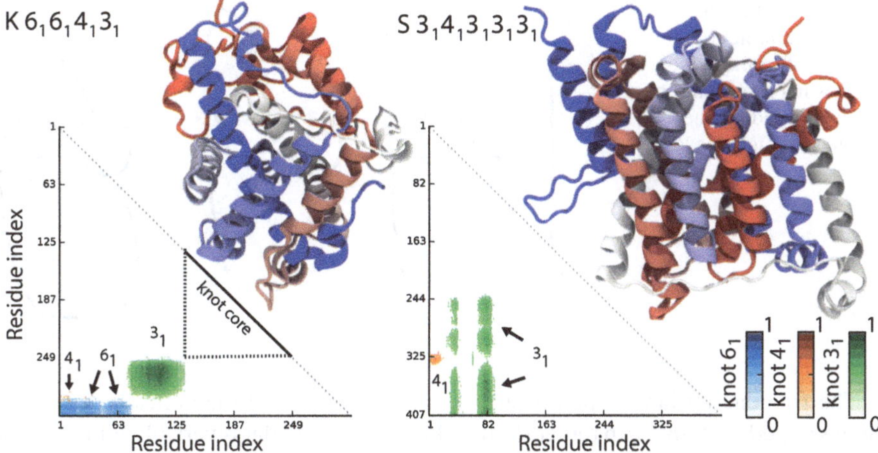

Fig. 8.7 Most complicated knotting fingerprints identified in the KnotProt database. The knotted protein (PDB code 3bjx chain A, left) forms 6_1 knot, and its various subchains form also another 6_1 knot and 4_1 and 3_1 knots – such a configuration is denoted schematically as $K6_16_14_13_1$. A protein with slipknot (PDB code 3qe7 chain A, right) is unknotted as a whole, however some of its subchains form 4_1 and various 3_1 knots. Each type of knot is denoted by different color (3_1 – green, 4_1 – orange, 6_1 – blue), whose intensity corresponds to probability of detection of this type of knot. Cartoon representations of these two proteins are also shown in each case, with colors changing from red at the N-terminus to blue at the C-terminus

activity and function, but also folding mechanisms which lead to such configurations. Even though some experiments with knotted proteins have been conducted, in recent years their folding mechanisms have been studied mainly theoretically, using various coarse grained models. Knots identified in proteins to date are of twist type, and they can be made in one step, by threading of a tail through a twisted loop (the number of twists determines the type of the resulting knot). Current results suggest that knots are not made spontaneously along the protein backbone (and then slight to the final native location), but they are created in two main steps [64]: the first is formation of a twisted native loop, and the second is threading a shorter knot tail through this loop, as shown in Fig. 8.8. In case of proteins with more complex structure, twisted native loop flips over a core of the protein creating e.g. the 6_1 knot also just in one step [5]. Flipping mechanism was also observed for proteins with slipknots [64].

One of the most commonly analysed proteins are Yibk and YebA, members of SPOUT family, which possess a deep trefoil knot. Theoretical studies with an unbiased structure based model revealed that these proteins can self tie, via formation of a twisted loop and threading shorter C-terminal tail [64]. Detailed analysis showed that the tail threads through the twisted loop in the slipknot configuration and this threading is a rate limiting step. The same pathway is also observed when non-native contacts are introduced [79]. Even though these proteins have been very extensively studied, still only kinetics pathways are known [41, 54, 74].

Fig. 8.8 Twist knots (which are found in proteins) are created by forming a loop and twisting it several times, and then threading one terminus through this loop

The free energy landscape however has been uncovered theoretically for a few small knotted proteins with rather shallow knots. In particular reversible folding via a slipknot conformation was observed for the smallest knotted protein (known to date), MJ0366 [48], with a structure based model, and it was further extensively studied e.g. in [4, 44]. Furthermore, explicit all atom simulations showed that this protein can self-tie and electrostatic interactions facilitate tying of a knot [49]. The self-tying mechanism of MJ0366 was also suggested experimentally [80]. Reversible folding was also observed for artificially designed proteins [31], where both theoretical and experimental studies suggested that topological constrains are responsible for their slower folding [37, 66].

Theoretical studies of proteins with more complex knots, such as 5_2 and 6_1, also suggest that knotting happens just in one step, which in this case corresponds to flipping the twisted loop over the core of the protein [5, 83]. Moreover folding of those proteins consists of parallel pathways and intermediates steps. Experimental studies of these proteins suggest that their folding mechanism involves at least one intermediate state, see e.g. [2, 36], and proteins are prone to misfold at the final state of folding [38, 81]. Influence of knot type on folding pathway based on lattice models was also investigated, e.g. in [20, 56].

Since proteins in vivo are surrounded by crowded environment and their folding could be supported by chaperons, or they could fold cotranslationally, it was also analyzed how these aspects affect knotting mechanism [8, 40, 57]. For example, when a chaperon is approximated by a cylindrical box, it was found that knotting is still a rate limiting step, however a chaperon significantly smooths the free energy landscape for smallest knotted proteins MJ0366, VirC2 and DndE [47], as well as proteins with complex fingerprint $5_2 3_1 3_1$ [83]. Experimental study on YibK, YebA shown that chaperon can significantly speed up the folding process [40]. Cotranslational knotting with theoretical methods was extensively studied e.g. in [8].

Even though in principle the folding pathways leading to knotted configurations have been proposed, the origin (type of interactions) of the driving force needed to overcome the topological barrier [12, 55], and the influence of chain stiffness [66], are still under investigation. Moreover it is important to mention that a certain exception in folding mechanism was observed based on a three domain protein from

T. Pallidum (PDB code 5jir chain A) where the middle domain is knotted. This protein probably folds via a shallow knot formed at random position [30]. Other approach to understand free knotting mechanism and reviews discussing folding pathways of knotted proteins can be found e.g. in [19, 28, 68, 77].

8.3.5 *Function of Knotted Proteins*

Another crucial challenge, apart from characterizing folding and geometry of entangled proteins, is to understand what is the role and the function of entanglement. This question is currently actively studied. A review of all knotted proteins shows that the topological fingerprint is well conserved in proteins separated even by million years of evolution and very low sequence similarity [67]. Moreover, more than 90% of known knotted proteins are enzymes, whose active sites are located inside the knotted core. These observations suggest that knots provide some advantages for the hosting organism.

Analysis of analogous proteins having the same function – methylation of tRNA [27] – but different topology also reveals some information about the role of a knot. An example of such a pair of proteins involves a bacterial one with a trefoil 3_1 knot (TrmD), and an eukaryotic one which is unknotted (Trm5). As showed in [7], the knot responds to the motions of the whole protein, even though it belongs to the rigid part of the structure. The trefoil knot in the TrmD protein is capable of transferring the signal coming from the substrates further through the protein, by its internal moves. Mutations of key residues in the area of the knot suppress this motion and make the protein unable to conduce the methyl transfer. The fact that the knot plays a role in the enzymatic function suggests that knots play more profound role than just stabilizing the structure. Stabilizing role of knots however should not be neglected, and it is important e.g. in phytochromes, which are red/far-red light photoreceptors that direct photosensory responses across the bacterial, fungal and plant kingdoms [17, 78]. The binding pocket of those proteins is stabilized by a deep figure-eight knot. Some other suggestions about the role of knots are presented in [14, 53].

Additional information about knots in proteins was obtained based on mechanical manipulations, via single molecule experiment both in vitro and in silico (by computer analysis) [51, 58]. First, molecular dynamics simulations with structure based model [59, 60] showed that a knotted protein is more resistant than an unknotted one [61]. Second, it was found in [6, 62] that upon pulling proteins by their termini, knots tighten along the backbone at deterministic locations. Conditions to untie knotted proteins were established in [65] and tested experimentally in [84]. Furthermore, it was shown that knotted proteins can block or be pulled through narrow channels, depending on the applied force [69, 70, 82]. A surprising tying of a knot on the protein backbone was also observed in thermal and chemical denaturation [3, 41]. Furthermore, mechanical manipulations of proteins with slipknots showed that they form certain metastable conformations [25, 26, 63].

8.4 Links in Proteins

In knot theory, apart from knots, also another class of entangled structures is considered, which are called links. A link consists of several closed loops (called components of a link), such that each of those loops may be knotted, and moreover pairs, triples, etc. of loops can be simultaneously interlinked. Any knot may be regarded as a link with only one component. The simplest example of a non-trivial link is the Hopf link, formed by two loops linked in the simplest possible way. Another link that can be made out of two loops is the Solomon link. Hopf link and Solomon link are shown in Fig. 8.9 (in the middle).

Links in proteins have been identified only very recently, and they can be considered from two perspectives [11]. First, links can be identified in a single protein chain if it has additional linkages, for example disulfide bridges. Each such bridge defines a closed loop, and if several such loops are present in one protein chain, then the whole structure forms a link (which is nontrivial if those loops are interlinked). Importantly, links defined in such a way are properly defined from mathematical perspective and there is no need to consider probabilistic methods to identify them – we call such links deterministic. Once orientation of component loops is introduced – e.g. induced from the orientation of the backbone from N to C terminus – the notion of chirality of links can be defined, see Fig. 8.9.

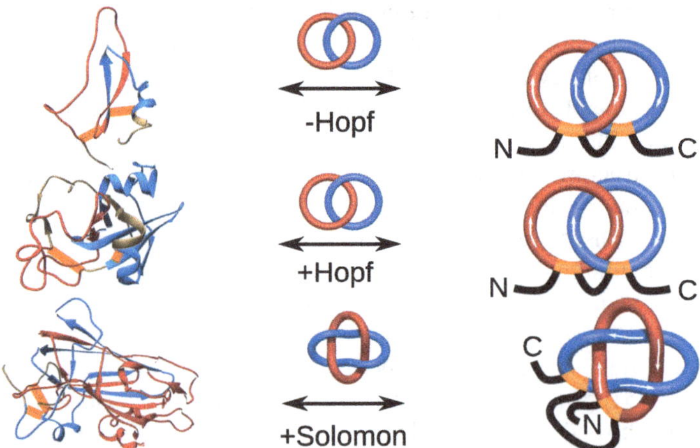

Fig. 8.9 Deterministic links identified in proteins (following [11]): the Hopf link and the Solomon link (whose mathematical structure is shown in the middle). In the left a cartoon representation of a protein forming a given link is shown. A schematic representation of protein backbone is shown in the right, with two link components in blue and red, the linkages in orange, and remaining parts of the protein chain in black. The orientation of each component loop can be induced from the orientation of the backbone from N to C terminus – once it is taken into account, two types of the Hopf link can be considered

Table 8.3 Families of proteins (following [11]) with the Hopf link (with positive and negative orientation, respectively in the upper and lower part of the table, as indicated in the last column). For each family the function and a representative chain are given respectively in the second and the third column. L1 and L2 denote the size of the two loops of the Hopf link, P1 and P2 is the signed index of a residue piercing through each of those loops. "# of hom." is the number of homologs for a given structure, "# of loops" is the number of disulfide-based covalent loops in the structure (e.g. if 4, there are 2 covalent loops forming the Hopf link and two trivial covalent loops), "Size" is the number of residues in the structure, "Loop sep." is the sequential distance between the loops. The dashed line separates the humanly modified protein with PDB code 3T93. Proteins are ordered according to the size of the first pierced loop

Protein family	Classification	PDB code	L1	P1	L2	P2	# of hom.	# of loops	Size	Loop sep.
Barwin	Lectin	$1bw3_A$	22	82	57	49	4	3	125	2
Cerato-platanin	Toxin	$2kqa_A$	23	75	55	45	1	2	129	2
Expansin-B1	Allergen	$2hcz_X$	28	94	67	58	2	3	245	2
Endoglucanase	Hydrolase	$3 \times 2g_A$	28	84	85	36	11	5	180	2
Cerato-platanin	–	$3suk_A$	37	97	59	64	4	2	125	2
Cerato-platanin	–	$3sum_A$	37	97	62	72	1	3	136	2
Endoglucanase	Hydrolase	$1wc2_A$	39	100	85	51	1	6	181	2
Endoglucanase	Hydrolase	$1hd5_A$	70	109	113	74	7	7	213	0
Tryptase inhibitor	Hydrolase	$2lfk_A$	27	−57	17	−45	4	4	57	0
Reelin	Signaling protein	$2e26_A$	39	−2473	166	−2380	2	12	725	5
BGP-1	Cell adhesion	$3r4d_B$	137	−240	79	−126	2	3	288	6
Ectonucleotide NPP1	Hydrolase	$4b56_A$	212	−529	387	−230	25	16	820	64
Growth arrest-specific protein	–	$1h30_A$	287	−668	27	−477	3	4	422	72
Glutamate receptor	Transport protein	$3t93_B$	77	−219	55	−109	243	2	258	65

Table 8.4 Families of proteins (following [11]) with the Solomon link. For each family the function and a representative chain are given respectively in the second and the third column. L1 and L2 denote the size of the two loops of the Hopf link, P1 and P2 is the signed index of a residue piercing through each of those loops. "# of hom." is the number of homologs for a given structure, "# of loops" is the number of disulfide-based covalent loops in the structure (e.g. if 4, there are 2 covalent loops forming the Solomon link and two trivial covalent loops), "Size" is the number of residues in the structure, "Loop sep." is the sequential distance between the loops

Protein family	Classification	PDB code	L1	P1	L2	P2	# of hom.	# of loops	Size	Loop sep.
Flocculation protein	Cell adhesion	$2xjp_A$	146	−218	87	−142	19	4	258	0
				−252		−165				
Epa1A	Cell adhesion	$4asl_A$	129	−221	82	−147	15	3	229	0
				−253		−170				

On the other hand, links can also be made of several separate chains, which form dimers, trimers, etc. [15]. To identify such links, each component chain must be closed (e.g. by connecting its termini on a large sphere, similarly as in the case of knots), and then the HOMFLY-PT polynomial is calculated. A type of such a link may depend on details how termini of all chains are connected, and so we call such links as probabilistic.

All proteins with links identified in the Protein Data Bank are presented in the LinkProt database [15], available at http://linkprot.cent.uw.edu.pl. As of spring 2017, 124 deterministic and 8456 probabilistic links (1071 linked proteins with 30% of sequence similarity) have been identified in this database. This database is also regularly updated, so that it always contains an up to date list of proteins with links. Deterministic links identified in proteins to date form one of two simplest links – Hopf link and Solomon link – as shown in examples in Fig. 8.9. Once the orientation of two component loops is introduced, one can consider two versions of the Hopf link, and both of them are found in proteins (in this case there is a natural orientation of each loop induced by the ordering of the protein from N to C terminus). In Fig. 8.9 in the left a cartoon representation of a given protein is shown, and in the right the protein backbone is shown in a simplified way, with link components colored in blue and red, linkages (closing the loops) shown in orange, and remaining parts of the protein chain in black. Furthermore, all non-redundant structures (found to date) that form deterministic Hopf link, together with their geometric properties and the function, are listed in Table 8.3. Analogous data for non-redundant proteins that form deterministic Solomon link is shown in Table 8.4.

Probabilistic chains identified in LinkProt database form more configurations, which involve two, three or four components. One of the most interesting configurations is 6^3_3 (probabilistic) link, which involves three symmetrically and mutually interlinked components (note that it should not be confused with the Borromean rings). As there are many more protein configurations forming probabilistic links, to learn about their properties we encourage a reader to browse the LinkProt database.

Currently the function of links in proteins is not well understood, however they definitely provide additional topological stability [11] (in addition to stabilizing role of disulfide bridges and closed loops). Link topology is also responsible for misfolding of a hosted protein [11].

8.5 Proteins with Lassos

The last class of entangled proteins that we discuss are proteins with lassos, introduced systematically in [45]. Lassos can be identified in structures with disulfide (or other types of) bridges. Similarly as in (deterministic) links, such a bridge defines a closed loop. By a lasso we mean a configurations which consists of such a loop, through which some other part (or parts) of the backbone chain is threaded. The pattern of such threading may be quite complicated: a protein chain may pierce the loop several times, it can wind around the backbone chain forming a loop, etc. Note that configurations similar to lassos were studied in [75]; we also stress that lassos should not be confused with cystein knots [9, 10, 16].

In order to define lassos unambiguously, we propose to span an auxiliary surface of minimal area (analogous to a soap bubble), called the minimal surface, on the closed loop. To determine such a surface, or more precisely a triangulated approximation to such a surface, we use tools and algorithms from computer graphics [45]. The orientation of the loop (from N to C terminal) induces the orientation of the surface spanned on this loop, which enables to identify a direction of piercing of this surface by a protein terminus, as shown in Fig. 8.10.

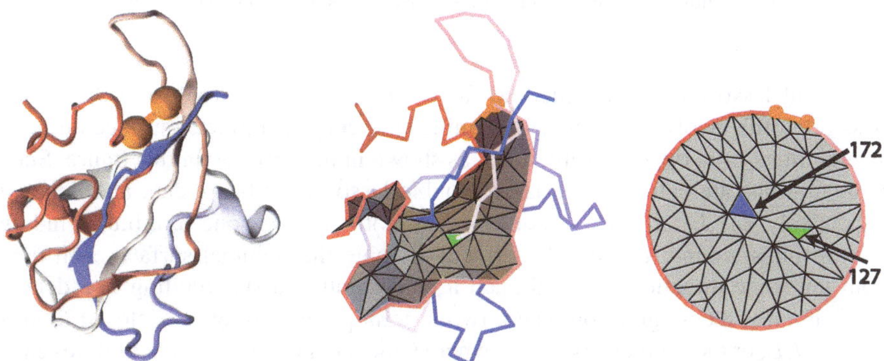

Fig. 8.10 An oxidoreductase protein (PDB code 2oiz) forming a lasso (left), with the disulfide bridge shown in orange (following [45]). In the middle a schematic representation of the protein chain and the triangulated minimal surface spanned on the loop (closed by the disulfide bridge) are shown; the surface is pierced, respectively by 127th and 172nd chain segment, at two triangles (in green and blue color, which indicates the direction of piercing). A baricentric representation of the minimal surface is shown in the right

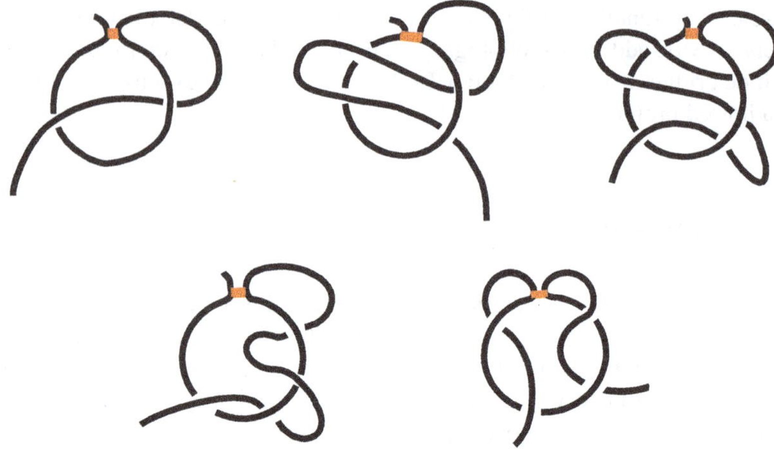

Fig. 8.11 Various types of lassos identified in proteins (following [45]). The orange segment denotes a bridge/linkage forming a closed loop. Top row: single lasso, double lasso, and triple lasso (denoted respectively L_1, L_2 and L_3). Bottom row: supercoiled lasso (denoted LS), and lasso involving two termini piercing the loop (denoted LL)

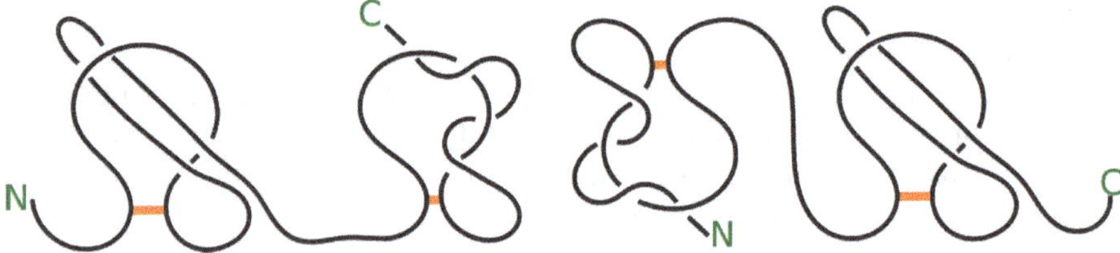

Fig. 8.12 More complicated lasso configurations, involving several loops (following [13])

Several lasso motifs identified so far in proteins are shown in Fig. 8.11. One class of motifs involves one protein terminus piercing the minimal surface once, or several times from opposite directions, as shown in the first row in this figure. Such configurations are called single lasso, double lasso, triple lasso, etc., and denoted respectively L_1, L_2, L_3, etc. Another lasso motif involves one terminus winding several times around the closed loop and piercing the minimal surface each time from the same direction; we call such a configuration a supercoiling and denote it LS. Finally, a configuration where two termini pierce through the closed loop is denoted LL, or more precisely $LL_{i,j}$, where i and j denote the number of times each terminus pierces the minimal surface. For each of those motifs an information about the direction of piercings and the piercing terminus can also be provided. More complicated lasso configurations, involving several loops, also exist, for example such as shown in Fig. 8.12.

An example of a protein forming a lasso of type L_2 – an oxidoreductase protein (PDB code 2oiz) – is shown in Fig. 8.10. Cartoon representation of the protein is

Table 8.5 Families of proteins with various types of lassos (following [45]). In the first bullet groups of enzymes are listed in order of decreasing number of occurrences; in the second bullet PDB classes of non-enzymatic proteins are listed in order of decreasing number of occurrences. In case of L_1 lasso only groups with more than 9 elements are listed

Lasso	Protein families
L_1	• Hydrolases (85), transferases (15), oxidoreductases (14), lyases (3), isomerase (1) • Binding protein (26), antimicrobial proteins (16), viral proteins (15), immune system related (12), transport proteins (12), toxines (11), cytokines (10), membrane proteins (9) …
L_2	• Hydrolases (9), oxidoreductases (9), transferases (4) • Cytokines (11), immune system, related (3), signaling proteins (3), viral proteins (3), other (4)
L_3	• Hydrolases (2), isomerases (1), oxidoreductases (1); • Transport proteins (10), allergens (3), immune system related (2), viral proteins (2), other (2)
L_6	• Oxidoreductase (1)
LS	• Lyases (3), hydrolases (1); • Cell adhesion related (5), metal binding protein (2), structural proteins (1), transport protein (1), GAS(1)
LL	• Hydrolases (2) • Cell adhesion related (2), membrane proteins (2), toxin (1), structural protein (1), cytokine (1), transport protein (1)

shown in the left, with the disulfide bridge shown in orange. A simplified protein structure is shown in the middle, with the grey triangulated minimal surface spanned on the closed loop, pierced by one terminus at two triangles (green and blue color, indicating the direction of piercing). A baricentric planar representation of the minimal surface with pierced triangles is shown in the right. The numbers 127 and 172 indicate the numbers of segments connecting consecutive C_α atoms that pierce the surface.

All proteins with lassos identified to date are collected in the LassoProt database [13], available at http://lassoprot.cent.uw.edu.pl. As of spring 2017, there are 6446 structures with lassos in this database, which have been identified in all kingdoms of life, in globular and membrane proteins. We list families of these proteins in Table 8.5, together with the number of lasso structures in each family, and grouped into enzymes and non-enzymes. It truns out that around 18%, i.e. 376 out of 2021 protein structures with disulfide bonds (data from the year 2015), form lassos in the nonredundant set (i.e. proteins with sequence similarity lower than 30%). Furthermore, geometric properties of lassos can be analyzed using the online PyLasso server [46], available at http://pylasso.cent.uw.edu.pl.

Apart from their geometry, also functions and other properties of proteins with lassos have been studied. In [1, 52] it was shown that the lasso motif is well preserved in antimicrobial proteins, where the lasso stabilizes the entire fold. Lassos also have a therapeutic potential, similarly as proteins with cystein knots [16]. Even though the influence of lassos on proteins' stability and biological activity have been analyzed systematically only recently, it is already known that e.g. in leptin the presence of a lasso slows down folding, but facilitates receptor binding [22–24].

8.6 Conclusions

We have presented various types of entangled structures in proteins: knots, slipknots, links and lassos. Their crucial feature is the fact that the whole protein chain needs to be considered to identify its type of entanglement; in this sense entanglement is a global property of a protein chain. Some years ago it was believed that entangled proteins, in particular knots, cannot exist, due to the complexity of their structure. However more recently many such structures have been found, and it has become clear that their existence is not accidental and they must play certain biological and physical role.

We have also presented databases and servers that assemble and regularly update information about entangled structures, as well as a plugin that facilitates analysis of lassos:

- KnotProt, http://knotprot.cent.uw.edu.pl
- LinkProt, http://linkprot.cent.uw.edu.pl
- LassoProt, http://lassoprot.cent.uw.edu.pl
- PyLasso http://pylasso.cent.uw.edu.pl

Apart from geometric structure of all protein chains, also an information on their function, various classifications, a list of similar chains, etc., are provided in the above databases. In these websites one can also upload other polymer-like structures, not necessarily proteins, and analyze if they contain knots, slipknots, links or lassos.

In last few years the studies of entangled proteins have grown into a new, rapidly developing and interdisciplinary field, which involves methods and techniques from biophysics, biochemistry, computer science, branches of mathematics such as topology and knot theory, etc. There are plenty of opportunities and outstanding questions in studying entangled proteins, which involve understanding their function, evolution, folding mechanisms, and other features. We encourage a reader to try to answer some of these questions too.

Acknowledgements This work has been financed from the budget for science in the years 2016-2019 [#0003/ID3/2016/64 MNiSW – Ideas Plus], EMBO [#2057 Installation Grant], and National Science Centre [#2012/07/E/NZ1/01900] to J.I.S. The work of P.S. has been supported by the ERC Starting Grant no. 335739 *"Quantum fields and knot homologies"* funded by the European Research Council under the European Union's Seventh Framework Programme.

References

1. C.D. Allen, M.Y. Chen, A.Y. Trick, D.T. Le, A.L. Ferguson, A.J. Link, Thermal unthreading of the lasso peptides astexin-2 and astexin-3. ACS Chem. Biol. (2016)
2. F.I. Andersson, D.G. Pina, A.L. Mallam, G. Blaser, S.E. Jackson, Untangling the folding mechanism of the 52-knotted protein uch-l3. FEBS J. **276**(9), 2625–2635 (2009)
3. B.T. Andrews, D.T. Capraro, J.I. Sulkowska, J.N. Onuchic, P.A. Jennings, Hysteresis as a marker for complex, overlapping landscapes in proteins. J. Phys. Chem. Lett. **4**(1), 180–188 (2012)
4. S.A. Beccara, T. Škrbić, R. Covino, C. Micheletti, P. Faccioli, Folding pathways of a knotted protein with a realistic atomistic force field. PLoS Comput. Biol. **9**(3), e1003002 (2013)
5. D. Bölinger, J.I. Sułkowska, H.-P. Hsu, L.A. Mirny, M. Kardar, J.N. Onuchic, P. Virnau, A Stevedore's protein knot. PLoS Comput. Biol. **6**(4), e1000731–e1000731 (2010)
6. T. Bornschlögl, D.M. Anstrom, E. Mey, J. Dzubiella, M. Rief, K.T. Forest, Tightening the knot in phytochrome by single-molecule atomic force microscopy. Biophys. J. **96**(4), 1508–1514 (2009)
7. T. Christian, R. Sakaguchi, A.P. Perlinska, G. Lahoud, T. Ito, E.A. Taylor, S. Yokoyama, J.I. Sulkowska, Y-M. Hou, Methyl transfer by substrate signaling from a knotted protein fold. Nat. Struct. Mol. Biol. (2016)
8. M. Chwastyk, M. Cieplak, Cotranslational folding of deeply knotted proteins. J. Phys. Condens. Matter **27**(35), 354105 (2015)
9. D.J. Craik, N.L. Daly, T. Bond, C. Waine, Plant cyclotides: a unique family of cyclic and knotted proteins that defines the cyclic cystine knot structural motif. J. Mol. Biol. **294**(5), 1327–1336 (1999)
10. D.J. Craik, M. Čemažar, C.K.L. Wang, N.L. Daly, The cyclotide family of circular miniproteins: nature's combinatorial peptide template. Pept. Sci. **84**(3), 250–266 (2006)
11. P. Dabrowski-Tumanski, J.I. Sulkowska, Topological knots and links in proteins. Proc. Natl. Acad. Sci. **114**(13), 3415–3420 (2017)
12. P. Dabrowski-Tumanski, A.I. Jarmolinska, J.I. Sulkowska, Prediction of the optimal set of contacts to fold the smallest knotted protein. J. Phys. Condens. Matter **27**(35), 354109 (2015)
13. P. Dabrowski-Tumanski, W. Niemyska, P. Pasznik, J.I. Sulkowska, Lassoprot: server to analyze biopolymers with lassos. Nucleic Acids Res. **44**(W1), W383–W389, 2016
14. P. Dabrowski-Tumanski, A. Stasiak, J.I. Sulkowska, In search of functional advantages of knots in proteins. PloS one, **11**(11), e0165986 (2016)
15. P. Dabrowski-Tumanski, A.I. Jarmolinska, W. Niemyska, E.J. Rawdon, K.C. Millett, J.I. Sulkowska, Linkprot: a database collecting information about biological links. Nucleic Acids Res. **45**(D1), D243 (2017)
16. N.L. Daly, D.J. Craik. Bioactive cystine knot proteins. Curr. Opin. Chem. Biol. **15**(3), 362–368 (2011)
17. L.-O. Essen, J. Mailliet, J. Hughes, The structure of a complete phytochrome sensory module in the Pr ground state. Proc. Natl. Acad. Sci. **105**(38), 14709–14714 (2008)
18. B. Ewing, K.C. Millett, Computational algorithms and the complexity of link polynomials. Prog. Knot Theory Relat. Top. **56**, 51–68 (1997)
19. P.F.N. Faísca, Knotted proteins: a tangled tale of structural biology. Comput. Struct. Biotechnol. J. **13**, 459–468 (2015)
20. P.F.N. Faísca, R.D.M. Travasso, T. Charters, A. Nunes, M. Cieplak, The folding of knotted proteins: insights from lattice simulations. Phys. Biol. **7**(1), 016009 (2010)
21. P. Freyd, D. Yetter, J. Hoste, W.B.R. Lickorish, K. Millett, A. Ocneanu, A new polynomial invariant of knots and links. Bull. Am. Math. Soc. **12**(2), 239–246 (1985)
22. E. Haglund, J.I. Sułkowska, Z. He, G-S. Feng, P.A. Jennings, J.N. Onuchic, The unique cysteine knot regulates the pleotropic hormone leptin. PloS one **7**(9), e45654 (2012)
23. E. Haglund, J.I. Sulkowska, J.K. Noel, H. Lammert, J.N. Onuchic, P.A. Jennings, Pierced lasso bundles are a new class of knot-like motifs. PLoS Comput. Biol. **10**(6), e1003613 (2014)

24. E. Haglund, A. Pilko, R. Wollman, P.A. Jennings, J.N. Onuchic, Pierced lasso topology controls function in leptin. J. Phys. Chem. B **121**(4), 706–718 (2017)
25. C. He, G.Z. Genchev, H. Lu, H. Li, Mechanically untying a protein slipknot: multiple pathways revealed by force spectroscopy and steered molecular dynamics simulations. J. Am. Chem. Soc. **134**(25), 10428–10435 (2012)
26. C. He, G. Lamour, A. Xiao, J. Gsponer, H. Li, Mechanically tightening a protein slipknot into a trefoil knot. J. Am. Chem. Soc. **136**(34), 11946–11955 (2014)
27. Y.M. Hou, R. Matsubara, R. Takase, I. Masuda, J.I. Sulkowska. TrmD: a methyl transferase for tRNA methylation with m1G37. The Enzymes (2017)
28. S.E. Jackson, A. Suma, C. Micheletti, How to fold intricately: using theory and experiments to unravel the properties of knotted proteins. Curr. Opin. Struct. Biol. **42**, 6–14 (2017)
29. M. Jamroz, W. Niemyska, E.J. Rawdon, A. Stasiak, K.C. Millett, P. Sułkowski, J.I. Sulkowska, Knotprot: a database of proteins with knots and slipknots. Nucleic Acids Res. **43**(D1), D306–D314 (2015)
30. A.I. Jarmolinska, A.P. Perlinska, R. Runkel, B. Trefz, P. Virnau, J.I. Sulkowska, Proteins? knotty problems (2017) (under review)
31. N.P. King, A.W. Jacobitz, M.R. Sawaya, L. Goldschmidt, T.O. Yeates, Structure and folding of a designed knotted protein. Proc. Natl. Acad. Sci. **107**(48), 20732–20737 (2010)
32. N.P. King, E.O. Yeates, T.O. Yeates, Identification of rare slipknots in proteins and their implications for stability and folding. J. Mol. Biol. **373**(1), 153–166 (2007)
33. G. Kolesov, P. Virnau, M. Kardar, L.A. Mirny, Protein knot server: detection of knots in protein structures. Nucleic Acids Res. **35**, W425–8 (2007)
34. K. Koniaris, M. Muthukumar, Self-entanglement in ring polymers. J. Chem. Phys. **95**(4), 2873–2881 (1991)
35. Y.-L. Lai, S.-C. Yen, Y. Sung-Huan, J.-K. Hwang, pknot: the protein knot web server. Nucleic Acids Res. **35**(2), W420–W424 (2007)
36. Y.-T.C. Lee, C-Y. Chang, S.-Y. Chen, Y.-R. Pan, M.-R. Ho, S.T.D. Hsu, Entropic stabilization of a deubiquitinase provides conformational plasticity and slow unfolding kinetics beneficial for functioning on the proteasome. Sci. Rep. **7**, 45174 (2017)
37. W. Li, T. Terakawa, W. Wang, S. Takada, Energy landscape and multiroute folding of topologically complex proteins adenylate kinase and 2ouf-knot. Proc. Natl. Acad. Sci. **109**(44), 17789–17794 (2012)
38. S.-C. Lou, S. Wetzel, H. Zhang, E.W. Crone, Y.-T. Lee, S.E. Jackson, S.-T.D. Hsu, The knotted protein UCh-L1 exhibits partially unfolded forms under native conditions that share common structural features with its kinetic folding intermediates. J. Mol. Biol. **428**(11), 2507–2520 (2016)
39. R.C. Lua, Pyknot, a pymol tool for the discovery and analysis of knots in proteins. Bioinformatics **28**(15), 2069–2071 (2012)
40. A.L. Mallam, S.E. Jackson, Knot formation in newly translated proteins is spontaneous and accelerated by chaperonins. Nat. Chem. Biol. **8**(2), 147–153 (2012)
41. A.L. Mallam, J.M. Rogers, S.E. Jackson, Experimental detection of knotted conformations in denatured proteins. Proc. Natl. Acad. Sci. **107**(18), 8189–8194 (2010)
42. M.L. Mansfield, Are there knots in proteins? Nat. Struct. Mol. Biol. **1**(4), 213–214 (1994)
43. K.C. Millett, E.J. Rawdon, A. Stasiak, J.I. Sułkowska, Identifying knots in proteins. Biochem. Soc. Trans. **41**(2), 533–537 (2013)
44. S. Najafi, R. Potestio, Folding of small knotted proteins: insights from a mean field coarse-grained model. J. Chem. Phys. 143(24):12B606_1 (2015)
45. W. Niemyska, P. Dabrowski-Tumanski, M. Kadlof, E. Haglund, P. Sułkowski, J.I. Sulkowska, Complex lasso: new entangled motifs in proteins. Sci. Rep. **6**, 36895 (2016)
46. W. Niemyska, A.M. Gierut, P. Sulkowski, P. Dabrowski-Tumanski, J.I. Sulkowska, Pylasso a pymol plugin to identify lassos (2017) (under review)
47. S. Niewieczerzal, J.I. Sulkowska, Knotting and unknotting proteins in the chaperonin cage: effects of the excluded volume. PloS one **12**(5), e0176744 (2017)

48. J.K. Noel, J.I. Sułkowska, J.N. Onuchic, Slipknotting upon native-like loop formation in a trefoil knot protein. Proc. Natl. Acad. Sci. **107**(35), 15403–15408 (2010)
49. J.K. Noel, J.N. Onuchic, J.I. Sulkowska, Knotting a protein in explicit solvent. J. Phys. Chem. Lett. **4**(21), 3570–3573 (2013)
50. J.H. Przytycki, P. Traczyk, Invariants of links of conway type. Kobe J. Math. **4**, 115–139 (1988)
51. M. Rief, H. Grubmüller, Force spectroscopy of single biomolecules. Chem. Phys. Chem. **3**(3), 255–261 (2002)
52. K.J. Rosengren, R.J. Clark, N.L. Daly, U. Göransson, A. Jones, D.J. Craik, Microcin j25 has a threaded sidechain-to-backbone ring structure and not a head-to-tail cyclized backbone. J. Am. Chem. Soc. **125**(41), 12464–12474 (2003)
53. T.C. Sayre, T.M. Lee, N.P. King, T.O. Yeates, Protein stabilization in a highly knotted protein polymer. Protein Eng. Des. Select. **24**(8), 627–630 (2011)
54. E. Shakhnovich, Protein folding: to knot or not to knot? Nat. Mater. **10**(2), 84–86 (2011)
55. T. Škrbić, C. Micheletti, P. Faccioli, The role of non-native interactions in the folding of knotted proteins. PLoS Comput. Biol. **8**(6), e1002504 (2012)
56. M.A. Soler, A. Nunes, P.F.N. Faísca, Effects of knot type in the folding of topologically complex lattice proteins. J. Chem. Phys. **141**(2), 07B607_1 (2014)
57. M.A. Soler, A. Rey, P.F.N. Faísca, Steric confinement and enhanced local flexibility assist knotting in simple models of protein folding. Phys. Chem. Chem. Phys. **18**(38), 26391–26403 (2016)
58. M. Sotomayor, K. Schulten, Single-molecule experiments in vitro and in silico. Science **316**(5828), 1144–1148 (2007)
59. J.I. Sułkowska, M. Cieplak, Mechanical stretching of proteins—a theoretical survey of the protein data bank. J. Phys. Condens. Matter **19**(28), 283201 (2007)
60. J.I. Sułkowska, M. Cieplak, Selection of optimal variants of gō-like models of proteins through studies of stretching. Biophys. J. **95**(7), 3174–3191 (2008)
61. J.I. Sułkowska, P. Sułkowski, P. Szymczak, M. Cieplak, Stabilizing effect of knots on proteins. Proc. Natl. Acad. Sci. **105**(50), 19714–19719 (2008)
62. J.I. Sułkowska, P. Sułkowski, P. Szymczak, M. Cieplak, Tightening of knots in proteins. Phys. Rev. Lett. **100**(5), 058106 (2008)
63. J.I. Sułkowska, P. Sułkowski, J.N. Onuchic, Jamming proteins with slipknots and their free energy landscape. Phys. Rev. Lett. **103**(26), 268103 (2009)
64. J.I. Sułkowska, P. Sułkowski, J. Onuchic, Dodging the crisis of folding proteins with knots. Proc. Natl. Acad. Sci. **106**(9), 3119–3124 (2009)
65. J.I. Sułkowska, P. Sułkowski, P. Szymczak, M. Cieplak, Untying knots in proteins. J. Am. Chem. Soc. **132**(40), 13954–13956 (2010)
66. J.I. Sułkowska, J.K. Noel, J.N. Onuchic, Energy landscape of knotted protein folding. Proc. Natl. Acad. Sci. **109**(44), 17783–17788 (2012)
67. J.I. Sułkowska, E.J. Rawdon, K.C. Millett, J.N. Onuchic, A. Stasiak, Conservation of complex knotting and slipknotting patterns in proteins. Proc. Natl. Acad. Sci. **109**(26), E1715–E1723 (2012)
68. J.I. Sułkowska, J.K. Noel, C.A. Ramírez-Sarmiento, E.J. Rawdon, K.C. Millett, J.N. Onuchic, Knotting pathways in proteins. Biochem. Soc. Trans. **41**(2), 523–527 (2013)
69. P. Szymczak, Tight knots in proteins: can they block the mitochondrial pores? Biochem. Soc. Trans. **41**(2), 620–624 (2013)
70. P. Szymczak, Periodic forces trigger knot untying during translocation of knotted proteins. Sci. Rep. **6** (2016)
71. W.R. Taylor, A deeply knotted protein structure and how it might fold. Nature **406**(6798), 916–919 (2000)
72. K.L. Tkaczuk, S. Dunin-Horkawicz, E. Purta, J.M. Bujnicki, Structural and evolutionary bioinformatics of the spout superfamily of methyltransferases. BMC Bioinform. **8**(1), 73 (2007)
73. L. Tubiana, E. Orlandini, C. Micheletti, Probing the entanglement and locating knots in ring polymers: a comparative study of different arc closure schemes. Prog. Theor. Phys. Suppl. **191**, 192–204 (2011)

74. I. Tuszynska, J.M. Bujnicki, Predicting atomic details of the unfolding pathway for yibk, a knotted protein from the spout superfamily. J. Biomol. Struct. Dyn. **27**(4), 511–520 (2010)
75. E. Uehara, T. Deguchi, Statistical and hydrodynamic properties of topological polymers for various graphs showing enhanced short-range correlation. J. Chem. Phys. **145**(16), 164905 (2016)
76. P. Virnau, L.A. Mirny, M. Kardar, Intricate knots in proteins: function and evolution. PLoS Comput. Biol. **2**(9), e122 (2006)
77. P. Virnau, A. Mallam, S. Jackson, Structures and folding pathways of topologically knotted proteins. J. Phys. Condens. Matter **23**(3), 033101 (2010)
78. J.R. Wagner, J.S. Brunzelle, K.T. Forest, R.D. Vierstra, A light-sensing knot revealed by the structure of the chromophore-binding domain of phytochrome. Nature **438**(7066), 325–331 (2005)
79. S. Wallin, K.B. Zeldovich, E.I. Shakhnovich, The folding mechanics of a knotted protein. J. Mol. Biol. **368**(3), 884–893 (2007)
80. I. Wang, S.-Y. Chen, S.-T.D. Hsu, Unraveling the folding mechanism of the smallest knotted protein, mj0366. J. Phys. Chem. B **119**(12), 4359–4370 (2015)
81. I. Wang, S.-Y. Chen, S.-T.D. Hsu, Folding analysis of the most complex stevedore's protein knot. Sci. Rep. **6** (2016)
82. M. Wojciechowski, À. Gómez-Sicilia, M. Carrión-Vázquez, M. Cieplak, Unfolding knots by proteasome-like systems: simulations of the behaviour of folded and neurotoxic proteins. Mol. Biosyst. **12**(9), 2700–2712 (2016)
83. Y. Zhao, S. Niewieczerzal, P. Dabrowski-Tumanski, J.I. Sulkowska, The exclusive effects of chaperonin on the behavior of the 52 knotted proteins (under review)
84. F. Ziegler, N.C. Lim, S.S. Mandal, B. Pelz, W.-P. Ng, M. Schlierf, S.E. Jackson, M. Rief, Knotting and unknotting of a protein in single molecule experiments. Proc. Natl. Acad. Sci. 201600614 (2016)

Part IV
Soft Matter and Biophotonics

Chapter 9
Topology in Liquid Crystal Phases

Gareth P. Alexander

Abstract Liquid crystals exhibit a rich set of phenomena with a geometric and topological flavour. We provide a survey of this in the nematic, smectic and cholesteric mesophases. Starting with Schlieren textures in nematics, we introduce the topological methods used to identify and classify point and line defects as well as non-singular textures. Particular attention is given to the characterisation of disclination loops and the use of the Pontryagin–Thom construction as a method of visualising complex three-dimensional textures. We illustrate these with examples drawn from recent experiments. Smectics have non-uniform ground states, which adds geometrical constraints to their topological properties. We describe the implications this has for two-dimensional smectics with the aid of a three-dimensional visualisation in terms of 'height' functions. Next, we describe geometrical methods for analysing line fields in terms of their curvatures and the degeneracies in them, which we call umbilics. These methods are applied to identify defects in the pitch axis in cholesterics. Finally, we provide a brief overview of the topological characterisation of knotted defect lines in nematics as well as an explicit method for constructing a director field with arbitrary knotted disclination line.

9.1 Introduction

Liquid crystals are as beautiful and inspiring as they are varied. The term encompasses the enormous range of soft materials whose properties are in some way intermediate between those of an isotropic fluid and a crystalline solid. They are at once soft and flow, and also structured and elastic. Optical microscopy paints them as colourful and often dramatic textures, whose features include flexible threads, distinctive polarisation brushes, fingerprint patterns, brightly coloured platelets and exquisite arrays of

G. P. Alexander (✉)
Department of Physics and Centre for Complexity Science,
University of Warwick, Coventry CV4 7AL, UK
e-mail: G.P.Alexander@warwick.ac.uk

S. Gupta and A. Saxena (eds.), *The Role of Topology in Materials*,
Springer Series in Solid-State Sciences 189,
https://doi.org/10.1007/978-3-319-76596-9_9

confocal conic sections. Many, or even most, of these are reflections of the topology inherent in liquid crystals.

Mesophases of liquid crystals have been prominent in the introduction of topological methods into the characterisation of physical phenomena. The fundamental defects were identified and classified in topological terms by Frank in his seminal paper [1]. This was formalised and generalised through the homotopy theory of the 1970s, many of the more subtle aspects of which derive from liquid crystals [2–4]. Over the past 20 years experimental advances have continued to foster the advancement of topological concepts, stimulated by the properties of emulsions and colloidal inclusions [5–8], and including realisations of the celebrated Hopf map [9, 10] and the creation of arbitrary knotted defect lines [11, 12].

Topology pervades the physical properties of liquid crystals. Many of their phase transitions involve the proliferation of defects; the two-dimensional isotropic–nematic transition is of Berezinskii–Kosterlitz–Thouless type and the cholesteric–smectic transition is analogous to the Abrikosov transition of type II superconductors, to give only two examples. Defects and textures can be used to control self-assembly through the elastic interactions they create [7, 13, 14] and provide novel prospects for soft photonics and metamaterials [15, 16]. For the most part, however, this survey will not address these aspects directly and instead focus on the underlying topology.

The basic liquid crystal phases are the nematic, smectic and cholesteric phases [17, 18]. These are illustrated schematically in Fig. 9.1a. Nematics are orientationally ordered fluids and the simplest of the liquid crystal mesophases. They are formed from compounds with an elongated molecular shape; a thin rod, or piece of chalk, in physicists' cartoons. Models of rod-like hard particles demonstrate an entropic drive to a state of coherent alignment, but without any positional order. A fundamental feature of this nematic alignment is that it is line-like rather than vectorial. Even if the molecules themselves have a 'head' and a 'tail' the macroscopic alignment that they create does not. This has dramatic consequences for the topology of liquid crystals.

The direction of alignment in a nematic is called the director field. Usually it is denoted by a unit magnitude vector $\mathbf{n}$, with the understanding that the nematic symmetry implies that $\mathbf{n}$ and $-\mathbf{n}$ should be considered physically identical. The set of all possible orientations for the director at any point is the set of directions in space at that point, the unit 2-sphere S^2, with exactly opposite directions identified. This is known as the real projective plane $\mathbb{RP}^2$, one of the simplest non-orientable spaces. In many respects the textures of liquid crystals furnish visual representations of it.

Both the smectic and cholesteric phases share the orientational order of nematics but also have their own unique characteristics. Smectics possess translational order in which the positions of the molecules are correlated as well as their orientations. This is not the crystalline arrangement of atoms as in a solid but an intermediate arrangement where the positions are ordered in one dimension but remain uncorrelated, or fluid-like, in the transverse directions. Smectics are a hybrid of a one-dimensional crystal and a two-dimensional fluid, resulting in a structure consisting of a family of regularly spaced fluid layers. It is not unlike the pattern of ridges that form our fingertips, the grain in a piece of wood, or the contour lines in a topographical map. We shall restrict

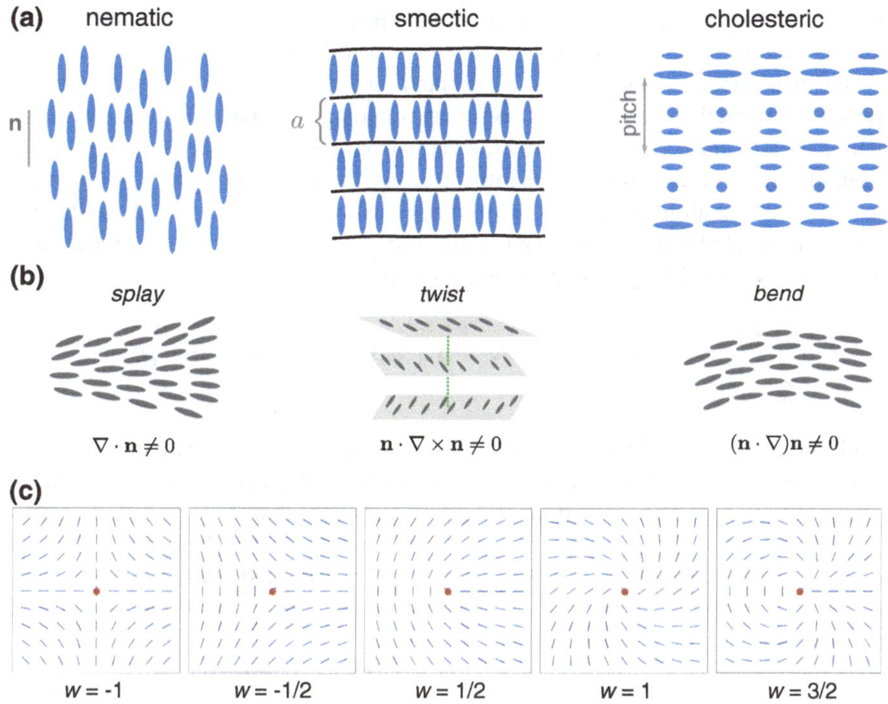

Fig. 9.1 **a** The main phases of liquid crystals: nematics, smectics and cholesterics. **b** Curvature elasticity: splay, twist and bend distortions. **c** Disclinations, topological point defects, in two-dimensional nematics

ourselves to the smectic-A phase where the direction of positional order – the layer normal – coincides with the director field of molecular orientational alignment.

Cholesterics are not really a distinct phase from nematics. Notionally, they differ only in that the nematic mesogens possess an intrinsic chirality or that a small amount ($\lesssim$ few % weight) of a chiral dopant has been added to an achiral nematic. However, in terms of their textures, and topology, the difference can be quite striking. They organise into a non-uniform ground state in which the director field regularly rotates about a spontaneously chosen direction, called the pitch axis. The regular rotation gives cholesterics a one-dimensional periodicity with a length scale, called the pitch, that is anywhere from several hundred nanometres to tens of microns. As a result they can Bragg scatter light in the optical range, leading to photonic applications in tunable lasers [19, 20] as well as the basis for structural colour in certain jewelled beetles [21, 22].

Cholesterics are similar to smectics but they are not the same. In particular, their periodicity is only in the molecular alignment and they remain full three-dimensional fluids. Smectics, by contrast, have a modulated density corresponding to their translational order but the molecules are uniformly aligned in the ground state. Nonetheless, the feature of a ground state that is a function of position gives to both materials geometric and topological properties that are still not fully understood [23–26].

The theoretical foundation for understanding liquid crystals was cemented in Frank's influential paper [1]. His prescient treatment emphasised both geometric methods – *curvature elasticity* – and concepts from topology – *disinclinations*. The Frank elasticity derives from an identification of the basic curvature distortions of the liquid crystal alignment, known as splay, twist and bend, see Fig. 9.1b. Each of these is associated with the non-vanishing of certain gradients of the director field and taken together they give rise to the Frank free energy

$$F = \int_{\Omega} \left\{ \frac{K_1}{2} (\nabla \cdot \mathbf{n})^2 + \frac{K_2}{2} (\mathbf{n} \cdot \nabla \times \mathbf{n} + q_0)^2 + \frac{K_3}{2} ((\mathbf{n} \cdot \nabla)\mathbf{n})^2 \right\} dV, \qquad (9.1)$$

for elastic distortions of a liquid crystal. The K_i are known as elastic constants and the parameter q_0 is called the chirality; it is non-zero only in cholesteric materials.

In addition to elastic distortions, liquid crystals also display singularities where the director is discontinuous in an essential way. The simplest of these were identified and classified by Frank, see Fig. 9.1c, and called by him *disinclinations*; nowadays this has transmuted to the term disclination. We take them as a starting point for our survey of topology in liquid crystals.

9.2 Schlieren Textures and Two-Dimensional Nematics

Two-dimensional nematics serve as something of a prototype for the description of topology in ordered media. They are typically realised in thin cells with parallel anchoring conditions, or in films at an oil-water interface. The entire texture can be visualised directly and one of the principal methods for doing so – under crossed polarisers with light microscopy – embodies some core topological concepts. These experimental images are known as Schlieren textures, illustrative examples of which are shown in Fig. 9.2. It is a pattern of dark brushes where the light does not pass through the crossed polarisers and marked by isolated points where the brushes appear to 'come together' or 'pinch off'.

At most points of the sample there is a local orientation, which varies smoothly from one point to the next. However, there are also certain singular points, called disclinations, at which the orientation is discontinuous, or ill-defined. These are the 'pinch' points of the brushes. They may be characterised by the change in orientation around any closed loop that encircles the defect, known as a winding number. Let us represent the orientation in the texture by a unit magnitude vector field $\mathbf{n} = (\cos(\theta(\mathbf{x})), \sin(\theta(\mathbf{x})), 0)$ with the understanding that the directions $\mathbf{n}$ and $-\mathbf{n}$ are to be considered the same. We have

$$d\mathbf{n} = \big(-\sin(\theta), \cos(\theta), 0\big)\, d\theta \qquad \text{and} \qquad \mathbf{n} \times d\mathbf{n} = \big(0, 0, 1\big)\, d\theta, \qquad (9.2)$$

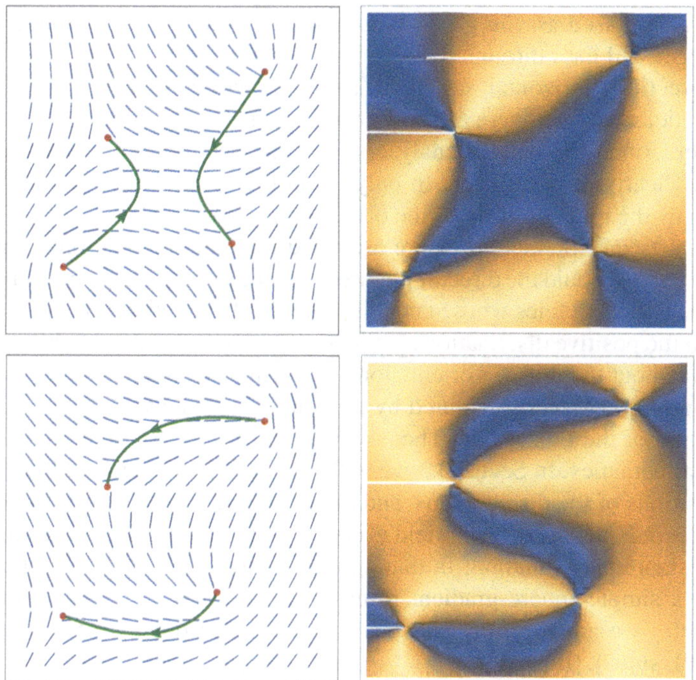

Fig. 9.2 Examples of Schlieren textures in nematic liquid crystals. Left: director field with four $\pm 1/2$ disclinations. A Pontryagin–Thom representation for each texture is overlaid. Right: visualisation of brushes seen under crossed polarisers. Regions of the material where the orientation is either vertical or horizontal appear dark, producing a characteristic pattern of 'brushes'

or, recast in a purely two-dimensional notation, $d\theta = \epsilon_{ij} n_i dn_j$, where ϵ_{ij} is the antisymmetric symbol, $\epsilon_{ij} = -\epsilon_{ji}$, with $\epsilon_{12} = +1$. It then follows that the winding number around any closed loop γ can be computed from the formula

$$w = \frac{1}{2\pi} \int_\gamma d\theta = \frac{1}{2\pi} \int_\gamma \epsilon_{ij} n_i \partial_k n_j \, dx_k. \tag{9.3}$$

Since the angle θ can only change by integer multiples of π, the winding number is always a half-integer, $w \in \frac{1}{2}\mathbb{Z}$. It is easy to 'see' this half-integer from the Schlieren texture. At each disclination in Fig. 9.2 two dark brushes come together; the winding number is this number of brushes divided by four. The sign of the winding number is determined experimentally by rotating the crossed polarisers. If the pattern of brushes co-rotates the disclination has positive winding, while if the pattern counter-rotates it has negative winding.

Let us spell out a little more why the simple counting works. The brushes correspond to those places in the sample where the local orientation is parallel to either the polariser or the analyser, so counting them counts the number of times that those directions are crossed as you go once around the defect. If the orientation rotates by a full 2π it will pass through the polariser direction (vertical, say) twice and the analyser direction (horizontal, say) twice and there will be four brushes in total that meet at the defect. When there are only two brushes the director passes through the polariser direction once and the analyser direction once, so that the total rotation is only by π. Clearly, it is sufficient to only count the number of 'analyser brushes', which tells us the number of times the orientation is horizontal.

The brushes in a Schlieren texture are vivid illustrations of inverse images. In Fig. 9.2 we have overlaid the director pictures with the inverse image of the horizontal orientation. These are lines whose endpoints are the disclinations. They are oriented to run from the positive disclinations to the negative ones. Evidently, the topological information about the director orientation – the winding numbers about each of the disclinations – is succinctly captured by these inverse images. This correspondence is an example of the Pontryagin–Thom construction [27, 28], here applied to two-dimensional nematics. In Sect. 9.3.3 we will extend it to three dimensions.

The positions of the disclinations, and their winding numbers, is sufficient information to reconstruct the director field everywhere in the sample, at least in the one-elastic-constant approximation. In that approximation, the Frank free energy (9.1) reduces to a simple square gradient form, $F = \int_{\Omega} \frac{K}{2} |\nabla \mathbf{n}|^2 \, dV$, which in the thin film geometry we are currently considering with a two-dimensional director field $\mathbf{n} = (\cos(\theta), \sin(\theta), 0)$ becomes

$$F = \frac{Kh}{2} \int |\nabla \theta|^2 \, d^2x, \tag{9.4}$$

where h is the thickness of the film. The Euler-Lagrange equation then gives that θ is a harmonic function, $\nabla^2 \theta = 0$. It is convenient to write the solution as

$$\theta = \frac{1}{2} \mathrm{Im} \ln \Theta(x + iy), \tag{9.5}$$

where Θ is a meromorphic function of $x + iy$ and the prefactor of $\frac{1}{2}$ accounts for the nematic symmetry. Meromorphic functions can be represented in terms of their zeros and poles, together with their orders or multiplicities, so that we can write

$$\theta = \frac{1}{2} \mathrm{Im} \ln \prod_j \big(x - x_j + i(y - y_j)\big)^{m_j} = \sum_j \frac{m_j}{2} \arctan \frac{y - y_j}{x - x_j}. \tag{9.6}$$

It is clear that the winding number at each of the defects is $w_j = m_j/2$.

The utility of this explicit expression is not that it represents any real texture with high precision, as real materials have distinct elastic constants. Still, the texture in a real material will not be drastically different and, in particular, will be related to the one-elastic-constant formula by an everywhere smooth rotation of the local orientation. What (9.6) illustrates is that the topological properties in any texture allow for its complete reconstruction up to continuous deformation, or homotopy. In other words, the relationship between textures and their topological properties is a two-way street. Given an explicit texture, such as the Schlieren textures in Fig. 9.2, we can identify the defects and compute their winding numbers. And conversely, given only the green lines in the left panels of Fig. 9.2 we can reconstruct the entire texture, up to a smooth rotation. We will see how to extend these constructions to three-dimensional textures in Sect. 9.3.3.

9.3 The Homotopy Theory of Defects

Winding numbers are homotopy invariants, meaning they are characteristics of the texture that are preserved under continuous changes to the director field. The study of properties that are preserved under continuous transformations is known as homotopy theory.

In three dimensions nematics possess line defects, again called disclinations. Prototypes for them are obtained by simply extending the two-dimensional disclinations of Fig. 9.1c uniformly into the third dimension. The analogue of winding numbers is to identify the line defects of three-dimensional nematics with the distinct homotopy classes of maps $S^1 \to \mathbb{RP}^2$, which is the two element set $\mathbb{Z}_2$. Given the infinite number of disclinations in two dimensions it can come as a surprise to find that there is a unique homotopy class in three dimensions. A homotopy connecting the standard $+1/2$ and $-1/2$ profiles can be given as

$$\mathbf{n}_t(r, \phi, z) = \Big(\cos\big(\tfrac{1}{2}\phi\big), \cos(\pi t)\sin\big(\tfrac{1}{2}\phi\big), \sin(\pi t)\sin\big(\tfrac{1}{2}\phi\big)\Big), \tag{9.7}$$

using cylindrical coordinates (r, ϕ, z). $\mathbf{n}_0$ is the $+1/2$ profile, while $\mathbf{n}_1$ is the $-1/2$, and the homotopy is simply a rotation of each molecule by angle π about the x-direction. The homotopy that eliminates the $+1$ line defect entirely is known as 'escape in the third dimension' and is realised, for instance, in cylindrical capillaries with normal anchoring [29, 30].

These principles of the classification of disclinations in nematics were encapsulated into a homotopy classification of defects developed in the mid 1970s to apply to any kind of ordered material. The general approach was given as an extension of Landau's symmetry breaking theory of phase transitions in which an ordered mesophase is characterised by the spontaneous reduction of the symmetry group G of the high temperature (isotropic) phase to a subgroup H of residual symmetries of the low temperature (ordered) phase. In the case of a nematic we can take the high temperature group to be the rotation group $SO(3)$ of Euclidean space, while the symmetry of the low temperature nematic phase is the subgroup D_∞ of symmetries of a cylinder, or piece of chalk. The ground state manifold is the coset space $G/H = SO(3)/D_\infty = \mathbb{RP}^2$, the real projective plane, and any measurement of the texture on a measuring loop is a map $S^1 \to \mathbb{RP}^2 = G/H$ so that the defects are classified by the (conjugacy classes) of the fundamental group $\pi_1(G/H)$. The formalism is excellently summarised in a number of review articles [31–34].

The formalism of homotopy groups, especially as applied in the general context of symmetry breaking, provides powerful tools for the description of defects in ordered media, all of which flow freely from the machinery of algebraic topology. Without going into any details we mention a couple of examples. First, the homotopy groups $\pi_2(\mathbb{RP}^2) \cong \mathbb{Z}$ and $\pi_3(\mathbb{RP}^2) \cong \mathbb{Z}$ inform us that there are an infinite number of distinct point defects in nematics and also an infinite number of distinct defect free textures, respectively. We shall describe each of these shortly. Second, the action of π_1 on the other homotopy groups suggests a non-commutativity of defect crossing [35], characterised by Whitehead products [36]. For instance, this is predicted to arise in biaxial nematics and cholesterics, but has not yet been observed in either material.

Despite the powerful formal machinery, it is still instructive to approach the topological description of defects in more physical terms. An example of this is provided by disclination loops, the understanding of which benefited greatly by the development of new experiments focused on them [8, 11, 37, 38]. A second cautionary tale comes from smectics. Two-dimensional smectics can be described in terms of the symmetry breaking of the high temperature Euclidean group $G = \mathbb{R}^2 \rtimes SO(2)$ to the low temperature group $H = (\mathbb{R} \times \mathbb{Z}) \rtimes \mathbb{Z}_2$ corresponding to the symmetries of the smectic phase. Applying the formal machinery leads to the prediction that the defects are classified by the fundamental group of the Klein bottle, which turns out to be erroneous [24]. We shall describe this in Sect. 9.5.

9.3.1 Point Defects: Hedgehogs

In addition to line defects, nematics in three dimensions also exhibit singularities at isolated points. These are referred to colloquially as hedgehogs because the prototype for such fields is the purely radial texture $\mathbf{n} = \mathbf{x}/|\mathbf{x}|$. These defects are characterised, in part, by the behaviour of the director field on any spherical surface Σ enclosing the singular point. This constitutes a map $\Sigma \to \mathbb{RP}^2$ and the different defects correspond to the different homotopy classes of such maps.

A feature of hedgehogs that contrasts them with disclinations is that they can be oriented. For instance, the radial hedgehog is conventionally oriented to point outwards ($\mathbf{n} = \mathbf{x}/|\mathbf{x}|$) rather than inwards ($\mathbf{n} = -\mathbf{x}/|\mathbf{x}|$). With a choice of orientation, the behaviour of the director field on a spherical surface Σ enclosing the singular point yields a map $\Sigma \to S^2$, and these are known to be classified by degree. There are a number of equivalent ways of computing the degree of a map. One of the most common representations is as the integral

$$\deg(\mathbf{n}) = \frac{1}{4\pi} \int_{\Sigma} \mathbf{n} \cdot \partial_{x_1} \mathbf{n} \times \partial_{x_2} \mathbf{n} \, dx_1 dx_2, \tag{9.8}$$

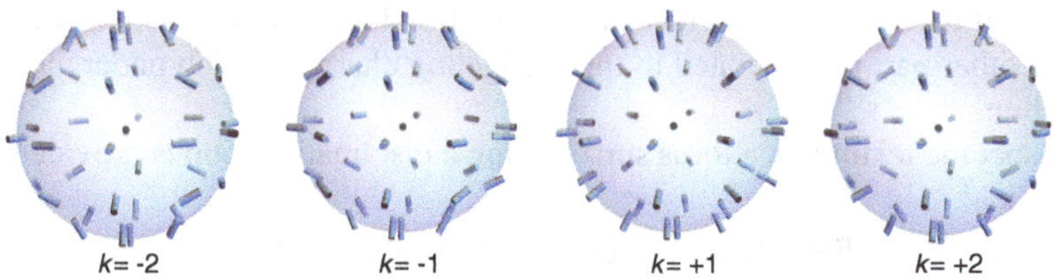

Fig. 9.3 Representative director fields on spherical surfaces enclosing hedgehogs with different integer degrees. The $k = +1$ configuration is called a 'radial hedgehog' and $k = -1$ a 'hyperbolic hedgehog'

where (x_1, x_2) are local coordinates for Σ. Note that $\deg(\mathbf{n})$ is not invariant under $\mathbf{n} \to -\mathbf{n}$ so that it depends upon a choice of orientation for the director field. As a result the sign of nematic hedgehogs is subject to ambiguity [38].

The degree can take any integer value, corresponding to the homotopy group $\pi_2(\mathbb{RP}^2) \cong \mathbb{Z}$ and giving an integers worth of topologically distinct point defects in nematics. Representative director fields for each of these hedgehogs can be given as

$$\mathbf{n}_k(r, \theta, \phi) = \sin(\theta)\big[\cos(k\phi)\mathbf{e}_x + \sin(k\phi)\mathbf{e}_y\big] + \cos(\theta)\mathbf{e}_z, \qquad (9.9)$$

where (r, θ, ϕ) are the standard spherical polar coordinates and k is the integer degree. A selection of these are illustrated in Fig. 9.3. Of course, for these representatives there is no pretence that they coincide with free energy minimisers, as was the case in two-dimensions, (9.6), only that they capture the appropriate topology.

9.3.2 Disclination Loops

The interplay between point and line defects yields one of the most interesting aspects of the topological theory of defects in nematics. At its heart is the observation that a disclination loop may shrink to become a point, or conversely a point may 'open up' into a loop. Thus disclination loops must somehow carry topological characteristics pertaining to both line defects and point defects. For instance, suppose a point defect of degree k opens up to form a loop. A naive expectation is that the 'hedgehog charge' should somehow be conserved. Indeed, any spherical measuring surface Σ that continues to enclose the entire loop will still record the same degree k, as expected.

A different characterisation is provided by looking at the behaviour of the director field on a small tubular neighbourhood of the disclination. This surface is a torus rather than a sphere and so has different properties. It constitutes a map $T^2 \to \mathbb{RP}^2$ that is non-trivial along the meridian of the torus. It turns out that there are only four homotopy classes of such maps [39], rather than an infinite number. A part of this comes from whether or not the director field is also non-orientable along the longitude of the torus, i.e. whether the disclination loop is linked with any others or not. The other part is associated to the 'hedgehog charge', which undergoes a mod 2 reduction.

One specific texture on a torus surrounding a disclination loop is provided by the director field

$$\mathbf{n}(x_1, x_2) = \cos\bigl(\pi x_2/\ell\bigr)\mathbf{e}_x + \sin\bigl(\pi x_2/\ell\bigr)\mathbf{e}_y, \tag{9.10}$$

where ℓ is the 'meridional circumference'. It is non-orientable around the meridian but does not depend upon the longitudinal direction at all. We use it as a starting point from which we can generate other textures. The basic way in which (9.10) can be modified is to pick any point in the torus, take a small disc around it, and replace the director field on the interior of that disc with a configuration representing a degree 1 map. Such modifications generate all of the homotopy classes of maps that have the same behaviour as the starting one along the longitudinal and meridional cycles. However, not all of these modifications yield distinct textures. Indeed, performing two such modifications produces a texture that is homotopic to the original one.

A way of seeing this is to double cover the torus around the meridian, as shown in Fig. 9.4. It is then possible to orient the director field on the double cover. There is an obvious 2:1 projection that takes the oriented doubled version back to the proper director field and everything needs to be consistent with respect to this projection. This is called equivariance. Modifications must be done consistently between the two sheets of the cover; that is, a modification by a degree +1 map in one sheet must be accompanied by an equal and opposite modification by a degree −1 map at the same location in the other sheet. As shown in Fig. 9.4, there is an equivariant homotopy, indicated by the dashed red lines, that allows modifications to be removed in pairs and shows that there are only two homotopy classes of director fields with the prescribed behaviour around the meridian and longitude.

This is the basic 'local' characterisation of disclination loops, but as we shall describe in Sect. 9.8 there is a lot more to their global properties.

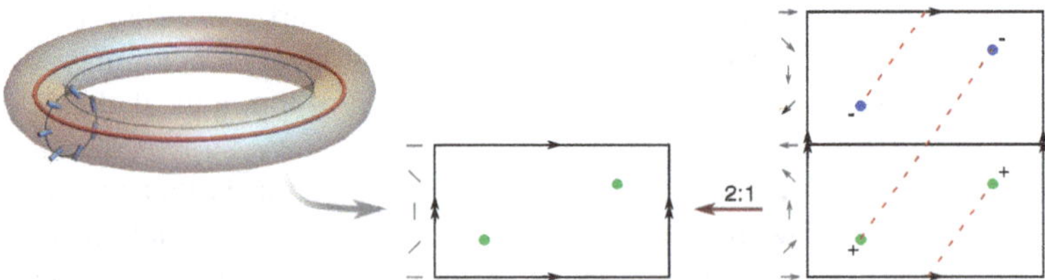

Fig. 9.4 Schematic illustration of the homotopy characterisation of nematic textures on a torus enclosing a disclination loop. Such textures can be analysed by lifting to a double cover over the meridional direction that goes around the disclination. Modifications of the texture by degree 1 maps (indicated by coloured dots) appear in pairs in the two sheets of the cover, with opposite degree. An equivariant homotopy illustrating that two such modifications are homotopic to none at all is indicated by the dashed red lines

9.3.3 The Pontryagin–Thom Construction

Schlieren textures present a vivid depiction of the topology in a liquid crystal. Part of what makes them so effective is that they are two-dimensional, so that everything can be seen at once on a single screen. Typically, a dense forest of lines showing the orientation of every molecule in some three-dimensional volume is uninformative and unilluminating. So there is great value in obtaining an analogous vivid and accessible depiction of the structure and topology in three dimensions. One method that is founded on deep results in topology is the Pontryagin–Thom construction [9, 40].

At its core, a Schlieren texture is about inverse images; it highlights those parts of the domain that are 'inverse images', or preimages, of a particular orientation, for instance horizontal. The Pontryagin–Thom construction extends this to a more general setting. We first describe briefly orientable director fields before addressing the general case. For a more thorough account the reader is especially directed to the excellent description given by Milnor [27].

For maps into a sphere (polar order) the general construction is really exactly the same as the two-dimensional Schlieren textures; one looks at the inverse image of any particular orientation and these preimages convey the topology. Again, they are one-dimensional curves that end on point defects. The analogue of the orientation (from plus to minus) of the inverse image curves in two dimensions is that now they should carry a framing. To give an example of how the inverse image curves convey the topology, consider the degree of a point defect. As with computing the winding number in two dimensions, the degree of a point defect can be calculated from a signed count of preimages. For any $y \in S^2$ we count

$$\deg(\mathbf{n}) = \sum_{\mathbf{x} \in \mathbf{n}^{-1}(y)} \epsilon_{\mathbf{x}}, \tag{9.11}$$

where $\epsilon_{\mathbf{x}}$ is equal to $+1$ if the map is orientation preserving at the point $\mathbf{x}$ where the preimage intersects a spherical measuring surface, and is equal to -1 if it is orientation reversing. The reduction of a three-dimensional texture to a set of framed curves represents a tremendous compression of information and provides a significantly simplified visual representation.

The general case of non-orientable director fields is a little more subtle and I describe it here only in practical terms and also only for situations in which the orientation at large distances is vertical – for the mathematical details see [27, 28, 40]. In this case the appropriate inverse image to display is everywhere that the orientation is horizontal, the entire equator in $\mathbb{RP}^2$. This will be a two-dimensional surface, or collection of two-dimensional surfaces, whose boundaries are the point and line defects in the director field. We will refer to it as a 'PT surface'. The PT surface is further decorated in two ways. First, we may colour it by the horizontal orientation at each point, running from red to green to blue and back to red again. Second, there is an assignment of 'normal data' to the surface. The normal neighbourhood of the

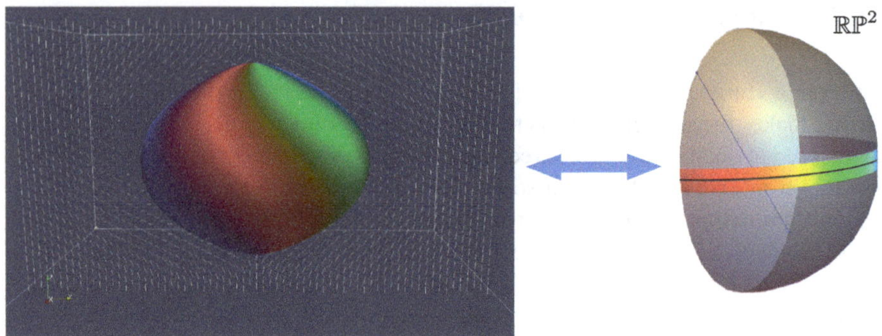

Fig. 9.5 The Pontryagin–Thom construction for visualisation of three-dimensional nematic textures. It is the set of points in the texture where the director is horizontal, or the inverse image of the equator in $\mathbb{RP}^2$. This is a surface that is then coloured according to the different horizontal directions. The example shown is a toron, and described in Sect. 9.4.3

PT surface, an orientable manifold, maps into a non-orientable Möbius strip around the equator in $\mathbb{RP}^2$ and the additional data encodes the structure of this map. In practical terms what this entails is that there are an even number of colour windings around any orientable cycle of the PT surface and an odd number around any non-orientable cycle. The examples I consider here involve only orientable PT surfaces, which effectively renders this second point moot. The PT construction is illustrated in Fig. 9.5 and employed throughout the remainder of this chapter.

9.4 Illustrations in Liquid Crystals

In this section we describe a selection of textures in liquid crystals that illustrate or embody some of the topological concepts we have just described.

9.4.1 Skyrmions

Consider a thin film geometry where the domain is quasi-two-dimensional but the director orientation is unconstrained. A uniform film, with the director aligned along the normal directions, $\mathbf{n} = \mathbf{e}_z$ say, can be modified on a disc, of radius R, by replacing the uniform texture inside it with the director field

$$\mathbf{n}(r, \phi) = \sin(\pi r/R)\big[-\sin(\phi)\mathbf{e}_x + \cos(\phi)\mathbf{e}_y\big] - \cos(\pi r/R)\mathbf{e}_z, \tag{9.12}$$

where (r, ϕ) are polar coordinates for the disc. This modification represents a degree 1 map of the disc to S^2, which as long as the boundary conditions, $\mathbf{n} \to \mathbf{e}_z$ as $r \to \infty$,

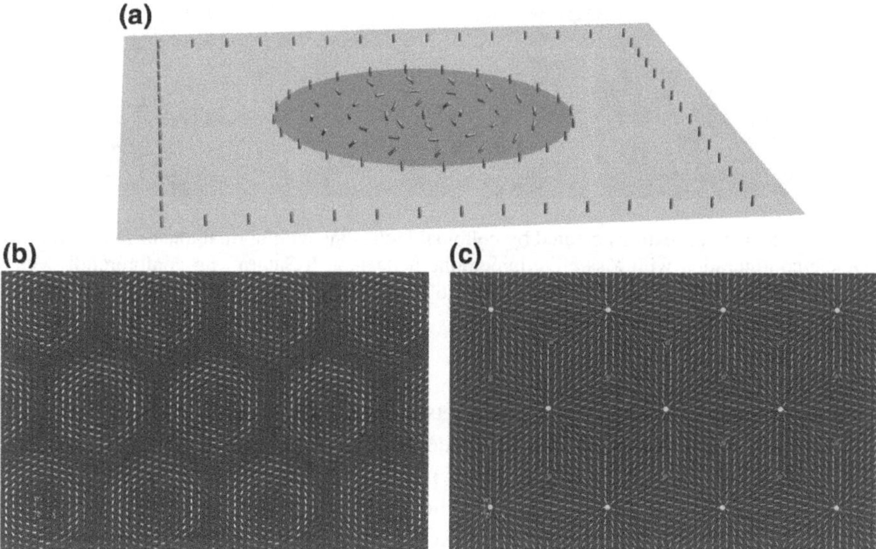

Fig. 9.6 **a** Schematic of a single Skyrmion inserted into a background uniform texture. This is a degree $+1$ map on a disc that is constant along the boundary. **b** Hexagonal lattice of Skyrmions. These configurations are energetically stable in some chiral materials. **c** Pitch axis of the hexagonal lattice (see Sect. 9.7) with defects highlighted. The red dots are λ^- lines and the yellow dots are the axes of double twist cylinders

are maintained is topologically robust. These textures have come to be known as Skyrmions, originally in the context of chiral ferromagnets [41–47] but now also in the liquid crystal literature [48–50].

In situations where they are energetically favourable they typically rapidly proliferate, leading to phases in which individual Skyrmions pack into hexagonal lattice arrangements. An example of the director field for a hexagonal lattice of Skyrmions is shown in Fig. 9.6b. Since the sample geometry is that of a thin film it is possible to identify the location of each Skyrmion by those points where the director is exactly parallel to the surface normal direction. Although this method is often used it is not really topological. One can instead identify the location of Skyrmions by plotting the maxima of the 'charge density' – the integrand of (9.8) – which yields the same positions and is founded on both geometric and topological concepts. In Sect. 9.6.2 we will describe a dual of this that is shown in Fig. 9.6c.

9.4.2 Colloids

The drive behind much of the recent advances in topology in liquid crystals has come from the properties of colloidal inclusions [5–8, 11, 37, 38, 51–53]. Initially

Fig. 9.7 Examples of textures created by colloidal inclusions with normal anchoring conditions. **a** Dipole configuration with a satellite hyperbolic hedgehog. **b** Saturn ring configuration with a disclination loop encircling the colloid. **c** Two colloids entangled by a single disclination loop in the 'figure Omega' configuration

these were spherical colloids but platelets, tori, handlebodies up to genus 5 and even Möbius strips have also been considered. Colloids provide internal boundary conditions for the liquid crystal with the main point being that these are typically topologically non-trivial. For instance, when the molecules align along the surface normal direction then the director field on the colloidal boundary carries a degree. Spherical colloids have degree $+1$ and in general the degree is $1 - g$, where g is the genus of the surface [51].

If at large distances the director is uniform then the total degree of all the defects in the liquid crystal must compensate that presented by the colloidal inclusions. However, there is large variety in how the liquid crystal achieves this. Even in the simplest case of a single spherical colloid the degree represented by the surface may be compensated either by an accompanying hyperbolic hedgehog, as in Fig. 9.7a, or by a disclination loop, as in Fig. 9.7b, a configuration known as the Saturn ring. These have different symmetries and create different elastic distortions in the liquid crystal, the former dipolar and the latter quadrupolar, leading to differences in the interactions between colloids and the chains and lattice structures that form from them [54–56].

From the topological perspective, these examples illustrate the interconversion that is possible between point defects and disclination loops – one slides the Saturn ring towards one of the poles and shrinks it to a point – a feature that is central to the more global aspects of nematic topology. With two or more colloids, there arises also the possibility that a single disclination loop will wrap itself around them all. Several configurations are possible [8, 57, 58]; we show in Fig. 9.7c one of the original examples involving two colloids called the 'figure Omega'. These shapes can be purely geometric, as in the figure Omega example, or they can involve the formation of non-trivial knots, links and braids [11, 12, 37]. The topological implications of such defects will be considered in Sect. 9.8.

9.4.3 Torons and Hopf Textures

Another class of experiments that has proved especially fruitful in providing new examples of topology in liquid crystals is from frustrated cholesteric cells. In thin cells, where the cell gap is comparable to the cholesteric pitch, normal anchoring conditions can frustrate the natural cholesteric order by making the director align uniformly along the vertical direction. However, suitable stimulation, with laser tweezers for instance, can prompt the creation of localised twisted structures to relieve the frustration [9, 10, 59, 60].

One example of these was dubbed the toron [59]. It is a toroidal region of twisted director field (a circular variation on the double twist cylinder) that is flanked above and below by hyperbolic hedgehogs. This is the example that we used to illustrate the Pontryagin–Thom construction in Fig. 9.5. The PT surface is a sphere that 'pinches' at the top and bottom where the hedgehogs sit. It separates an interior region where the director field is tilted downward to some degree from the exterior where it has an upwards tilt and asymptotically matches the uniformly vertical alignment. The surface itself is coloured with two full turns of the 'colour wheel' conveying a full 2π rotation of the horizontal direction field as you go around the PT surface. Each colour, for instance the preimage of red ($\mathbf{n} = \mathbf{e}_x$), forms a line with endpoints on the two point defects. That they are not straight lines of longitude reflects the twisting of the director. With these points being said, the Pontryagin–Thom construction allows for an effective reconstruction of the director field throughout the sample.

A second texture that can be created under the same conditions is the non-singular Hopf texture [9]. Schematically it can be thought of as what is obtained if the two point defects can be made to come together and mutually annihilate to leave behind a non-trivial smooth texture. The result is shown in Fig. 9.8. It can be thought of as an illustration of the homotopy group $\pi_3(\mathbb{RP}^2) \cong \mathbb{Z}$, which expresses that there are an infinite number of distinct everywhere smooth director fields that are different from the state of uniform alignment. The prototype for these is the celebrated Hopf map, the generator of $\pi_3(S^2)$.

The Hopf map is an assignment of a unit vector to each point in $\mathbb{R}^3$ with boundary conditions that the vector takes a fixed orientation at large distances. First, we represent the point $(x, y, z) \in \mathbb{R}^3$ in terms of a pair of complex numbers (z_1, z_2), with $|z_1|^2 + |z_2|^2 = 1$, by stereographic projection to S^3

$$z_1 = \frac{2(x+iy)}{x^2+y^2+z^2+1}, \qquad z_2 = \frac{2z + i(x^2+y^2+z^2-1)}{x^2+y^2+z^2+1}. \tag{9.13}$$

The Hopf map $S^3 \to S^2$ can be represented in terms of a complex coordinate for S^2 by the simple formula $w = z_1/z_2$. Converting this to a unit vector using stereographic projection gives

$$n_x + i n_y = 2 z_1 \bar{z}_2, \qquad n_z = |z_1|^2 - |z_2|^2. \tag{9.14}$$

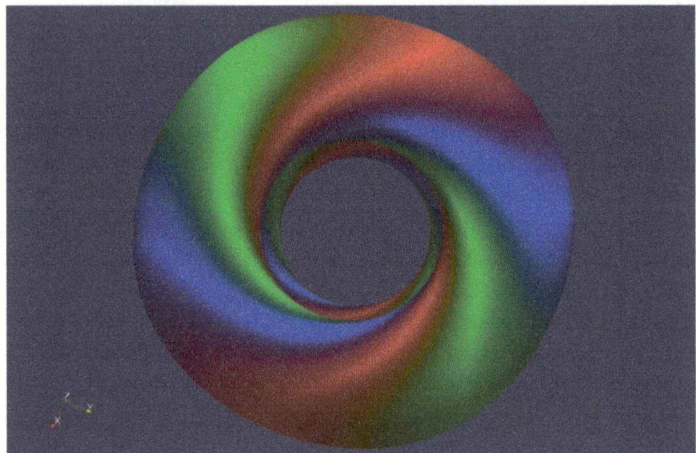

Fig. 9.8 The Hopf texture visualised using the Pontryagin–Thom construction. The PT surface shows all points in the material where the director is horizontal and is coloured according to the horizontal direction. There are two full colour windings corresponding to a full 2π variation of the horizontal orientation. The surface is a torus and it can be seen that every orientation describes a circle on the torus that links with every other orientation exactly once

It is easy to read off from this that $n_z = 1$ everywhere that z_2 vanishes, which is the unit circle in the xy-plane. Similarly, $n_z = -1$ everywhere that z_1 vanishes, which is the z-axis. So the inverse images of these two orientations are linked with each other, precisely once. This is the characteristic feature of the Hopf map; the inverse image of any orientation is a circle of points in $\mathbb{R}^3$ and the inverse images of any two distinct orientations is a pair of circles that are linked precisely once. The linking of inverse images is a spectacular and instantly recognisable feature of Hopf textures.

9.5 Smectics

Smectics share the orientational order of nematics and have in addition positional order associated to breaking translational symmetry in one dimension only. They are described by one-dimensional mass density waves, of the form $\rho(\mathbf{x}) = \rho_0 + \delta\rho\cos\big(\frac{2\pi}{a}\phi(\mathbf{x})\big)$, where $\phi(\mathbf{x})$ is a phase field whose level sets at integer multiples of a give the loci of the maxima in the mass density, or, more simply, the positions of the molecules. The distance between adjacent maxima is the layer spacing, a, typically equal to the molecular length. In the ground state, the phase field is a linear function of position, say $\phi = z$, and the layers are equally spaced and flat. Of course, there are many equivalent, but distinct, ground states, obtained from this one by uniform translations and rotations. We arrive back at precisely the same ground state if the translation is by the layer spacing a, or the rotation is by π. This pattern of

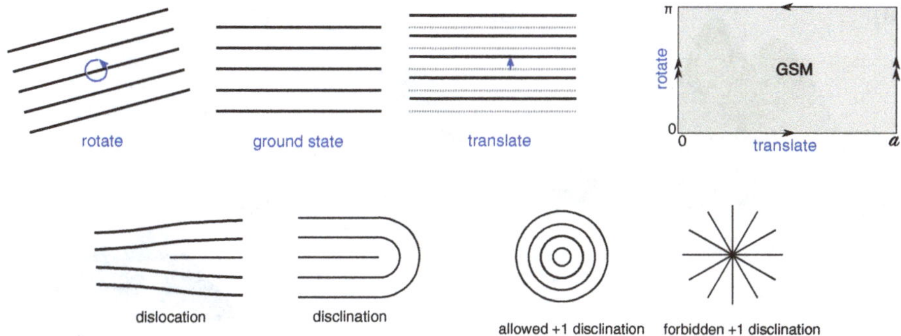

Fig. 9.9 The ground state manifold (GSM) for two-dimensional smectics. Both translational and rotational symmetries are broken so that both translations and rotations yield equivalent but distinct ground states. The identifications under translations by the layer spacing, a, and rotations by π give the topology of the Klein bottle, a non-orientable manifold with non-Abelian fundamental group. There are two types of defects; disclinations, associated to the rotational freedom, and dislocations, associated to the translation freedom

identifications leads to a ground state manifold with the topology of the Klein bottle, see Fig. 9.9.

Following the general rubric would lead to a classification of defects in terms of the fundamental group of the Klein bottle. This is a non-Abelian group generated by translations by the lattice spacing, which we shall call 'shifts' S, and rotations by π, which we shall call 'flips' F. The defects that correspond to the shifts are called dislocations, while those that correspond to the flips are the disclinations. The flips act on the shifts according to $FSF^{-1} = S^{-1}$, but this is the only relation and the fundamental group has the presentation $\langle S, F | FSF^{-1} = S^{-1} \rangle$.

In the classical formulation of the homotopy theory of defects, the allowed dislocations and disclinations in two dimensional smectics (and their properties) would correspond to this fundamental group. The main point is to understand that this structure is not correct. Mermin was the first to point out that disclinations of arbitrary negative strength can occur in smectics, but positive strength disclinations are restricted to winding number $+1/2$ or $+1$ [31]. In fact more than this is true, the disclinations are constrained in their geometry in a manner that they are not in a nematic. To illustrate, the $+1$ disclination has layers that are concentric circles and a radial director field; the converse configuration of a circular director field and radial layers entails uncontrolled variation in the layer spacing and so does not correspond to any physical smectic texture. This restriction on smectic disclinations was proved formally by Poénaru [61] as a theorem on measured foliations. We shall restrict ourselves here to a heuristic pictorial description introduced by Chen [24].

Smectics do not correspond to maps from the domain into the ground state manifold, as is the case for nematics. A way of thinking about this is in terms of Goldstone modes. As is well known, the translations and rotations in a smectic are not described by independent Goldstone modes; rather, both are captured by the Eulerian

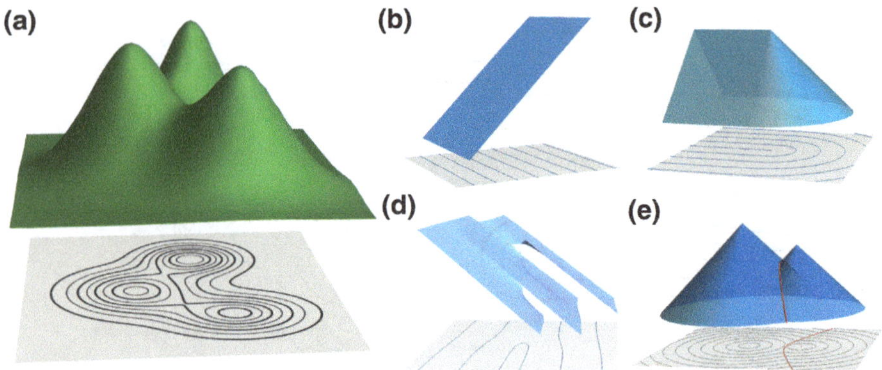

Fig. 9.10 Height functions and smectics. **a** A mountain range and its contour map. The contour map shares the properties of smectic textures. **b** The smectic ground state is a plane at 45° to the vertical. **c** A $+1/2$ disclination and its 'height function'. **d** The surface for a smectic dislocation is a tilted helicoid, or 'spiral staircase'. **e** Focal conics arise naturally from the intersections of mountains with constant slope

displacement field $u(\mathbf{x})$. Indeed smectic order is described entirely by a single function, the phase field $\phi(\mathbf{x})$ that enters the mass density wave. Locally, this is a linear function of position. The smectic layers are its level sets. Although it is common to depict the smectic only using the layers, as in Fig. 9.9, it is important to keep in mind that between the layers the phase field still exists and needs to be well-defined. The global structure can be represented by displaying the graph of the phase field. This is a surface in $\mathbb{R}^3$ with coordinates $(x, y, \phi(x, y))$; the level sets of ϕ – the smectic layers – are contours of the surface at fixed heights, equal to integer multiples of the layer spacing.

From this perspective, pictures of smectic layers are seen to share many of the same characteristics as topographical maps, which proves insightful in understanding disclinations in smectics, see Fig. 9.10. Mountain tops are high points that the contours of a topographical map loop around; they can be identified with $+1$ disclinations in smectics. Between the mountain tops are mountain passes and the contours around them have the saddle-like structure of -1 disclinations in smectics. Defects with higher negative winding are possible – they correspond to degenerate mountain passes that simultaneously connect three or more mountain peaks. However, disclinations with higher positive winding are not possible since combining two mountains does not produce anything new, only a bigger mountain. Or, what is the same, between any two mountain peaks there is always a mountain pass. It simply is not possible to combine two $+1$ disclinations only in a smectic as there is always a -1 disclination between them.

The height function picture provides visual intuition for the defects in two-dimensional smectics. The smectic ground state is a plane with 45° slope, whose contours at $\phi = na$ – the smectic layers – are equally spaced. This plane is not unique. The density ρ is unchanged by the replacements $\phi \to \phi + na$, for any integer n, and $\phi \to -\phi$, which therefore correspond to equivalent ground states. Defects in the smectic order correspond to connections between these equivalent ground states. Some examples of how this appears in the height function picture are shown in Fig. 9.10.

Immediate extension to three dimensions is hampered by the need to visualise a complicated three-dimensional surface in $\mathbb{R}^4$. Nonetheless, some progress has been made in two directions: screw dislocations [26, 62] and focal conics [63–65]. The latter correspond to the special class of surfaces where the slope is always 45°, giving equally spaced layers. Their geometry turns out to be the same as the geometry of null surfaces in Minkowski space-time [63].

9.6 Geometry of Line Fields

As remarked in the introduction, liquid crystals are highly geometrical and there is much to be gained from studying their geometry, including additional insight into topological properties [25, 66, 67]. For the most part, the geometry that we will describe now is that of vector fields, or more properly of vector bundles. Any liquid crystal texture defines two natural vector bundles as follows. At every point in the material the director picks out a preferred direction and so gives a canonical splitting of the tangent space to $\mathbb{R}^3$ into the line parallel to the director (a rank 1 vector space) and the 2-plane of all directions perpendicular to it (a rank 2 vector space). These two vector spaces vary smoothly so long as the director field does, so we have a pair of vector bundles over the set of points Ω where the director field is well-defined, i.e. a canonical splitting $T\mathbb{R}^3|_\Omega \cong L \oplus \xi$, where L is the line bundle and ξ the orthogonal 2-plane field. Some examples of this structure are shown in Fig. 9.11.

Gradients of the director field may be separated into those parallel to the local orientation and those along perpendicular directions, $\nabla \mathbf{n} = \nabla_{\parallel}\mathbf{n} + \nabla_{\perp}\mathbf{n}$. More concretely, as liquid crystals are uniaxial there is a local subgroup of the rotation group isomorphic to $SO(2)$ that preserves the director at any given point and the director gradients can be decomposed with respect to the action of this group

$$\partial_i n_j = n_i(n_k\partial_k)n_j + \frac{\nabla \cdot \mathbf{n}}{2}\big(\delta_{ij} - n_i n_j\big) + \frac{\mathbf{n} \cdot \nabla \times \mathbf{n}}{2}\epsilon_{ijk}n_k + \Delta_{ij}, \tag{9.15}$$

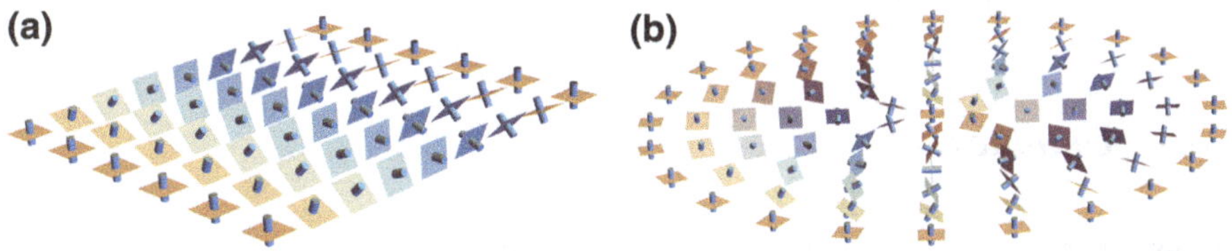

Fig. 9.11 Illustrations of plane fields and the geometry of liquid crystals. The examples shown are **a** the cholesteric ground state texture, **b** a Skyrmion

where the deviatoric part Δ_{ij} has the explicit form

$$\Delta_{ij} = \frac{1}{2}\big(\delta_{ik} - n_i n_k\big)\partial_k n_j + \frac{1}{2}\big(\delta_{jk} - n_j n_k\big)\partial_k n_i - \frac{\nabla \cdot \mathbf{n}}{2}\big(\delta_{ij} - n_i n_j\big). \quad (9.16)$$

The four pieces of the director gradients each have their own geometrical significance. The first is equivalent to the vector $\mathbf{b} = (\mathbf{n} \cdot \nabla)\mathbf{n}$ representing the bend distortions of the liquid crystal. It is a vector that is everywhere orthogonal to the director field and so always lies in the orthogonal 2-planes, i.e. it is a section of ξ. This means it is not a typical vector field. In particular, because it is an element of a rank 2 vector space it vanishes generically on sets of codimension 2, so that zeros of bend occur along one-dimensional lines. A typical vector field only vanishes at isolated points. Nonetheless, these zeros encode topological information about the nematic texture. For a regular (tangent) vector field, the zeros encode the Euler characteristic of the space by the celebrated Poincaré–Hopf index theorem [68]. The zeros of the bend vector realise the more general result that the zero locus of any section represents the Poincaré dual of the Euler class of the vector bundle [69, 70]. We do not develop this further here but focus instead on the orthogonal part of the director gradients where similar structures are present.

The last three terms in (9.15) constitute the 'shape operator' for the director field. The terminology is taken from the differential geometry of surfaces. The gradients of the surface normal as you move around on a surface are known as the shape operator, as they encode the 'shape' of the surface. It is a linear transformation on the tangent planes of the surface. In a smectic, $\mathbf{n}$ is the unit normal to a family of surfaces – the smectic layers – and the shape operator encodes the 'shape' of the layers. The expression (9.15) is the general case for arbitrary $\mathbf{n}$ and not just smectics. It is a linear transformation on the orthogonal 2-planes ξ.

The first part of the shape operator is isotropic in the orthogonal planes. Its magnitude is the splay of the director field, or its mean curvature. The second is also isotropic, but a pseudoscalar rather than a scalar. Its magnitude is the twist of the director field, or its mean torsion. As an operator, it acts on the orthogonal planes by a $\pi/2$ rotation. It is called a complex structure. The final term is the deviatoric part of the orthogonal director gradients and is a spin-2 object. Locally it is a 2×2 matrix of the form $\begin{bmatrix} \Delta_1 & \Delta_2 \\ \Delta_2 & -\Delta_1 \end{bmatrix}$ whose eigenvectors define the principal directions of curvature. Again, the terminology extends that of the differential geometry of surfaces, since in a smectic where $\mathbf{n}$ is the normal to a family of layers, these eigenvectors are precisely the directions of principal curvature of the smectic layers.

9.6.1 Umbilics

Points where the spin 2 field Δ vanishes are places where the shape operator is locally isotropic. In the classical differential geometry of surfaces such points are called umbilics and we adopt the same terminology here [66].

The umbilics of liquid crystal textures are codimension 2 objects – they require the vanishing of two real numbers Δ_1 and Δ_2 – and hence occur along extended one-dimensional curves, which endows them with some additional features compared to the umbilic points of surfaces. Near an umbilic Δ has the form $\Delta = |\Delta|\left[\begin{smallmatrix} \cos(\theta) & \sin(\theta) \\ \sin(\theta) & -\cos(\theta) \end{smallmatrix}\right]$, where $|\Delta|$ generically vanishes linearly. A loop that goes around the umbilic gives a map $\theta : S^1 \to S^1$, which carries an integer winding number. This may be visualised in terms of the eigenvectors of Δ, as is done for the umbilic points of surfaces. It is easy to see that the eigenvectors wind by half as much as Δ itself, so that around generic umbilics the profile is the same as the $\pm\frac{1}{2}$ disclinations of Fig. 9.1c. We show examples of this in Sect. 9.7.1.

Finally, we mention only in passing that if the umbilic forms a closed loop then the variation of θ around a longitude will convey a second integer winding number [66], which is related in part to the linking of umbilics.

9.6.2 Chirality Pseudotensor

One may compose the shape operator with the complex structure $(\epsilon_{ijk}n_k)$ to obtain another geometric linear transformation on ξ of some interest and importance. This is known as the chirality pseudotensor [25, 71]

$$(\nabla_\perp \mathbf{n})_{il}\epsilon_{ljk}n_k \equiv C_{ij} = -\frac{\mathbf{n}\cdot\nabla\times\mathbf{n}}{2}\big(\delta_{ij} - n_i n_j\big) + \frac{\nabla\cdot\mathbf{n}}{2}\epsilon_{ijk}n_k + \Pi_{ij}. \tag{9.17}$$

Its deviatoric part has the explicit expression

$$\begin{aligned}\Pi_{ij} = &\frac{1}{4}\epsilon_{ilk}\big[n_l\partial_k n_j + n_l\partial_j n_k - n_j n_l n_m \partial_m n_k\big] \\ &+ \frac{1}{4}\epsilon_{jlk}\big[n_l\partial_k n_i + n_l\partial_i n_k - n_i n_l n_m \partial_m n_k\big],\end{aligned} \tag{9.18}$$

and has the same basic properties as Δ_{ij}. For instance they both vanish in the same places – the umbilics – but nonetheless Π_{ij} is an independent quantity. Note in particular, that Π_{ij} is invariant under the nematic symmetry $\mathbf{n} \to -\mathbf{n}$ whereas Δ_{ij} is not, so that it is globally defined for line fields even though Δ_{ij} is not.

Insight into the geometric significance of the chirality pseudotensor comes from its eigenvectors. The eigenvectors of Δ_{ij} define directions of principal curvature for the director field. Those of Π_{ij} define directions of principal torsion, or twist. These are of course related since each defines a basis for the orthogonal planes ξ. The relation is a simple $\pi/4$ rotation; it is beautifully illustrated in the experiments of Armon et al. [72] on the opening of chiral seed pods, and of Efrati and Irvine [71] on the shapes of chiral elastic strips. The eigenvectors of Π_{ij} also serve as one definition of the pitch axis in cholesteric materials [25, 66, 71], as we now describe.

9.7 Cholesterics

Cholesterics differ from nematics in that they have an intrinsic handedness and propensity to twist. In terms of the Frank free energy (9.1), cholesterics are materials for which $q_0 \neq 0$. The minimiser, or cholesteric ground state, is the director field

$$\mathbf{n} = \cos(q_0 z)\,\mathbf{e}_x + \sin(q_0 z)\,\mathbf{e}_y, \tag{9.19}$$

or any equivalent to this. Its main feature, like the smectic ground state, is that it is not uniform; the director lies always in the xy-plane with its orientation rotating at a uniform rate along the z-direction. The axis of rotation is called the pitch axis and is everywhere orthogonal to the director.

One way of presenting the local order at any point in a cholesteric is to specify the director field $\mathbf{n}$ and also the direction of the pitch axis $\mathbf{p}$. Since these are orthogonal we also have the direction $\mathbf{n}_\perp = \mathbf{p} \times \mathbf{n}$ so that the local order in a cholesteric is an orthonormal frame. This is really an unoriented frame since the director is properly a line field. In this picture there are three distinct types of disclination. Defects in the director but not the pitch axis are called χ lines; defects in the pitch axis but not the director are called λ lines; and defects in both the director and the pitch are called τ lines [4, 73]. This triptych corresponds well with observations of defects in cholesterics [74, 75].

However, the description in terms of a local orthonormal frame has some shortcomings, for both in the form of the free energy and in performing numerical simulations, only the director field needs to be used; separate mention of the pitch axis is not needed, nor ever given. It would appear that all the necessary information is already present in the director field. This comes from the geometry of Sect. 9.6. For instance, the pitch axis can be identified with the positive eigenvector of the deviatoric part Π_{ij} of the chirality pseudotensor. To see this, consider the cholesteric texture (9.19) and introduce the basis $\mathbf{d}_1 = -\sin(q_0 z)\,\mathbf{e}_x + \cos(q_0 z)\,\mathbf{e}_y$, $\mathbf{d}_2 = \mathbf{e}_z$ for the orthogonal planes. We then have

$$\nabla\mathbf{n} = q_0 \mathbf{d}_2 \otimes \mathbf{d}_1 = -\frac{q_0}{2}\big[\mathbf{d}_1 \otimes \mathbf{d}_2 - \mathbf{d}_2 \otimes \mathbf{d}_1\big] + \frac{q_0}{2}\big[\mathbf{d}_1 \otimes \mathbf{d}_2 + \mathbf{d}_2 \otimes \mathbf{d}_1\big]. \tag{9.20}$$

Comparing with the general decomposition (9.15), we see that there is no bend, no splay and the twist is $-q_0$. The final term gives the deviatoric transformation Δ describing the principal curvatures. Transforming to its torsional counterpart we find

$$\Pi = \frac{q_0}{2}\big[\mathbf{d}_2 \otimes \mathbf{d}_2 - \mathbf{d}_1 \otimes \mathbf{d}_1\big], \tag{9.21}$$

which is diagonal, allowing the eigenvectors, and in particular the positive eigenvector $\mathbf{d}_2 = \mathbf{e}_z$, to be read off directly. We see that the positive eigenvector corresponds with the pitch axis.

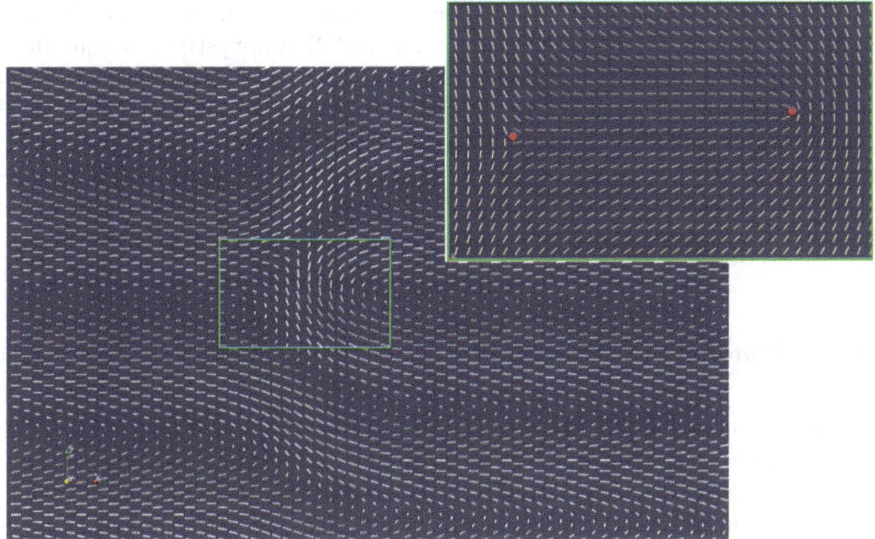

Fig. 9.12 λ lines in cholesterics. Main panel: director field of a 'dislocation' in a cholesteric. Inset: pitch axis, with the positions of the two λ lines indicated by the red dots

9.7.1 λ Lines: Defects in the Pitch

The pitch axis is defined everywhere that the eigenvectors of Π are defined, namely away from the umbilic lines. This gives a physical interpretation of umbilic lines in cholesterics as the λ defects in the pitch axis, features that are directly visible in optical microscopy. We show an example of a $\lambda^{\pm}$ pair, corresponding to a cholesteric 'dislocation', in Fig. 9.12. A second example is provided by the Skyrmion lattice of Fig. 9.6. Each Skyrmion is surrounded by six $-\frac{1}{2}$ profile λ lines and has at its centre a defect in the pitch with winding number $+1$, corresponding to a non-generic umbilic. In the liquid crystal literature this is called the axis of a double twist cylinder.

9.8 Knotted Fields

The experimental construction of knotted disclination lines in liquid crystals [11] opened up a new field in nematic topology. These were experiments with arrays of colloids in a twisted nematic, or cholesteric, cell. The disclinations that entangle the colloids can be manipulated with a laser tweezer to create any knot [12]. Subsequently knots have also been created in cholesteric droplets [76], handlebody droplets [77] and with Möbius strip colloids [53].

Knots are fascinating and almost endlessly varied, and bring with them the whole gamut of low-dimensional topology. This is a burgeoning field across many disciplines with examples of knotted continuous fields also arising in optics [78], fluid flows [79], superfluids [80, 81] and excitable media [82]. Amongst the basic questions that we might like to address are: What aspects of knot theory are directly pertinent to the physics of nematic liquid crystals? What properties do knotted nematics have? And, can we give explicit constructions of director fields with knotted disclination lines in them? The first two questions are answered partially by the homotopy classification of knotted nematics. The last can be accomplished, in part, by adapting Maxwell's descriptions of magnetic fields.

9.8.1 Homotopy Classification

We make no attempt to describe the details of the homotopy classification [83, 84] and provide only a concise summary of the main results. Fix a knotted disclination line, K, and assume, for simplicity, that the boundary conditions on the director field are that it is uniform (and aligned vertically) at large distances. The homotopy classes of such director fields are given by

$$[S^3 - K, \mathbb{RP}^2] \cong H_1\big(\Sigma(K); \mathbb{Z}\big)/(x \sim -x), \tag{9.22}$$

where $\Sigma(K)$ is the branched double cover of the knot complement, and the equivalence is as sets, i.e. without the group structure. This is a modestly sophisticated result. Part of its utility is that the homology group $H_1(\Sigma(K); \mathbb{Z})$ is relatively easy to compute from, say, any standard projection diagram for the knot [85]. For instance, for the figure-eight knot it is $\mathbb{Z}_5$, for the 7_6 knot it is $\mathbb{Z}_{19}$ and for the Kinoshita–Teresaka knot it is the one element group 0, the same as the unknot.

For any knot, the group $H_1(\Sigma(K); \mathbb{Z})$ has finite order. Consequently, there are only a finite number of homotopically distinct nematic textures associated to any knotted disclination. Links bring something new and in some cases may support an infinite number of homotopically distinct textures; perhaps the simplest example where this happens is the $(4, 4)$ torus link, for which $H_1(\Sigma(K); \mathbb{Z}) = \mathbb{Z}^2 \oplus \mathbb{Z}_2$. As one other additional property possessed by links, consider a Pontryagin–Thom surface for any link. In every homotopy class there is a representative where this is an orientable surface. The PT surface then induces pairwise linking numbers between the link components (that are independent of any choice of orientation for the PT surface) and for some links this distinguishes the homotopy classes. For instance, for the Hopf link there are two possibilities for the pairwise linking, $+1$ and -1, are these are homotopically distinct. Now the complement of the Hopf link has the homotopy type of a torus, T^2, so that these two pairwise linking numbers give another interpretation to the $\mathbb{Z}_2$ classification of Sect. 9.3.2.

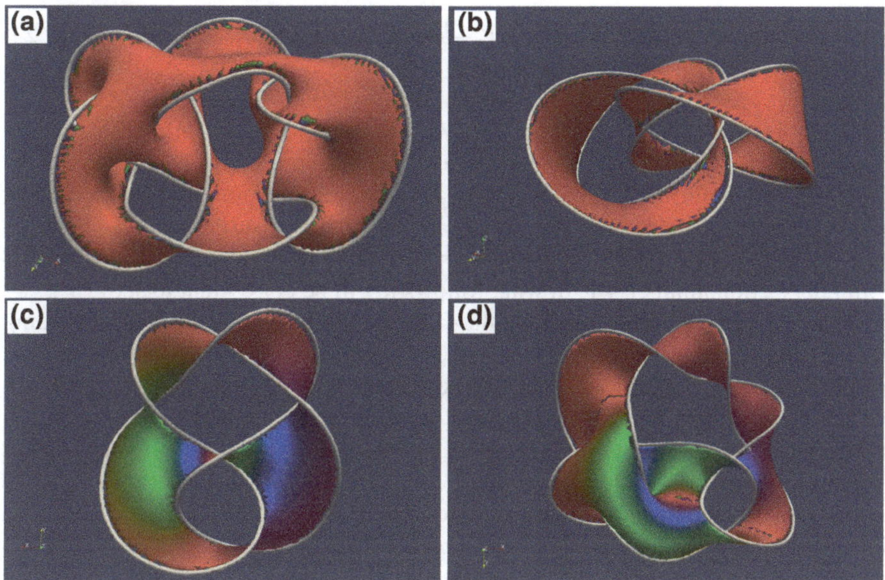

Fig. 9.13 Examples of knotted disclination lines in nematic liquid crystals. The Pontryagin–Thom surfaces are the level sets $\Phi = 0$ and are coloured according to the angle Θ. **a** Kinoshita–Teresaka knot, planar texture. **b** $(4, 4)$ torus link, planar texture. The PT surface consists of two interlocked Hopf link annuli. **c**, **d** Examples of non-planar textures on the complement of the 5_2 (**c**) and 8_8 (**d**) knots

9.8.2 Construction of Knots in Nematics

In his treatise on electricity and magnetism [86], Maxwell describes the magnetic fields created by current carrying wires. His constructions can be adapted to give explicit expressions for director fields containing knotted disclination lines. We refer to this as the Maxwell construction. Illustrations of it are given in Fig. 9.13.

A loop of wire K carrying a steady current I generates at any point $\mathbf{x}$ outside the wire a magnetic field given by the Biot–Savart formula

$$\mathbf{B}(\mathbf{x}) = \frac{I}{4\pi} \int_K \frac{\big(\mathbf{y}(s) - \mathbf{x}\big) \times d\mathbf{y}(s)}{|\mathbf{y}(s) - \mathbf{x}|^3}, \tag{9.23}$$

where s is arc length along the wire, whose position is denoted by $\mathbf{y}(s)$. Maxwell's equations imply that such a magnetic field is curl free and so may be represented locally as the gradient of a function, the magnetostatic potential. An application of Stokes' theorem to the Biot–Savart formula expresses this potential as an integral over any Seifert surface F for the wire

$$\mathbf{B}(\mathbf{x}) = \nabla\left(\frac{I}{4\pi}\int\limits_F \frac{\mathbf{x}-\mathbf{y}}{|\mathbf{x}-\mathbf{y}|^3}\cdot d\mathbf{S}\right) \equiv \nabla\Phi, \tag{9.24}$$

where $d\mathbf{S}$ is an element of surface area. We will refer to Φ as a phase field. It is a circle valued function ($\mathbb{R}/I\mathbb{Z}$) on the space outside the wire that is independent of the choice of Seifert surface, or, equivalently a real valued function on the complement of the Seifert surface that is independent of the choice of surface mod I. It winds by I on going once around any loop that encircles the wire, or pierces the surface F with intersection number $+1$. In what follows it will be convenient to set $I = 2\pi$.

We can use the phase field to orient the director in a liquid crystal texture according to

$$\mathbf{n}(\mathbf{x}) = \sin\bigl(\Phi(\mathbf{x})/2\bigr)\,\mathbf{e}_x + \cos\bigl(\Phi(\mathbf{x})/2\bigr)\,\mathbf{e}_z. \tag{9.25}$$

It then rotates by π along any loop that encircles the curve K once, which makes K a disclination line for the director field. This director field has several remarkable properties. It embeds an arbitrary knotted curve into a liquid crystal texture as a disclination loop. The texture is entirely planar, with an orientation that lies everywhere in the xz-plane and with boundary conditions that $\mathbf{n} \to \mathbf{e}_z$ at large distances from the knot, as in the construction we have given the phase field vanishes in the same limit. Finally, since $\nabla \cdot \mathbf{B} = 0$, Φ is a harmonic function and it follows that this director field is a critical point of the one-elastic-constant free energy, for a fixed defect set. Of course, in any situation in which the disclination loop is free to move it will do so, for instance to reduce its length, but if it is fixed then this director field is a minimiser. At the same time, the construction does present some limitations; for instance, it does not capture any geometry as the orientation of the liquid crystal is not directly correlated to the geometry of the curve. In particular, the choice of lying in the xz-plane is quite arbitrary.

We now describe briefly how to extend the construction to director fields that are not planar, and covering a representative of every homotopy class for any knot. Think of the Seifert surface F as a Pontryagin–Thom surface. In this planar representative it is monochrome. So what we need to do is describe how to colour it. The topologically interesting ways of colouring it are windings of the colour around the homology cycles of F. But the preceding Maxwell construction tells us how to generate such winding angles; we just need a current loop that is 'dual' to the winding of this angle. Maxwell might well have phrased the problem in the language of magnetic fields. If we wish to construct an angle that winds around some loop we need only thread that loop with a current carrying wire; the magnetic field that the wire generates will be the gradient of the angle that is desired.

In more formal terms, we have an instance of Alexander duality; to each homology cycle of F there corresponds a cycle in the complement of F that has linking number $+1$ with the given homology cycle of the surface and 0 with all others (in a basis for the homology).

So what we can do is choose a loop L, or collection of loops, in the complement of F, representing any homology cycle of the complement, and generate in the same manner as before a phase field Θ that winds by 2π around any closed curve that goes around the loop L. Then the director field

$$\mathbf{n} = \sin(\Phi/2)\big[\cos(\Theta)\,\mathbf{e}_x + \sin(\Theta)\,\mathbf{e}_y\big] + \cos(\Phi/2)\,\mathbf{e}_z, \tag{9.26}$$

corresponds to a texture with disclination loop K and Pontryagin–Thom surface F that is coloured by the angle Θ. It is known from the classification of knotted nematics [84] that this construction captures a representative of every distinct homotopy class. Not all such colour decorations generate topologically distinct textures. For instance, there is a unique homotopy class of nematic texture with a disclination loop in the form of the Kinoshita–Teresaka knot. This can be analysed in the same way as we did for maps $T^2 \to \mathbb{RP}^2$ in our discussion of disclination loops, by lifting to a double cover and considering equivariant homotopies. This process is described in [84], including with an explicit application to experiments on knotted defect lines in toroidal droplets [77].

Remark 1 For the director field (9.26) not to contain any artefacts the singularities in the angle Θ must lie in the surface $\Phi = 0$. However, for the purpose of initialising numerical simulations one does not need to be precise about this as any artefacts of the initialisation are removed at the first timestep of typical relaxation algorithms.

Remark 2 We have presented the phase field Φ in terms of an arbitrary choice of Seifert surface, F, as this proves convenient for our applications to liquid crystals. A representation that just uses the curve K was given by Maxwell in volume II of his treatise [86] and can be more convenient for other applications.

References

1. F.C. Frank, Discuss. Faraday Soc. **25**, 19 (1958)
2. G. Toulouse, M. Kléman, J. Phys. Lett. **37**, L149 (1976)
3. M. Kléman, L. Michel, G. Toulouse, J. Phys. Lett. **38**, L195 (1977)
4. Y. Bouligand, B. Derrida, V. Poénaru, Y. Pomeau, G. Toulouse, J. Phys. II Fr. **39**, 863 (1978)
5. E.M. Terentjev, Phys. Rev. E **51**, 1330 (1995)
6. P. Poulin, H. Stark, T.C. Lubensky, D.A. Weitz, Science **275**, 1770 (1997)
7. I. Muševič, M. Škarabot, U. Tkalec, M. Ravnik, S. Žumer, Science **313**, 954 (2006)
8. M. Ravnik, M. Škarabot, S. Žumer, U. Tkalec, I. Poberaj, D. Babič, N. Osterman, I. Muševič, Phys. Rev. Lett. **99**, 247801 (2007)
9. B.G. Chen, P.J. Ackerman, G.P. Alexander, R.D. Kamien, I.I. Smalyukh, Phys. Rev. Lett. **110**, 237801 (2013)
10. P.J. Ackerman, I.I. Smalyukh, Phys. Rev. X **7**, 011006 (2017)
11. U. Tkalec, M. Ravnik, S. Čopar, S. Žumer, I. Muševič, Science **333**, 62 (2011)
12. S. Čopar, U. Tkalec, I. Muševič, S. Žumer, Proc. Natl. Acad. Sci. U.S.A. **112**, 1675 (2015)
13. M. Ravnik, G.P. Alexander, J.M. Yeomans, S. Žumer, Proc. Natl. Acad. Sci. U.S.A. **108**, 5188 (2011)
14. A. Nych, U. Ognysta, M. Škarabot, M. Ravnik, S. Žumer, I. Muševič, Nat. Commun. **4**, 1489 (2013)
15. O.D. Lavrentovich, Proc. Natl. Acad. Sci. U.S.A. **108**, 5143 (2011)

16. I. Muševič, Philos. Trans. R. Soc. A **371**, 20120266 (2013)
17. P.G. de Gennes, J. Prost, *The Physics of Liquid Crystals*, 2nd edn. (Oxford University Press, Oxford, 1993)
18. P.M. Chaikin, T.C. Lubensky, *Principles of Condensed Matter Physics* (Cambridge University Press, Cambridge, 1995)
19. J. Schmidtke, W. Stille, H. Finkelmann, S.T. Kim, Adv. Mater. **14**, 746 (2002)
20. M. Humar, I. Muševič, Opt. Express **18**, 26995 (2011)
21. A.A. Michelson, Philos. Mag. **21**, 554 (1911)
22. V. Sharma, M. Crne, J.O. Park, M. Srinivasarao, Science **325**, 449 (2009)
23. J.P. Sethna, M. Huang, in *1991 Lectures in Complex Systems*, ed. by L. Nadal, D. Stein. SFI Studies in the Sciences of Complexity Proceedings, vol. XV (Addison Wesley, New York, 1992), p. 243
24. B.G. Chen, G.P. Alexander, R.D. Kamien, Proc. Natl. Acad. Sci. U.S.A. **106**, 15577 (2009)
25. D.A. Beller, T. Machon, S. Čopar, D.M. Sussman, G.P. Alexander, R.D. Kamien, R.A. Mosna, Phys. Rev. X **4**, 031050 (2014)
26. R.D. Kamien, R.A. Mosna, New J. Phys. **18**, 053012 (2016)
27. J. Milnor, *Topology from the Differentiable Viewpoint* (Princeton University Press, Princeton, 1965)
28. T. Tom Dieck, *Algebraic Topology* (American Mathematical Society, 2008)
29. R.B. Meyer, Philos. Mag. **27**, 405 (1973)
30. C. Williams, Y. Bouligand, J. Phys. Fr. **35**, 589 (1974)
31. N.D. Mermin, Rev. Mod. Phys. **51**, 591 (1979)
32. L. Michel, Rev. Mod. Phys. **52**, 617 (1980)
33. H.R. Trebin, Adv. Phys. **31**, 195 (1982)
34. M.V. Kurik, O.D. Lavrentovich, Sov. Phys. Usp. **31**, 196 (1988)
35. V. Poénaru, G. Toulouse, J. Phys. Fr. **38**, 887 (1977)
36. V. Poénaru, G. Toulouse, J. Math. Phys. **20**, 13 (1979)
37. S. Čopar, S. Žumer, Phys. Rev. Lett. **106**, 177801 (2011)
38. G.P. Alexander, B.G. Chen, E.A. Matsumoto, R.D. Kamien, Rev. Mod. Phys. **84**, 497 (2012)
39. K. Jänich, Acta Appl. Math. **8**, 65 (1987)
40. B.G. Chen, Dissertation, University of Pennsylvania (2012)
41. X.Z. Yu, Y. Onose, N. Kanazawa, J.H. Park, J.H. Han, Y. Matsui, N. Nagaosa, Y. Tokura, Nature **465**, 901 (2010)
42. X. Yu, M. Mostovoy, Y. Tokunaga, W. Zhang, K. Kimoto, Y. Matsui, Y. Kaneko, N. Nagaosa, Y. Tokura, Proc. Natl. Acad. Sci. U.S.A. **109**, 8856 (2012)
43. U.K. Rößler, A.N. Bogdanov, C. Pfleiderer, Nature **442**, 797 (2006)
44. S. Mühlbauer, B. Binz, F. Jonietz, C. Pfleiderer, A. Rosch, A. Neubauer, R. Georgii, P. Böni, Science **323**, 915 (2009)
45. A.N. Bogdanov, D.A. Yablonskii, Sov. Phys. JETP **68**, 101 (1989)
46. A.N. Bogdanov, A. Hubert, J. Magn. Magn. Mater. **138**, 255 (1994)
47. A.N. Bogdanov, JETP Lett. **62**, 247 (1995)
48. M.B. Pandey, T. Porenta, J. Brewer, A. Burkart, S. Čopar, S. Žumer, I.I. Smalyukh, Phys. Rev. E **89**, 060502 (2014)
49. P.J. Ackerman, R.P. Trivedi, B. Senyuk, J. van de Lagemaat, I.I. Smalyukh, Phys. Rev. E **90**, 012505 (2014)
50. A.O. Leonov, I.E. Dragunov, U.K. Rößler, A.N. Bogdanov, Phys. Rev. E **90**, 042502 (2014)
51. B. Senyuk, Q. Liu, S. He, R.D. Kamien, R.B. Kusner, T.C. Lubensky, I.I. Smalyukh, Nature **493**, 200 (2012)
52. M. Cavallaro Jr., M.A. Gharbi, D.A. Beller, S. Čopar, Z. Shi, R.D. Kamien, S. Yang, T. Baumbart, K.J. Stebe, Soft Matter **9**, 9099 (2013)
53. T. Machon, G.P. Alexander, Proc. Natl. Acad. Sci. U.S.A. **110**, 14174 (2013)
54. M. Škarabot, M. Ravnik, S. Žumer, U. Tkalec, I. Poberaj, D. Babič, N. Osterman, I. Muševič, Phys. Rev. E **76**, 051406 (2007)

55. M. Škarabot, M. Ravnik, S. Žumer, U. Tkalec, I. Poberaj, D. Babič, N. Osterman, I. Muševič, Phys. Rev. E **77**, 031705 (2008)
56. M. Ravnik, S. Žumer, Liq. Cryst. **36**, 1201 (2009)
57. U. Tkalec, M. Ravnik, S. Žumer, I. Muševič, Phys. Rev. Lett. **103**, 127801 (2009)
58. S. Čopar, T. Porenta, V.S.R. Jampani, I. Muševič, S. Žumer, Soft Matter **8**, 8595 (2012)
59. I.I. Smalyukh, Y. Lansac, N.A. Clark, R.P. Trivedi, Nat. Mater. **9**, 139 (2010)
60. P.J. Ackerman, I.I. Smalyukh, Nat. Mater. **16**, 426 (2016)
61. V. Poénaru, Commun. Math. Phys. **80**, 127 (1981)
62. H. Aharoni, T. Machon, R.D. Kamien, Phys. Rev. Lett. (2017). (to appear)
63. G.P. Alexander, B.G. Chen, E.A. Matsumoto, R.D. Kamien, Phys. Rev. Lett. **104**, 257802 (2010)
64. G.P. Alexander, R.D. Kamien, R.A. Mosna, Phys. Rev. E **85**, 050701(R) (2012)
65. D.B. Liarte, M. Bierbaum, R.A. Mosna, R.D. Kamien, J.P. Sethna, Phys. Rev. Lett. **116**, 147802 (2016)
66. T. Machon, G.P. Alexander, Phys. Rev. X **6**, 011033 (2016)
67. T.J. Machon, Dissertation, University of Warwick (2016)
68. H. Hopf, *Differential Geometry in the Large*, 2nd edn. (Springer, Berlin, 1989)
69. J.W. Milnor, J.D. Stasheff, *Characteristic Classes* (Princeton University Press, Princeton, 1974)
70. R. Bott, L.W. Tu, *Differential Forms in Algebraic Topology* (Springer, New York, 1982)
71. E. Efrati, W.T.M. Irvine, Phys. Rev. X **4**, 011003 (2014)
72. S. Armon, E. Efrati, R. Kupferman, E. Sharon, Science **333**, 1726 (2011)
73. M. Kléman, G. Toulouse, J. Phys. Colloq. Paris **30**, C4 (1969)
74. Y. Bouligand, J. Phys. II Fr. **35**, 959 (1974)
75. O.D. Lavrentovich, M. Kléman, in *Chirality in Liquid Crystals*, ed. by H.S. Kitzerow, C. Bahr (Springer, Berlin, 2001)
76. D. Seč, S. Čopar, S. Žumer, Nat. Commun. **5**, 3057 (2014)
77. M. Tasinkevych, M.G. Campbell, I.I. Smalyukh, Proc. Natl. Acad. Sci. U.S.A. **111**, 16268 (2014)
78. M.R. Dennis, R.P. King, B. Jack, K. O'Holleran, M.J. Padgett, Nat. Phys. **6**, 118 (2010)
79. D. Kleckner, W.T.M. Irvine, Nat. Phys. **9**, 253 (2013)
80. D. Proment, M. Onorato, C.F. Barenghi, Phys. Rev. E **85**, 036306 (2012)
81. H. Salman, Phys. Rev. Lett. **111**, 165301 (2013)
82. F. Maucher, P.M. Sutcliffe, Phys. Rev. Lett. **116**, 178101 (2016)
83. T. Machon, G.P. Alexander, Phys. Rev. Lett. **113**, 027801 (2014)
84. T. Machon, G.P. Alexander, Proc. R. Soc. A **472**, 20160265 (2016)
85. W.B.R. Lickorish, *An Introduction to Knot Theory* (Springer, New York, 1997)
86. J.C. Maxwell, *A Treatise on Electricity and Magnetism*, vol. I–II (Dover Publications Inc., USA, 1954)

Chapter 10
Topologically Complex Morphologies in Block Copolymer Melts

J. J. K. Kirkensgaard

Abstract Polymers are macromolecules built from chains of subunits. Most synthetic polymers are built from a single subunit, the monomer, and are termed homopolymers. The connection of two or more homopolymer chains into a larger macromolecule is termed a *block copolymer* and these can be made with multiple components connected into both linear or branched molecular architectures. Block copolymers remain a subject of significant research interest owing to the control and reproducibility of physical properties and the many fascinating nanoscale structures which can be obtained *via* self-assembly. The self-assembly behaviour of block copolymers originate from the tendency of the various polymer chains to undergo phase separation which is inherently constrained due to the molecular connectivity. This leads to the formation of ordered mesostructures with characteristic length scales on the order of the chain sizes, typically tens of nanometers. Here the focus is on the molecular architecture as a topological variable and how it influences the morphologies one finds in self-assembled block copolymer systems. We present a range of examples of morphologies with different and sometimes very complex mesoscale topology, i.e. patterns which emerges from the tendency of these molecules to undergo spatial phase separation.

10.1 Introduction

Polymers are macromolecules built from chains of subunits. Naturally occurring examples of polymers include DNA and proteins built from chains of nucleic and amino acids respectively or cellulose built from chains of connected glucose units. Most synthetic polymers are built from a single subunit, the monomer, and are termed

J. J. K. Kirkensgaard (✉)
Niels Bohr Institute, University of Copenhagen, Copenhagen, Denmark
e-mail: jjkk@nbi.dk

S. Gupta and A. Saxena (eds.), *The Role of Topology in Materials*,
Springer Series in Solid-State Sciences 189,
https://doi.org/10.1007/978-3-319-76596-9_10

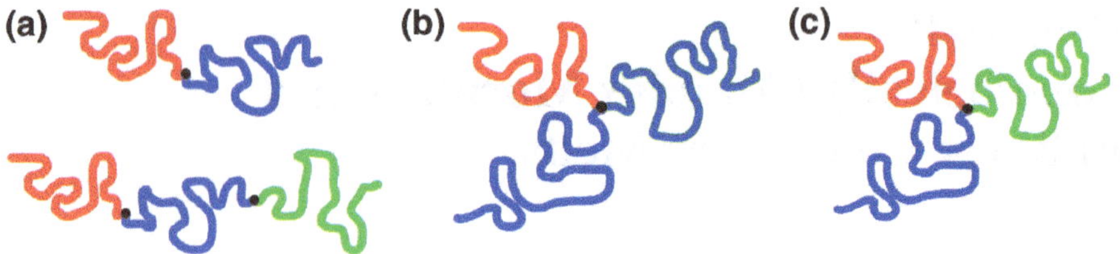

Fig. 10.1 Block copolymers. Covalent bonding of different polymer chains A, B, C, ...results in block copolymers of varying molecular architecture, here **a** linear AB diblock copolymers and ABC triblock terpolymers, **b** star-shaped AB_2 miktoarm copolymer and **c** ABC star miktoarm terpolymer

homopolymers. The connection of two or more homopolymer chains into a larger macromolecule is termed a *block copolymer* and these can be made with multiple components connected into both linear or branched molecular architectures as illustrated in Fig. 10.1. Block copolymers remain a subject of significant research interest owing to the control and reproducibility of physical properties and the many fascinating nanoscale structures which can be obtained [1]. The self-assembly behavior of block copolymers originate from the tendency of the various polymer chains to undergo phase separation which is however inherently constrained due to the molecular connectivity. This leads to the formation of ordered mesostructures with characteristic length scales on the order of the chain sizes, typically tens of nanometers. The formation of ordered structures in block copolymer systems stems from a competition between two effects: the entropic penalty associated with chain stretching and compression causing a preference for domains with constant thickness, and an enthalpic penalty associated with interfacial energy causing a preference for domain shapes which minimise surface area. In equilibrium a compromise between these factors is achieved by forming interfacial surfaces which tend to have approximately constant mean curvature. The nanostructures described below represent a tremendous potential for future technological applications because they provide a bottom-up route to materials with tailored optical, mechanical, electrical and photovoltaic properties (and combinations thereof), for example through phase selective chemistries or selective sequential removal of the blocks.

In this chapter two notions of topology will be relevant: First we will talk about the topology of the macromolecules, i.e. the arrangement of chains in the individual copolymers, but we will use the term 'molecular architecture' to describe this. We will focus in detail on the molecular architecture as a topological variable and how it influences the morphologies one finds in self-assembled block copolymer systems. Secondly, we present a range of examples of morphologies with different and sometimes very complex mesoscale topology, i.e. patterns which emerges from the tendency of these molecules to undergo spatial phase separation under the constraint of the molecular connectivity and which show periodicities on the length scale of the chain sizes as described above. We will restrict ourselves to looking at polymer

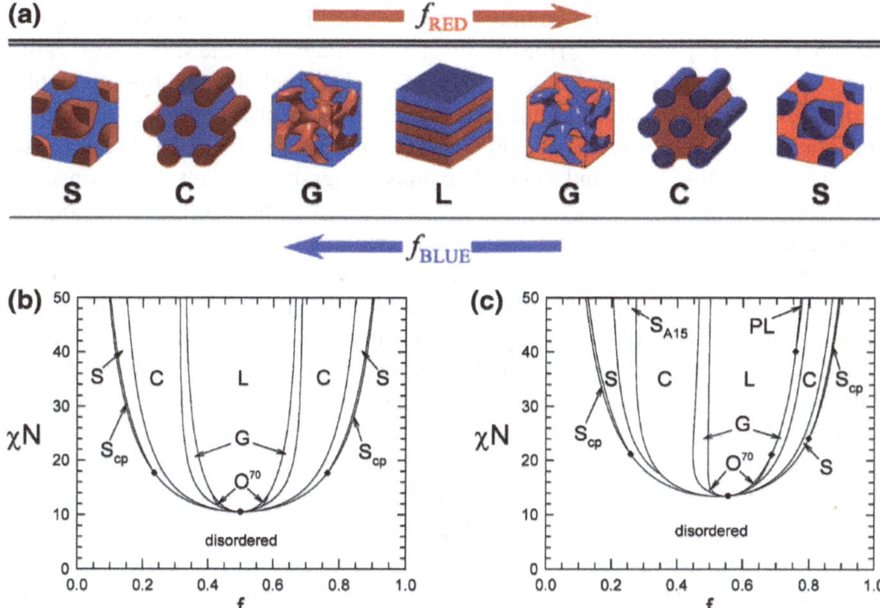

Fig. 10.2 Phase diagrams of AB-type block copolymers. **a** Morphologies found in AB diblock systems. S: cubic sphere packing, C: hexagonal cylinders, G: bicontinuous double gyroid, L: Lamellar. Figure from [2]. **b** Theoretical phase diagram for AB diblock copolymers. The O^{70} phase is an orthorhombic network phase described below. **c** Theoretical phase diagram for branched AB_2 miktoarm star copolymer. Phase diagrams from [3]

melts, i.e. polymers in a liquid state above the glass transition temperature without any added solvent.

10.2 AB Block Copolymers

The simplest block copolymers are AB diblocks where two polymer chains A and B are connected at a single junction point (see Fig. 10.1a). Diblock copolymers have been studied for decades and their self-assembly are generally well understood [3]. The structural phase behaviour of AB diblock copolymers is usually described as a function of two parameters: the composition, i.e. the relative volume fractions of the two components, and the degree of segregation described by the product χN where χ is the Flory-Huggins interaction parameter describing the chemical incompatibility between the different chains and N is the degree of polymerisation (the length of the polymer) [4, 5]. A characteristic feature of diblock copolymer self-assembly is that for increasing segregation, the phase behaviour becomes dominated by the composition. As a function of the composition described by the A component volume fraction $f = f_A$ ($f_B = 1 - f_A$) the universal phase diagram in the strong segregation limit consists of four ordered morphologies namely lamellar, bicontinuous double gyroid, hexagonally arranged cylinders and a cubic bcc sphere packing, all in principle appearing symmetrically around a 50/50 composition ($f = 0.5$). These phases are illustrated in Fig. 10.2a and a complete phase diagram based on a self-consistent field theory prediction is shown in Fig. 10.2b. The χ parameter for a given polymer pair is usually temperature dependent so that one can decrease the segregation by raising the temperature (and vice versa). This means that at $f = 0.4$ for example one should in principle be able to find phase transitions from L $\rightarrow$ G $\rightarrow$ C $\rightarrow$ disorder by increasing temperature (thus moving vertically in the phase diagram).[1] The latter is termed the order-disorder transition and the others order-order transitions. Inspection of the morphologies in Fig. 10.2a reveal that these phases are either based on simple topologies like planes, cylinders, spheres or in the case of the double gyroid, the more complex topology of networks whose midsurface is describable as a minimal surface [6]. Whichever it is, an order-order transition often necessitates a change of topology. Here we will not discuss such phase transitions in detail, but will rather focus on the effect of changing the molecular architecture and composition.

In Fig. 10.2c a phase diagram of two-component AB_2 miktoarm stars are shown. For a given composition quantified by f, the change in molecular architecture induces increased interfacial curvature leading to shifts in the phase diagram, increasing for example the regions of spherical packings by introducing the A15 phase (see Fig. 10.3) and also stabilising another new morphology, the perforated lamellae (PL) where lamellar sheets of the B-component is protruded by hexagonally arranged pillars of the majority matrix component A. Also, the O^{70} network phase exhibits a larger region on the $f > 0.5$ side of the diagram with the B-component forming the network and the A-component the matrix [3]. As we shall see below, changing the molecular architecture allows the formation of spectacular structures when increasing the number of components to more than two. The formation of low symmetry sphere packings is an ongoing topic in soft matter self-assembly [9] with a prominent example found in a block copolymer melt, namely the discovery of a large unit cell tetragonal structure known from metal alloys as the σ Frank–Kasper phase [8], see Fig. 10.3b, c. The low symmetry sphere packings like A15 and the σ phase are known as approximants to aperiodic quasicrystalline arrangements characterised by rotational symmetry, but not translational symmetry. The finding of the σ phase recently led to the discovery of a long-lived metastable dodecagonal quasicrystalline phase in a block copolymer system [10].

[1] There are a number of practical subtleties associated with this statement. First of all it requires that the molecular weight (or N) is not so large that the temperatures required to reach the transitions disintegrates the molecules, and second, the temperature range has to be above the glass transition temperature T_g which is a property of the specific chains. We will assume we are in a region of size and temperature where the notion of phase transitions makes sense. Note however that in structural studies of block copolymer morphologies one often utilises the glass transition of one or more of the chains to effectively 'freeze' a given structure by a rapid temperature quench.

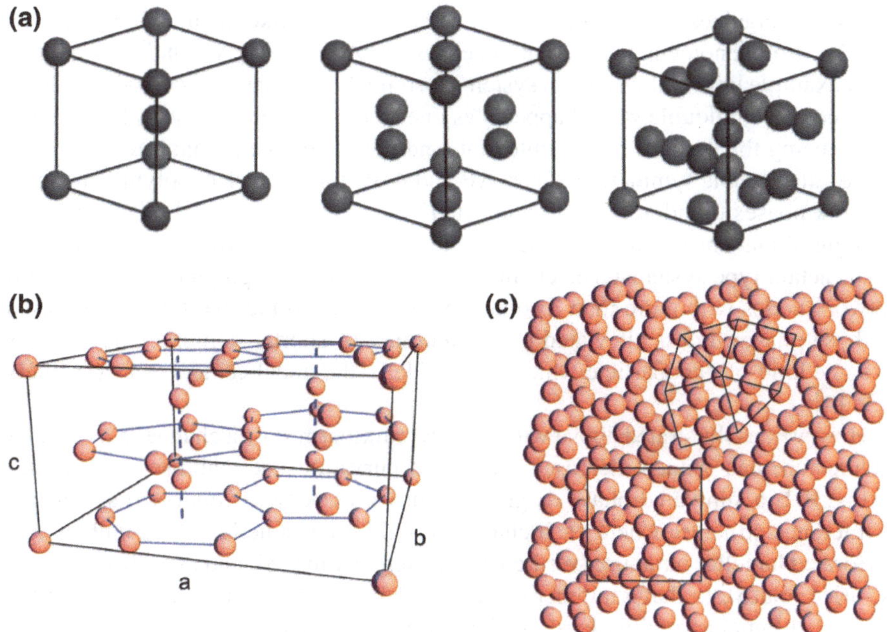

Fig. 10.3 Sphere packings found in block copolymer systems. **a** Cubic packings BCC, FCC and A15. Figure from [7]. **b, c** Frank–Kasper σ-phase. Figure from [8]

10.3 ABC Block Copolymers

The addition of a third component dramatically increases the morphological phase space as seen with linear ABC terpolymers where a large number of different structures have been found [1, 11]. A unifying feature between the structures formed in linear ABC systems and those found in simpler AB systems is that they are all characterised by the *interfaces* of each pair of polymer species. So speaking in terms of structural motifs, linear ABC block copolymers also explore variations of the surface topologies mentioned above: sphere, cylinder, plane, networks/minimal surface, but in multicolor versions, see Fig. 10.4 for a few examples. Nevertheless, the addition of a third component still allows more detailed control of the resulting morphologies and also opens up regions in the phase diagram of new and complex phases. An example is the poly(isoprene-b-styrene-b-ethylene oxide) linear triblock terpolymer system (ISO) studied experimentally and theoretically by Bates and colleagues [12] where several complex network phases are shown to form in the phase region between 2- and 3-colored lamellar structures, see Fig. 10.4b, c. The orthorhombic O^{70} phase is a single net structure while the Q^{230} and Q^{214} phases are both of the double gyroid type. The difference is the molecular packing: in the Q^{230} the nets are symmetric while in Q^{214} the two nets are built from different chemical species lowering the symmetry to the chiral subgroup since each of the double gyroid nets are chiral enantiomers (of opposite handedness). The formation of network phases like the double gyroid is a well-known phenomena in soft matter self-assembly where it is found ubiquitously in for example lipid and surfactant systems forming lyotropic liquid crystals [6]. Here the 3-connected double gyroid appears as one of typically three phases formed, the others being the 4-connected double diamond phase and the 6-connected primitive phase - all of cubic symmetry. In pure AB and ABC linear block copolymer melts all network phases found is of the 3-connected kind which is ultimately a result of the configurational entropy loss associated with chain stretching which is not penalised in surfactant type systems: The chain stretching required in polymer systems to fill network nodes with more than 3 connectors becomes prohibitive for the formation of those phases unless alleviated by the addition of shorter homopolymers chains or some kind of nanoparticle which can reside in the nodal centers as space fillers [13, 14].

However, as illustrated above with the AB_2 miktoarm stars a change in molecular architecture can induce increased interfacial curvature and stabilise new phases. In Fig. 10.5 a simulated phase diagram is shown of $A(BC)_2$ miktoarm star melts. A reference structure where each chain is roughly the same length assembles to a perforated lamellae structure (verified experimentally in [15]) and as each arm length is varied relative to that a number of new structures appear on the periphery of the perforated lamellae region. First, as the A arm is shortened, a single gyroid phase emerges (GL_{AB}) - note that this is a chiral structure - while for longer A arms a double diamond phase is stabilised. Increasing the middle B-block leads to a hybrid structure with the red A component forming a spherical packing while the green C species forms a 3-connected gyroid-like network. Note that because of the multicomponent nature of the molecules, the middle B blocks in these phases form topologically very complex continuous morphologies with channels and cavities.

As mentioned above, a unifying feature between the structures formed in the linear AB and ABC systems shown in Fig. 10.1 is that they are all characterised by the interfaces of each pair of polymer species. Although the $A(BC)_2$ miktoarm stars allow the formation of a range of topologically complex phases, they can still effectively be though of as a linear ABC molecule only with added splay due to the two diblock chains. This is because the topology of the star branch point only involves two of the chain species and so effectively acts in the same way as the connection of a linear molecule. However, another option when adding a third component is to make a star-like topology where all species meet at the branch point. Such molecules are called ABC miktoarm star terpolymers (Fig. 10.1). The self-assembly behaviour of ABC stars have been investigated experimentally [17] and theoretically [18, 21–23]. The generic theoretical phase diagram as described by these sources is shown in Fig. 10.6 under the compositional constraint of two components occupying equal volume fractions and of symmetric interaction parameters between the different polymer species. Compared with the self-assembly of linear block copolymers a fundamental result appears despite these severe constraints: dictated by the molecular star topology ABC lines are formed where the three different interfaces between AB, AC and BC meet [18–21]. As a consequence, a sequence of columnar structures with cross-sections following various polygonal tiling patterns appears. This sequence of

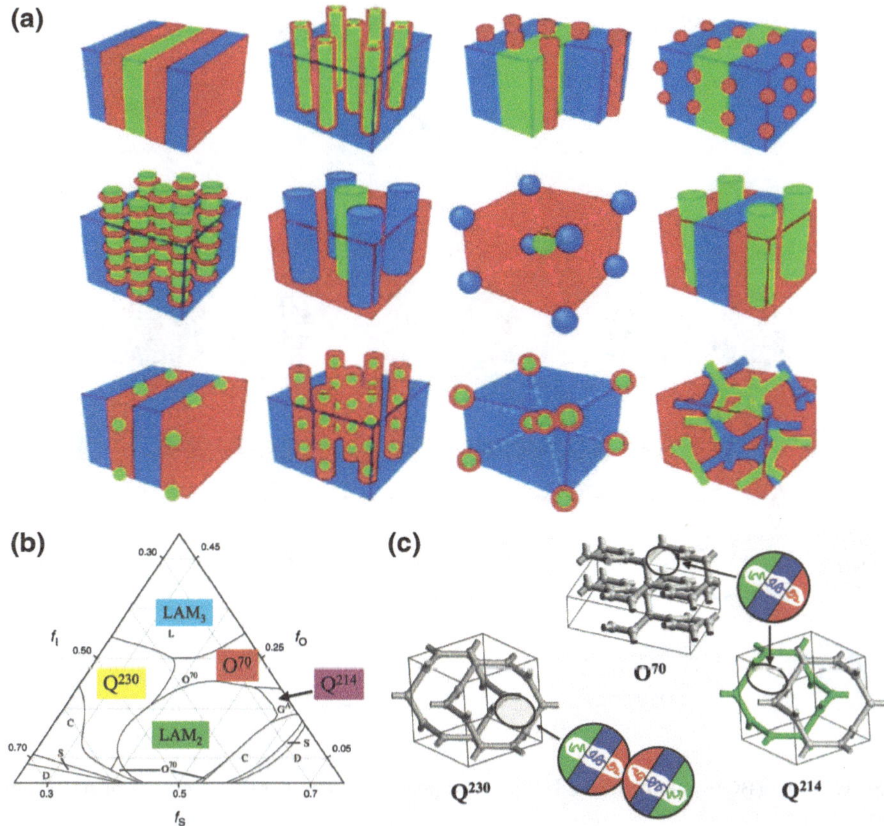

Fig. 10.4 Linear ABC triblock copolymer phases. **a** A selection of structures formed in linear ABC systems. Figure from [11]. **b** Theoretical ternary phase diagram showing the appearance of three different 3-connected network phases [12], the orthorhombic O^{70} phase and the two cubic double gyroid phases Q^{230} and Q^{214}. **c** Network representations and local molecular configurations of the networks from **b**. Figures (b, c) from [12]

tilings has been predicted from all the earlier mentioned theoretical studies and has been found in a number of experimental ABC 3-miktoarm star terpolymer systems [17].

Again, one can influence the interfacial curvature by altering the molecular architecture. In Fig. 10.7 the composition of ABC stars is altered by adding chains of equal length instead of increasing the length of one chain. Direct comparison with the ABC star phase diagram in Fig. 10.6 shows that the tiling patterns at $x = 2$ and $x = 3$ are now replaced with new decorations of the bicontinuous motifs of the diamond and gyroid network patterns. In these new structures one of the two nets are now built from alternating globular domains of the minority components. The striped double diamond has been found experimentally in a blend system as described below [24].

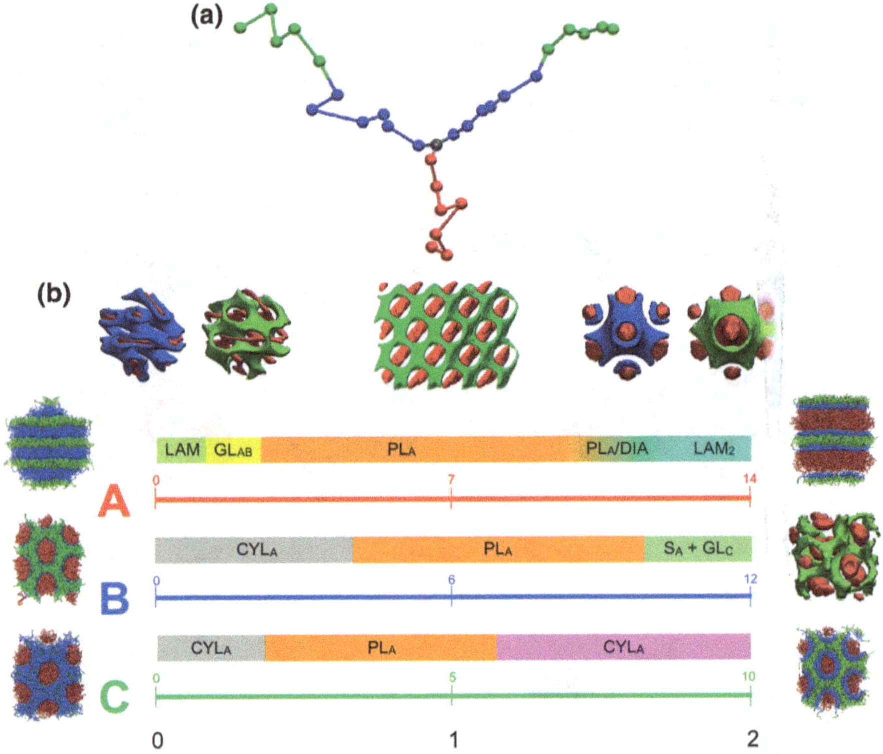

Fig. 10.5 **a** $A(BC)_2$ miktoarm star. **b** Simulated phase diagram of $A(BC)_2$ miktoarm stars from [16]

As mentioned above, the connectivity of ABC stars sets a topological constraint on the possible self-assembly morphologies but also leads to new possibilities. One very exciting option set forth by Hyde and co-workers is the formation of new tri- and polycontinuous patterns [20, 25]. They showed that such patterns were topologically consistent with the star molecular architecture and hypothesised their formation in star systems. A number of candidate structures were suggested (see Fig. 10.8) and evaluated in terms of energetics but any of them remains to be found experimentally in an ABC star system.[2] However, by suitably adjusting the molecular architecture and chemistry of ABC stars it turns out one can favour a thermodynamically stable tricontinuous structure based on three intertwined so-called ths-nets as was demonstrated using self-consistent field theory in [28] (see Fig. 10.9). This is a spectacular network structure effectively carving up space into three separated congruent labyrinths, each

[2]One of the predicted tricontinuous patterns have in fact been identified in both a hard and a soft matter context. In [26] such a pattern was found in a mesoporous silica and the same structure was later identified in a lyotropic liquid crystalline surfactant system [27]. However, in those cases the channels all contain the same material unlike the structures described here which have a different chemical species inside each channel.

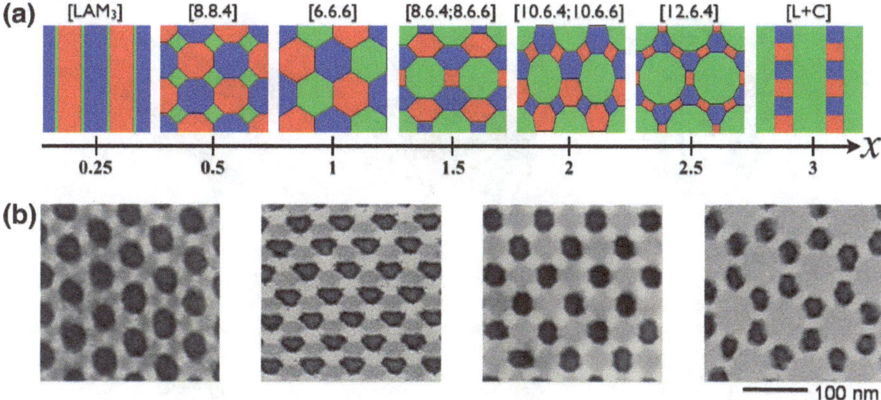

Fig. 10.6 **a** Generic phase diagram for ABC 3-miktoarm star terpolymers. The A and B components are constrained to occupy equal volume fractions and interactions between all unlike components are symmetric [18, 21–23]. The different phases are placed at their approximate compositional positions quantified by the parameter x, the volume ratio of the C and A (= B) components. The tilings are denoted by their Schläfli symbol [18, 21], a set of numbers $[k_1.k_2.\dots.k_l]$ indicating that a vertex is surrounded by a k_1-gon, a k_2-gon,…in cyclic order. Tilings with more than one topologically distinct vertex are named as $[k_1.k_2.k_3; k_4.k_5.k_6]$. Color code: A: red, B: blue, C: green. **b** Examples of tiling patterns from poly(isoprene-b-styrene-b-2-vinylpyridine) miktoarm star terpolymers visualized by transmission electron microscopy. Images from [17]

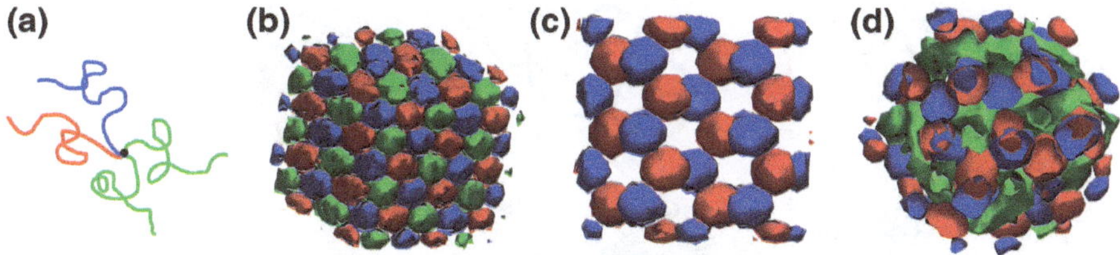

Fig. 10.7 Varying the composition by adding chains of equal length instead of increasing the length of one chain results in completely new structures [23]. **a** An ABC_2 star ($x = 2$). **b** For $x = 1$ the result is still the [6.6.6] tiling. **c** For $x = 2$ a double diamond network structure is found with one net built from alternating A and B domains. **d** For $x = 3$ the system also forms a striped network structure but now with the topology of the double gyroid structure with one net built from alternating red and blue domains. Figures from [23]

with a separate chemistry and thus properties. The key to stabilising this structure stems from the introduction of an extended core which effectively alters the balance between entropic and enthalpic free energy contributions, ultimately allowing this new pattern to outfavour the prismatic hexagonal honeycomb.

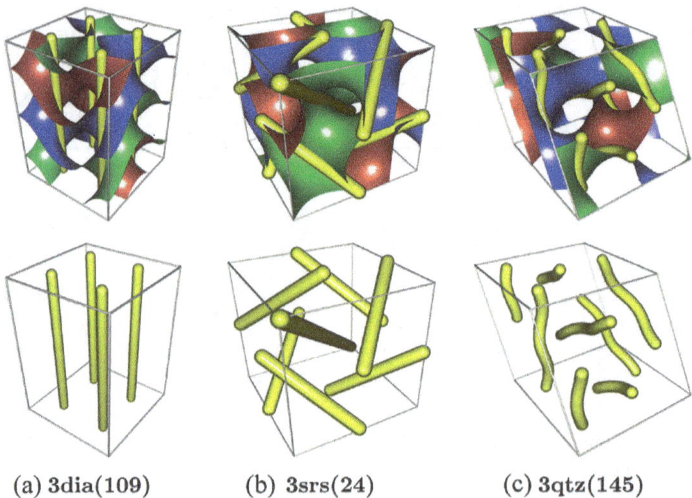

Fig. 10.8 Examples of tricontinuous candidate patterns topologically consistent with the star molecular architecture [20]. Yellow lines indicate the triple lines along which the star molecular cores pack. **a** 3 intertwined diamond nets. **b** 3 intertwined gyroidal srs-nets (all of same handedness, so a chiral structure). **c** 3 intertwined quartz qtz-nets (also chiral). Figure from [28]

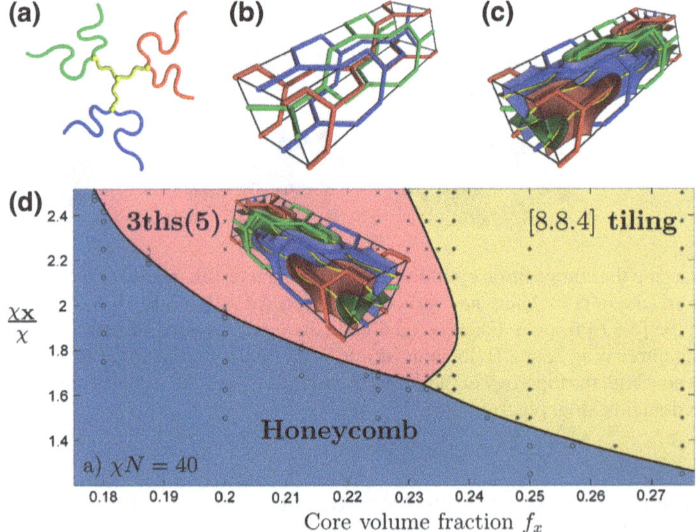

Fig. 10.9 **a** A dual chain core ABC star triblock copolymer. **b** Triply intergrown *ths*-nets. **c** As in **b** but showing the interfaces also. **d** Phase diagram from [28]: As a function of the core volume fraction and the interaction strength three structures dominate the phase diagram, one of which is a spectacular tricontinous network structure

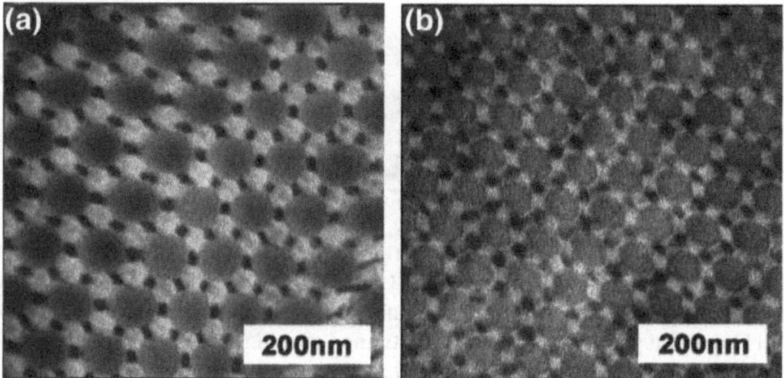

Fig. 10.10 Blending block copolymers. **a** An asymmetric polystyrene-b-polybutadiene-b-poly(2-vinylpyridine) ($S_{34}B_{11}V_{55}$) miktoarm star terpolymer forms a [12.6.4] tiling on its own (subscripts indicate molecular weight). The minority B component forms the dark 4-sided domains while the S component forms the white hexagonal domains in the tiling. **b** When blended with a $S_{45}V_{55}$ diblock copolymer the tiling pattern remains [12.6.4] but the chemical nature of the hexagons and squares swap so that B now forms hexagons and S forms squares. Figures from [29]

10.4 Blending Molecular Architectures

Blending polymers has been a heavily used strategy for many years to obtain a material with new properties akin to alloys in metallurgy. This also applies to block copolymers where one approach is to swell a particular domain with shorter homopolymer chains of the same species - in analogy to swelling in lyotropic liquid crystalline systems. For example, blending an ABC star which alone forms the [6.6.6] tiling pattern with C homopolymer chains can lead to a zinc-blende structure with alternating AB domains building up a diamond network [24] as the one illustrated in Fig. 10.7c above. If the swelling agent is not a homopolymer chain but another block copolymer, compatibility between pairs of blocks of both molecules becomes another control parameter for structure formation. An example is shown in Fig. 10.10 where blending an ABC star which forms the [12.6.4] tiling alone with AB diblock copolymers causes the A and B components to swap polygonal symmetry positions in the tiling pattern [29]. Thus, blending opens up the possibility of fine-tuning the structures found in the pure systems or allowing completely new patterns to appear. However, the possible phase space of a blend of different block copolymers stars is enormous. The main variables in play are (i) the molecular topology, i.e. the connectivity of the chains, (ii) the composition, i.e. the volume fractions of the different chains including the blend ratio, (iii) the chemical nature of the different polymer species, i.e. their mutual interaction parameters. Secondary variables can be polydispersity of the chains and chain flexibility for example [1].

In Fig. 10.11 an example is shown where a series of spectacular structures are predicted to form in blends of two different miktoarm stars [30]. Blending ABC and ABD star triblocks in a 50/50 ratio and with the majority domains (green C + yellow D) 2–7 times longer than the minority components (red A + blue B) a series

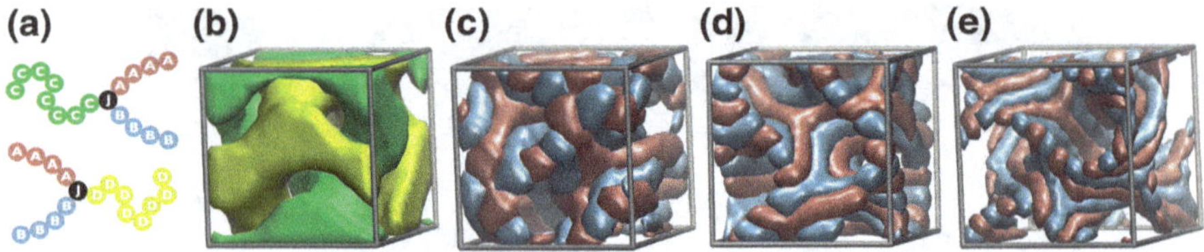

Fig. 10.11 **a** Blending ABC and ABD 3-miktoarm star terpolymers. All molecules contain equal sized A (red) and B (blue) arms, and longer C (green) and D (yellow) arms, but also of equal size. The parameter x (equal to $8/4 = 2$ in this image), corresponds to the number ratio of C to A beads. **b** C and D domain geometry, a pair of intertwined gyroid nets. **c–g** Single unit cell snapshots illustrating the curved striped pattern formed by the minority components A and B for varying x. **c** $x = 2$, **d** $x = 3.67$, **e** $x = 6$. Note the 3-fold branching for all values of x. Figures from [30]

of complex chiral network structures are predicted to form. The C and D species form two chiral enantiomeric nets separated by a membrane following the gyroid surface. As the C and D chains grow, the overall structure remains a chiral gyroid, but unlike all hitherto found gyroid(-like) structures, the channels in these new structures constitute majority domains resulting in the formation of a thin hyperbolic film.

Depending on the composition, interactions and molecular architectures in play, the minority components form remarkable structures on this hyperbolic curved film, in particular ordered branched structures which can be thought of as 'hyperbolic lamellae'. It is important to realise that if the film was flat, regular lamellae would form, but because of the curvature of the film branching has to occur. Remarkably, it turns out that these branched patterns are theoretically related to a family of tilings in the hyperbolic plane which when embedded on the gyroid surface yields distinct interwoven chiral nets, always of the same handedness and which changes topology systematically as a function of composition. The non-chiral nature of the block copolymer preclude a preferred handedness, which in the numerical simulations results in domains of opposite chirality forming, ultimately giving rise to defects, see Fig. 10.12. Nevertheless, the ideal mesostructures of these patterns in three-dimensional space are spectacular and extraordinarily complex. The minority components form multiple threaded chiral nets (all of equal handedness) whose topologies are that of the gyroid net. The number of disjoint nets making up the hyperbolic film can be up to 54 [30]. These intricate self-assemblies of liquid-like domains thus rival the complex interwoven networks found in synthetic metal-organic frameworks [31].

10.5 Concluding Remarks

The role of molecular architecture (chain topology) has been shown to be a defining variable influencing the resulting self-assembly morphologies in block copolymer melts and a number of patterns with complex topology has been presented from

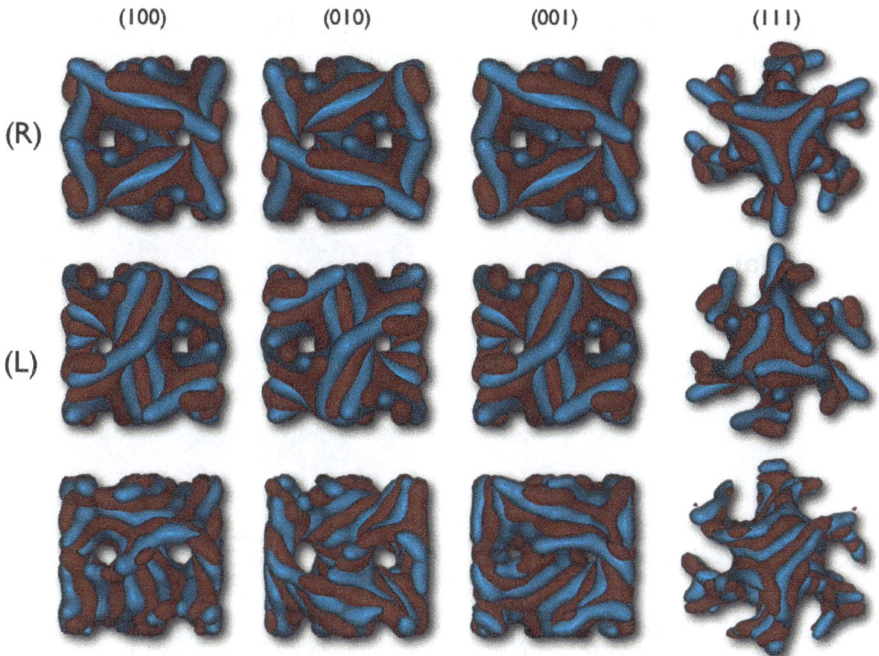

Fig. 10.12 Comparison of ideal left- (L) and right-handed (R) patterns (top and middle rows) viewed from various directions with a self-assembled morphology formed in a simulated mixture of terpolymers with $x = 4$ (bottom row). The simulated morphologies can be seen to match both of the ideal left- and right-handed patterns in distinct patches of the unit cell and are therefore of mixed chirality

experiments and simulations. From a topological perspective perhaps the simplest way to condense the message presented here is shown in Fig. 10.13. Here symmetric AB diblocks are compared as a 2-star with an ABC 3-star and a ABCD 4-star and it is clear that the interfacial dimensionality changes as the molecular topology is altered. Each pair of species defines a surface and as the surfaces have to meet in space we go from 2D interfaces to 1D line interfaces and finally to 0D point interfaces in a cellular packing. In the latter case, Monte Carlo simulations of symmetric ABCD 4-miktoarm stars by Dotera [32] showed that a 4-colored Kelvin foam was the optimal packing, and not for example a square 4-colored tiling pattern.

As the molecular composition is altered away from the symmetric case, the resulting structural response becomes a combination of these motifs as illustrated in Fig. 10.14. Here an asymmetric ABCD star is shown to form a hexagonal columnar structure with elements of a [12.6.4] tiling along the cylinder axis, but apart from the cylinders themselves remains a cellular structure. Many more structures are expected to be found in these and higher order block copolymer molecules as combinations of the topologies displayed here. From a theoretical perspective we can continue to add chains but the synthetic community is in some sense way ahead of theory:

In Fig. 10.15 a range of examples are shown of incredible molecular architectures synthesised in the group of A. Hirao. The potential morphologies and topologies to be found in these systems is unknown as these molecules are still largely unexplored structurally both theoretically and experimentally.

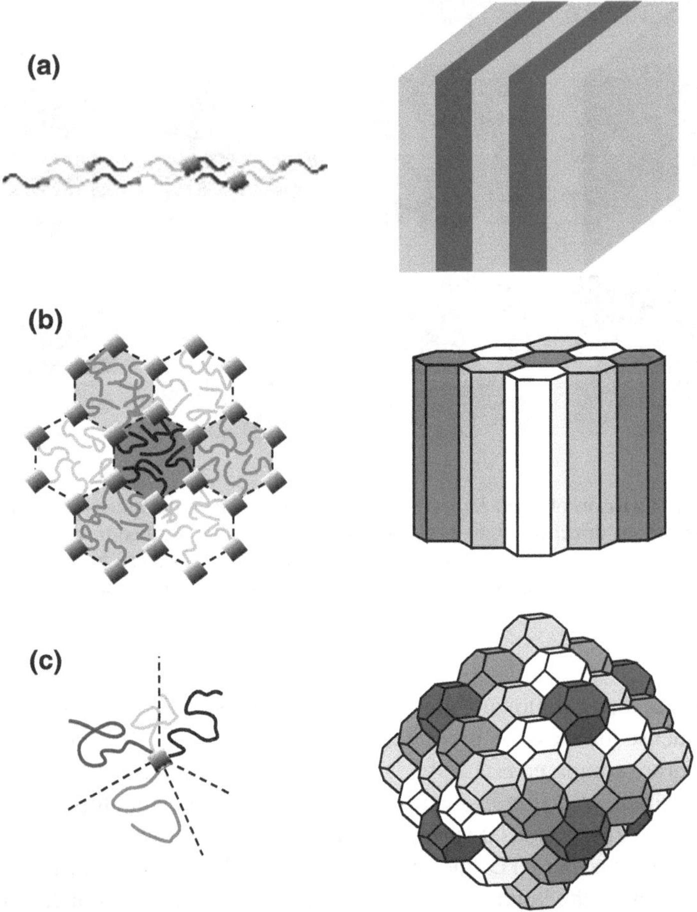

Fig. 10.13 The mutual interfacial dimensionality changes with the molecular architecture for symmetric block copolymers. **a** Diblocks form lamellae with 2D (surface) interfaces. **b** ABC star triblocks form tiling patterns with 2D and 1D (line) interfaces. **c** ABCD star tetrablocks form cellular packings with 2D, 1D and 0D (point) interfaces. This particular cellular packing is a 4-colored Kelvin foam, i.e. each component forms a closed cell in the shape of a truncated octahedron surrounded by 14 neighbors. This structure has been predicted to form in symmetric ABCD stars [32]. Figure from [32]

Fig. 10.14 Edge-on (left) and top-down (right) split views of morphology formed in an asymmetric four-armed ABCD miktoarm star with arm length ratios of 1:4:2:2. Color code: A (red), B (blue), C (green), D: (yellow)

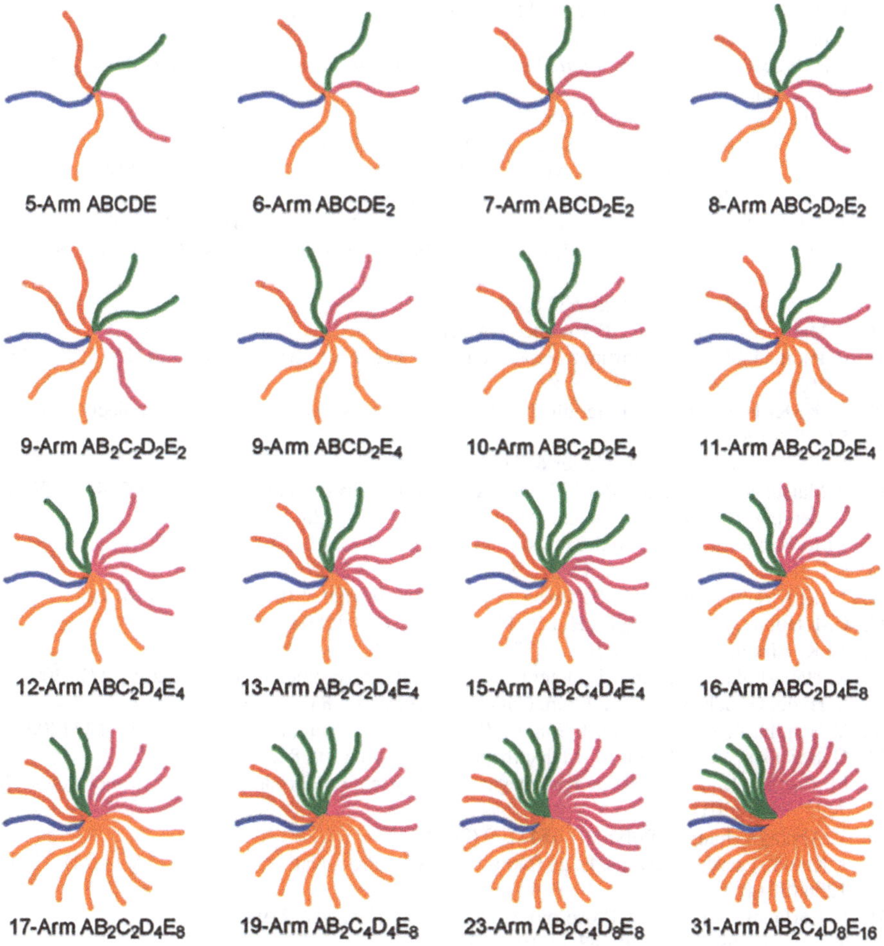

5-Arm ABCDE
6-Arm $ABCDE_2$
7-Arm $ABCD_2E_2$
8-Arm $ABC_2D_2E_2$
9-Arm $AB_2C_2D_2E_2$
9-Arm $ABCD_2E_4$
10-Arm $ABC_2D_2E_4$
11-Arm $AB_2C_2D_2E_4$
12-Arm $ABC_2D_4E_4$
13-Arm $AB_2C_2D_4E_4$
15-Arm $AB_2C_4D_4E_4$
16-Arm $ABC_2D_4E_8$
17-Arm $AB_2C_2D_4E_8$
19-Arm $AB_2C_4D_4E_8$
23-Arm $AB_2C_4D_8E_8$
31-Arm $AB_2C_4D_8E_{16}$

Fig. 10.15 A selection of possible structures of asymmetric star polymers obtained by Hirao and co-workers using advanced iterative synthetic techniques. Figure from [33]

Acknowledgements The author wishes to gratefully acknowledge colleagues and mutual co-authors of the authors own research presented in this chapter, in particular Stephen T. Hyde, Liliana de Campo, Myfanwy Evans, Martin C. Pedersen, Gerd E. Schröder-Turk, Michael G. Fischer, Panagiota Fragouli, Nikos Hadjichristidis and Kell Mortensen.

References

1. F. Bates, M. Hillmyer, T. Lodge, C. Bates, K. Delaney, G. Fredrickson, Science **336**, 434–440 (2012)
2. N.A. Lynd, A.J. Meuler, M.A. Hillmyer, Prog. Polym. Sci. **33**, 875–893 (2008)
3. M. Matsen, Macromolecules **45**, 2161–2165 (2012)
4. M. Huggins, J. Chem. Phys. **9**(5), 440 (1941)
5. P. Flory, J. Chem. Phys. **10**, 51 (1942)
6. S.T. Hyde, in *Handbook of Applied Surface and Colloid Chemistry*, ed. by K. Holmberg (Wiley, 2001); chapter 16
7. C. Huang, H. Yu, Polymer **48**, 4537–4546 (2007)
8. S. Lee, M. Bluemle, F. Bates, Science **330**, 349–353 (2010)
9. S. Kim, K. Jeong, A. Yethiraj, M. Mahanthappa, Proc. Natl. Acad. Sci. USA **114**(16), 4072–4077 (2017). https://doi.org/10.1073/pnas.1701608114
10. T. Gillard, S. Lee, F. Bates, Proc. Natl. Acad. Sci. USA **113**(19), 167–5172 (2016)
11. N. Hadjichristidis, H. Iatrou, M. Pitsikalis, S. Pispas, A. Avgeropoulos, Prog. Polym. Sci. **30**, 725–782 (2005)
12. F. Bates, MRS Bull. **30**, 525–532 (2005)
13. F. Martinez-Veracoechea, F. Escobedo, Macromolecules **40**, 7354–7365 (2007)
14. P. Padmanabhan, E. Martinez-Veracoechea, F. Escobedo, Macromolecules **49**, 5232–5243 (2016)
15. J.J.K. Kirkensgaard, P. Fragouli, N. Hadjichristidis, K. Mortensen, Macromolecules **44**(3), 575–582 (2011)
16. J.J.K. Kirkensgaard, Soft Matter **6**, 6102–6108 (2010)
17. Y. Matsushita, K. Hayashida, T. Dotera, A. Takano, J. Phys. Condens. Matter **23**, 284111 (2011)
18. T. Gemma, A. Hatano, T. Dotera, Macromolecules **35**, 3225–3227 (2002)
19. J.J.K. Kirkensgaard, S. Hyde, Phys. Chem. Chem. Phys. **11**, 2016–2022 (2009)
20. S.T. Hyde, L. de Campo, C. Oguey, Soft Matter **5**, 2782–2794 (2009)
21. J.J.K. Kirkensgaard, M.C. Pedersen, S.T. Hyde, Soft Matter **10**, 7182–7194 (2014)
22. C.-I. Huang, H.-K. Fang, C.-H. Lin, Phys. Rev. E **77**, 031804 (2008)
23. J.J.K. Kirkensgaard, Phys. Rev. E **85**, 031802 (2012)
24. K. Hayashida, A. Takano, T. Dotera, Y. Matsushita, Macromolecules **41**, 6269–6271 (2008)
25. S. Hyde, G. Schröder, Curr. Opin. Colloid Interface Sci. **8** (2003),
26. Y. Han, D. Zhang, L. Chng, J. Sun, L. Zhao, X. Zou, J. Ying, Nat. Chem. **1**, 123–127 (2009)
27. G. Sorenson, A. Schmitt, M. Mahanthappa, Soft Matter **10**, 8229–8235 (2014)
28. M. Fischer, L. de Campo, J. Kirkensgaard, S. Hyde, G. Schröder-Turk, Macromolecules **47**, 7424–7430 (2014)
29. V. Abetz, S. Jiang, e-Polymers **054**, 1–9 (2004)
30. J. Kirkensgaard, M. Evans, L. de Campo, S. Hyde, Proc. Natl. Acad. Sci. USA **111**(4), 1271–1276 (2014)
31. L. Carlucci, G. Ciani, D. Proserpio, Coord. Chem. Rev. 247–289 (2003)
32. T. Dotera, Phys. Rev. Lett. **82**(1), 105 (1999)
33. T. Higashihara, T. Sakurai, A. Hirao, Macromolecules **42**, 6006–6014 (2009)

Chapter 11
Topology of Minimal Surface Biophotonic Nanostructures in Arthropods

Vinodkumar Saranathan

Abstract Structural colors, including many vivid (ultra)violets, blues and greens are quite ubiquitous in animals and form an important aspect of their physical appearance. These colors, which have evolved over millions of years of evolution for a persistent color production function, are often used in camouflage and signaling, including in mate-choice and as warning coloration. By contrast to pigment or dye-based colors, these fade-proof colors are usually produced by the interference of light by a stunning diversity of nanostructures in the animal integument, none as spectacular as those found in certain arthropods. Recently, Saranathan et al. (Nano Lett 15:3735–3742, 2015 [1]) diagnosed the photonic nanostructure present in the cuticular scales and setae of 85 genera in 5 orders of terrestrial arthropods, using synchrotron small angle X-ray scattering (SAXS) and electron microscopy. We reported a rich diversity of nanostructures rivalling those seen in the phase behavior of amphiphilic surfactants, block copolymer, or lyotropic lipid-water systems, including ordered and disordered triply-periodic bicontinuous nanoporous networks, perforated lamellar, inverse hexagonal columnar and close-packed sphere morphologies. However, all these diverse nanostructures can be decomposed into their constituent topology, characterized by either negative, zero or positive Gaussian curvature, but constant mean curvature—namely, the saddle/hyperboloid (Schwarz's Primitive *P*, Schwarz's Diamond *D*, Schoen's Gyroid *G*) and lamellar-helicoid (Riemann's) surfaces, cylinder and sphere. Intriguingly, both the triply periodic saddle (*P*, *D*, and *G*) and the singly

V. Saranathan (✉)
Division of Science, Yale-NUS College, 10 College Avenue West, Singapore 138609, Singapore
e-mail: yncvks@nus.edu.sg

V. Saranathan
NUS Nanoscience and Nanotechnology Initiative, National University of Singapore, Singapore 117581, Singapore

V. Saranathan
Department of Biological Science, National University of Singapore, Singapore 117558, Singapore

V. Saranathan
Lee Kong Chian Natural History Museum, National University of Singapore, Singapore 117377, Singapore

S. Gupta and A. Saxena (eds.), *The Role of Topology in Materials*,
Springer Series in Solid-State Sciences 189,
https://doi.org/10.1007/978-3-319-76596-9_11

periodic Riemann surfaces are characterized by zero mean curvature, i.e., the building blocks of the corresponding nanostructures are all based on minimal surfaces. In light of this, I review the nanostructure, development, and the biomimetic and bioinspiration potential of these self-assembled minimal surface biophotonic nanostructures, when the large-scale, defect-free synthesis of mesoscopic nanostructures with the desired symmetry at optically relevant length scales is not currently facile.

11.1 Introduction

Colors in the biological world, as in the physical world, are produced either chemically via wavelength-selective absorption by pigments or physically as a result of light scattering from sub-micron features or nanostructures with compositional variation in the refractive index [2–4]. The latter class of colors called structural colors are produced as a result of either incoherent scattering from uncorrelated single particles (as in Rayleigh, Tyndall or Mie scattering) or coherent scattering as a result of constructive interference or diffraction of incident light due to periodic or quasi-periodic spatial variations in the refractive index, on the order of visible wavelengths of light [3, 5].

Structural colors are quite ubiquitous in animals and are known to be produced generally by the coherent scattering of light from a wide variety of underlying integumentary biophotonic nanostructures (100–350 nm) [1, 2, 5–12]. These colors constitute a very important part of the animal appearance, as they are hypothesized to function in camouflage, mate choice and as aposematic or warning coloration [10, 13–17], having evolved over millions of years of natural and sexual selection. Indeed many recent studies, even though not done with an explicit comparative evolutionary framework, suggest that these biophotonic nanostructures comprised of relatively low refractive index biomaterials (e.g., chitin ~ 1.56; feather β-keratin ~ 1.54 [18]), are perhaps optimized over evolutionary time for material and/or optical properties [19–24]. Unsurprisingly, a growing number of studies are turning to biophotonic nanostructures as a novel source of inspiration for advanced devices and technologies for use in photonics and sensing [6–9, 25–33].

However, the bioinspiration from and the biomimetics of biophotonic nanostructures have been hitherto hampered by a lack of accurate three-dimensional structural characterization, despite a long history of scientific inquiry, including by luminaries such as Newton, Lord Rayleigh, Michelson and Sir C. V. Raman (detailed in [8]). Considerable progress has been made only recently in diagnosing and elucidating the notoriously complex, three-dimensional symmetries of these biophotonic nanostructures, with the application of advanced materials characterization techniques such as small angle X-ray scattering (SAXS), X-ray tomography and ptychography [1, 16, 34–38]. Similarly, there has been a dearth of studies on the growth and development of these biophotonic nanostructures. While it has become abundantly clear that many of these nanostructures are self-assembled [1, 34, 35, 39–44], the precise details of their biological intra-cellular self-assembly processes are still uncertain. This knowledge is of considerable biomimetic and bioinspirational interest, given that the facile synthesis of large-scale, defect-free, three-dimensional photonic crystals (PCs) of desired symmetries at these rather large meso-scales (broadly 100–1000 nm) is rather challenging [22, 45–50].

Arthropods, especially insects, are the most abundant, colorful and ecologically diverse group of animals. Compared to birds [34, 51–54], lizards and frogs [55] and other land vertebrates [2, 4], cephalopods and other aquatic animals [10, 29, 56, 57], terrestrial arthropods unsurprisingly exhibit an unrivalled diversity of biophotonic nanostructures [1, 6–12] (Fig. 11.1). Recently, using SAXS, and EM, we structurally characterized the biophotonic nanostructures present inside cuticular scales and setae from an unprecedented number of taxa belonging to 85 genera in 5 orders of terrestrial arthropods [1]. In this chapter, I review from a topological point of view beyond just a language of shape [58–60], the nanostructure, development, and the potential of these self-assembled biophotonic nanostructures in arthropods, especially those whose fundamental interfacial topology is based on triply periodic minimal surfaces (TPMS), to perhaps bio-inspire the design and synthesis of next-generation multi-functionalities in photonics, sensing, and even energy storage/conversion.

11.2 Topology of Arthropod Biophotonic Nanostructures

In terrestrial animals generally and certainly in arthropods, structural colors especially iridescence (characterized by a change in hue with angle of observation or light incidence [3]) is quite frequently generated by the well-studied class of one-dimensional PCs, including thin-film, multi-layer reflectors and diffraction gratings [6–11, 61]. This is perhaps because they are the most easily evolvable and evolved biophotonic nanostructure, for all it takes to produce visible structural colors (think oil slick or soap bubble) is an optically relevant specification of the thickness of a single superficial layer that is optically different (refractive index) from the underlying matrix on the animal integument [5, 11, 62, 63]. Nevertheless, many beetles including scarab beetles (Coleoptera: Scarabaeidae), longhorn beetles (Coleoptera: Cerambycidae) and weevils (Coleoptera: Curculionidae), and certain butterflies (Lepidoptera: Lycaenidae, Papilionidae), bees (Hymenoptera: Apidae), jumping spiders (Araneomorphae: Salticidae), and tarantulas (Mygalomorphae: Theraphosidae) (see [1, 35] and references therein) have structurally colored scales or setae covering the integument on their elytra/wings, abdomen, and legs. These photonic scales and setae are known to have a variety of complex, three-dimensional biophotonic nanostructures within their interior lumen, made up of the polysaccharide chitin and air (see [1, 35] and references therein).

Amazingly, the diversity of arthropod biophotonic nanostructures spans the amphiphilic phase space [1], i.e., arthropod biophotonic scales and setae show a rich polymorphism comparable to those normally seen in the phase behavior of amphiphilic or lyotropic macromolecules, such as block copolymers [64–67], surfactants [68], and lipids [69, 70]. Nanostructural diagnoses based on an indexing of

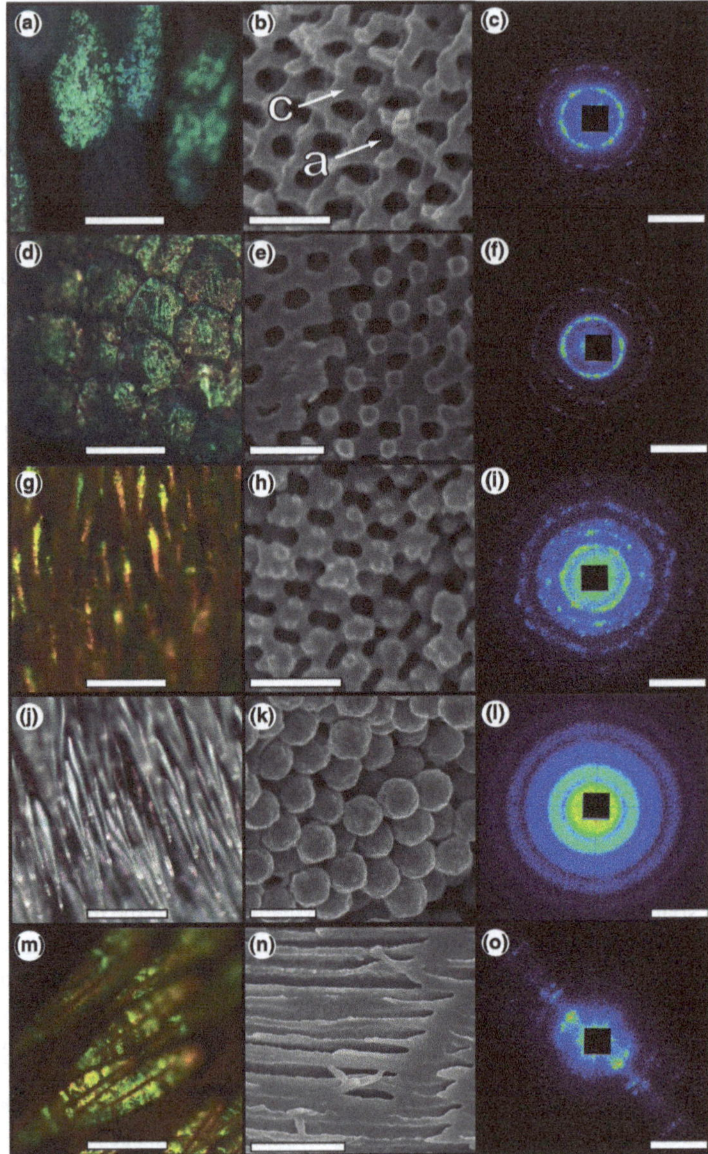

Fig. 11.1 Diversity of biophotonic arthropod cuticular nanostructures inside scales or setae. Representative light micrographs (first column), scanning electron micrographs (middle), and 2D SAXS patterns (last column) of: **a–c** *Rhinoscapha* sp. (Coleoptera: Curculionidae), single gyroid ($I4_132$); **d–f** *Pachyrrhynchus yamianus* (Coleoptera: Curculionidae), single diamond (*Fd*-3*m*); **g–i**) *Sternotomis mirabilis* (Coleoptera: Cerambycidae), simple or primitive cubic (*Pm*-3*m*); **j–l** *Anoplophora versteegi* (Coleoptera: Cerambycidae), quasi-ordered spheres; and **m–o** *Amegilla cingulata* (Hymenoptera: Apidae), inverse hexagonal columnar. SAXS patterns show the logarithm of scattering intensity as a function of the scattering wave vector, *q*, in false color. Scale Bars: **a**, **m**—50 μm; **d, g, j**—100 μm; **b**—600 nm; **e, n**—500 nm; **h, k**—250 nm; **c, f, i, l, o**—0.05 nm^{-1}. Abbreviations: **c**—chitin; **a**—air void

Bragg peaks [71] in the azimuthal SAXS profiles, together with real-space information from scanning and transmission EM images revealed that different families of arthropods have evolved specific classes of biophotonic nanostructure in a distinct, non-random fashion (Figs. 11.1 and 11.2) [1]. Triply periodic bicontinuous nanoporous networks with single gyroid (IUCr cubic space group 214, $I4_132$) and single diamond (IUCr cubic space group 227, *Fd*-3*m*) symmetries were found only in the iridescent scales of snout weevils (Coleoptera: Curculionidae; Figs. 11.1a–f, and 11.2a, b) [1]. Single gyroid PCs are also known to be present in the wing scales of certain lycaenid and papilionid butterflies [1, 11, 23, 35, 38, 72]. Triply periodic networks with simple or primitive cubic (IUCr cubic space group 221, *Pm*-3*m*) symmetry (Plumber's nightmare) were present only in the setae of the Sternotomini tribe of longhorn beetles (Coleoptera: Cerambycidae; Figs. 11.1g–i, and 11.2c) [1].

In general, these triply periodic networks exhibited several salient features quite remarkable for a biological soft-matter system [1]. First, the X-ray scattering from many such photonic scales exhibited 8 orders of Bragg peaks or more (Fig. 11.2), rivalling those typically seen, for instance, in synthetic polymer blends with a high degree of long-range orientational order [1, 73]. Overall, the degree of long-range order ranged, however, from monocrystal-like Bragg spots to various degree of poly-crystallinity (including concentric Debye-Scherrer rings at diagnosable spacings [73]), all the way to sponge-like or highly disordered versions of the networks exhibiting just the fundamental peak and at most one or two higher-order peaks (see Fig. S1 in [1]). In some Neotropical weevils colloquially called the "diamond" weevils (*Lamprocyphus* spp. see Figs. S1.72–80 in [1]; cf. [74]), both single diamond and single gyroid symmetries were present within the same scale assayed, suggesting a very interesting epitaxial relationship between these two co-existing phases [75, 76], albeit unexamined.

Whereas, the photonic scales of other longhorn beetles are comprised of close-packed arrays of chitin spheres with or without bridging necks (between neighboring spheres), occurring in ordered (face-centered cubic, *fcc*; body-centered cubic, *bcc*) and quasi-ordered arrangements (e.g., *Anoplophora versteegi*; Figs. 11.1j–l, and 11.2d) [1].

A two-dimensional, inverse hexagonal columnar (i.e., a honeycomb arrangement of cylindrical air holes in chitin) morphology was present in the photonic setae of the digger bee *Amegilla cingulata* (Hymenoptera: Apidae: Anthophorini; Figs. 11.1m–o, and 11.2e) and in a jumping spider (Araneae: Salticidae) [1]. A Bouligand-like [77] twisted inverse columnar (intermediate between a lamellar and an inverse hexagonal) morphology was present in the iridescent setae of the cuckoo bee *Thyreus nitidulus* (Hymenoptera: Apidae: Melectini), as well as in the scales of the *Hoplia* scarab beetles (Coleoptera: Scarabaeidae) [1]. Finally, a quasi-ordered or amorphous sponge-like photonic network was present in the setae of another cuckoo bee, *Thyreus pictus*, while the photonic nanostructures responsible for the purple to blue color in the setae of many tarantulas (Araneae: Theraphosidae) were found to be perforated lamellae [1], which are also known in many butterfly wing scales [11, 78–80].

Besides their corresponding space group symmetries, these diverse arthropod biophotonic nanostructures as well as their materials science analogs in block

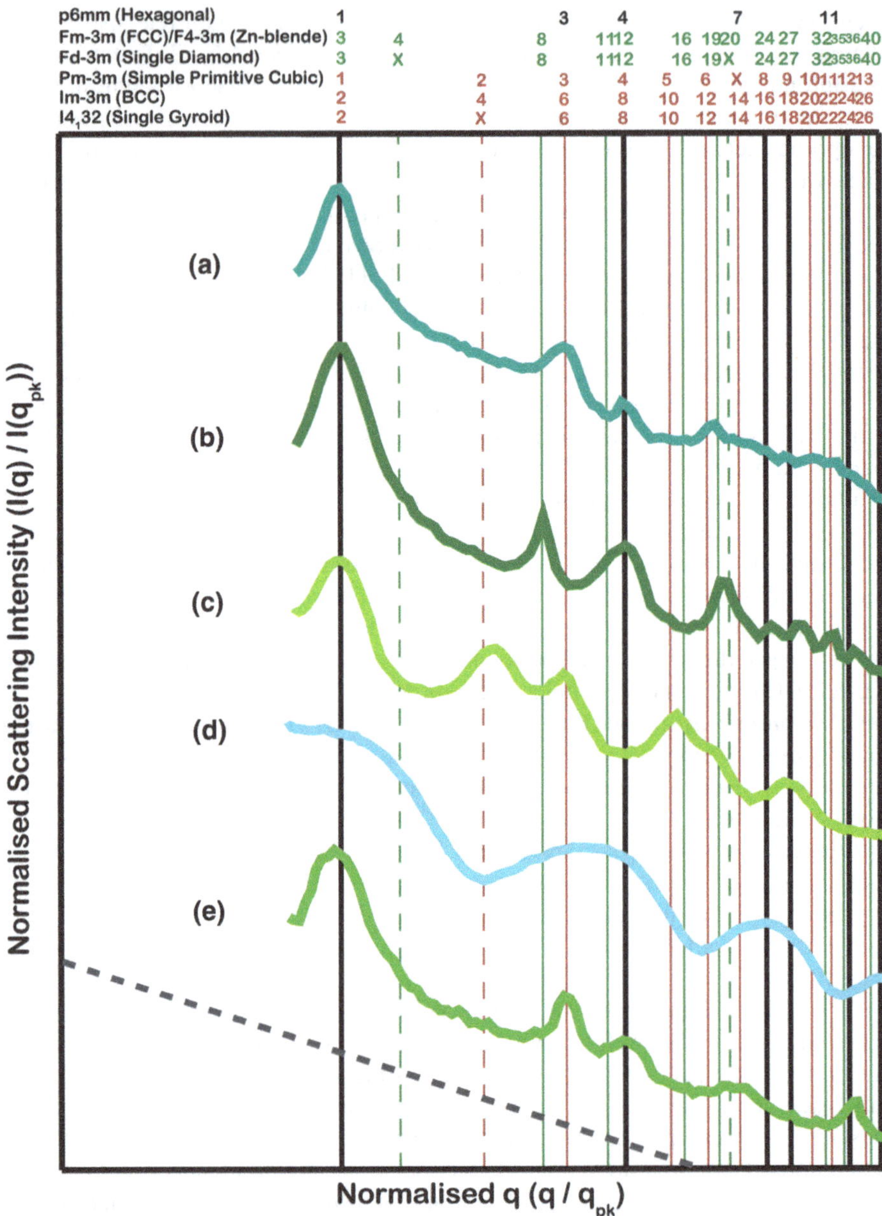

Fig. 11.2 Nanostructural diagnoses of representative SAXS patterns of arthropod cuticular photonic nanostructures presented in Fig. 11.1. The normalized, azimuthally-averaged SAXS profiles are shown on a log-log scale. Coleoptera: Curculionidae: **a** *Rhinoscapha* sp., single gyroid ($I4_132$), **b** *Pachyrrhynchus yamianus*, single diamond (*Fd*-3*m*); Coleoptera: Cerambycidae: **c** *Sternotomis mirabilis*, simple or primitive cubic (*Pm*-3*m*), **d** *Anoplophora versteegi*, amorphous or quasi-ordered spheres; Hymenoptera:Apidae: **e** *Amegilla cingulata*, inverse hexagonal columnar. The normalized Bragg peak positional ratios are indexed based on the predictions (colored vertical lines with corresponding squares of the moduli of the Miller indices *hkl* presented above) of specific space group symmetries, as per IUCr conventions [1, 71]

copolymer, lipid-water, and surfactant systems can all be classified based on their interfacial curvature [59, 69, 81, 82] into constituent topologies—namely, the plane, cylinder, sphere, saddle or hyperboloid (Schwarz's Primitive P, Schwarz's Diamond D and Schoen's Gyroid G) and lamellar-helicoid (Riemann's) surfaces [83–85] (Fig. 11.3). Interestingly, these topologies all have a *constant* mean curvature ($H = (\kappa_1 + \kappa_2)/2$, where κ_1 and κ_2 are the principal curvatures along orthogonal planes normal to the surface) [86, 87], but distinct (0 for cylinder and plane, constant positive for sphere and negative for saddle) saddle-splay or Gaussian curvature ($K = \kappa_1.\kappa_2$). However, the plane (trivially so), the singly periodic Riemann's surface and the triply periodic (i.e., translationally invariant in $\mathbf{R}^3$) P, D and G surfaces are *minimal* surfaces with zero mean curvature ($H = 0$). The P, D and G TPMS with double primitive (IUCr space group 229, *Im*-3*m*), double diamond (IUCr space group 224, *Pn*-3*m*) and double gyroid (IUCr space group 230, *Ia*-3*d*) symmetries respectively divide the unit cell into two equivalent bicontinuous labyrinths or networks on either side, each with the corresponding subgroup symmetry when considered alone (*Pm*-3*m*, *Fd*-3*m* and $I4_132$) and 6, 4 and 3-fold connectivity respectively [88, 89] (Fig. 11.3). The lowest attainable *genus* (i.e., the unit cell has a hole or handle along each of the three axes) for these TPMS is 3. The P, D, and G TPMS happen to belong to the same *associate* or *Bonnet* family as they are topologically related to each other by a Bonnet transformation [89–92], suggesting a topological link (for e.g., in order-order transitions) between these three phases. Let us now turn to the question of whether we can go beyond just cataloging and elucidate the origin and self-assembly of these arthropod biophotonic nanostructures, specifically the triply periodic bicontinuous minimal surface networks.

11.3 Self-assembly of Minimal Surface Biophotonic Nanostructures

Biological phospholipid bilayer membranes are generally known to self-assemble into a number of *inverse* (type-II) lyotropic liquid crystalline phases in aqueous media [59, 81, 93, 94], analogous to the diversity of biophotonic nanostructures reported in arthropods [1]. In vivo, the so-called *cubosomes* or cubic membrane morphologies with double primitive, double diamond and double gyroid symmetries, in addition to disordered sponge-like networks, tubular networks and lamellar phases, as well as reversible transition between phases and coexisting phases, have all been reported within membrane-bound organelles of both plant and animal cells [85, 94–103]. The self-organization of biological membranes thus appears to be an innate property of phospholipid bilayers, thought to be regulated by the energetics of membrane curvature, specifically by the interplay between the minimization of interfacial energy and chain-stretching energy [59, 82, 100, 104]. These cubic membranes have also been postulated to develop via a topology preserving intersection-free membrane folding or invagination as opposed to a topologically non-equivalent membrane

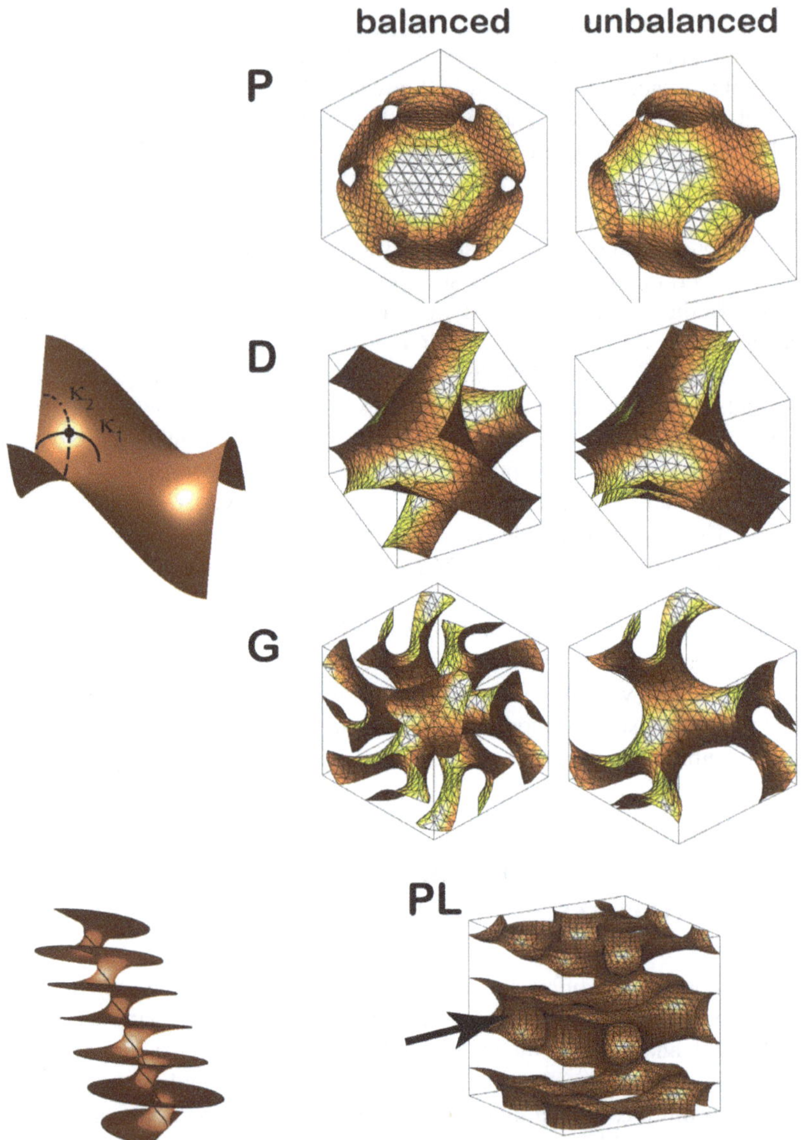

Fig. 11.3 Topology of arthropod minimal surface photonic nanostructures. Top panel: Triply periodic saddle or hyperboloid surfaces (Schwarz's Primitive *P* with double primitive *Im*-3*m*, Schwarz's Diamond *D* with double diamond *Pn*-3*m*, and Schoen's Gyroid *G* with double gyroid *Ia*-3*d* space-group symmetries). Pairs of parallel minimal surfaces (level set) are illustrated in both "balanced" and supergroup symmetry-breaking "unbalanced" configurations. Bottom panel: singly periodic Riemann's (lamellar-helicoid) surface (Riemann's surface, image credit: Matthias Weber). The pores (black arrow) in a perforated lamellar (PL) or a sponge (L_3, not shown) morphology can be modelled as a Riemann's minimal surface with helicoid-like bridges connecting asymptotic lamellar planes [83–85]

fusion process [98, 100, 101, 105]. Nonetheless, these morphologies occur with notably large lattice parameters (50–500 nm) than is typical for lipid-water systems [85, 94–103]. At a simple reading, it appears then that the in vivo self-assembly and observed polymorphism of biological bilayer lipid phases is not completely congruent with the thermodynamic phase behavior of lipids seen in bulk systems [94]. However, living cells can apparently generate spontaneous membrane curvature and induce membrane invagination by physically changing the shape of the membranes upon binding by certain classes of proteins that vary in shape, molecular weight, and electrostatic residues [82, 100–102, 104, 106], and perhaps also accounting for the unique biological length scales of self-assembly. One such membrane inclusion is the highly-conserved BAR domain protein superfamily [107–110]. Such membrane inclusions may not only play a regulatory role in the self-assembly of various biological lipid phases but may offer a powerful novel paradigm for material synthesis at the sought after meso-scales by controlling the local conformation/topology of membranes. Therefore, the membrane-binding proteins can perhaps be thought of as the topological "operator" for choosing among the states of the bilayer polymorphism.

Given that each arthropod family examined has evolved to occupy only specific regions of the lyotropic phase space [1] (see Sect. 11.2, and Figs. 11.1 and 11.2), it is tempting to hypothesize that different families or lineages of arthropods have evolved to independently co-opt this already existing intracellular self-assembly machinery, along with the controlled expression of lineage-specific membrane-binding proteins with characteristic bilayer bending (κ) and saddle-splay or Gaussian ($\bar{\kappa}$) moduli [81, 82], to template the lyotropic precursors of biophotonic nanostructures with these rather large lattice constants [1]. However, there is a pressing need for comparative developmental studies of the biophotonic nanostructure in arthropods in order to elucidate the precise details of their self-assembly. These studies could take either a descriptive, observational approach utilizing fluorescence assays of scales in a time series from developing pupae [111], or perhaps utilize the latest interference techniques (RNAi, CRISPR-Cas9) to knockdown [112, 113] candidate BAR-like proteins that could play a putative role in the self-assembly of biophotonic templates, or ideally opt for a materio-genomic approach integrating high-throughput RNA sequencing, combined with proteomics to identify the precise regulatory pathways [114].

Nonetheless, given at least the few developmental studies of biophotonic nanostructures in butterflies [40, 79, 111, 115, 116], here I outline the current state of knowledge, specifically on the intra-cellular self-assembly of the triply periodic minimal surface PCs in beetles and butterflies. Given that arthropod scales and setae are evolutionarily homologous, being derived from trichogen (shaft-forming) cells in the integument [117], it is reasonable to suppose that they share or at least have largely conserved elements of the intra-cellular mechanism of nanostructure development within the lumen.

Single gyroid PCs in certain papilionid and lycaenid butterfly wing scale cells are thought to develop via a process eerily reminiscent and perhaps pre-emptive of a materials engineering approach [35]. Inferring from a structural analogy of motifs in EM images of developing butterfly scale cells to the self-assembled, core-shell

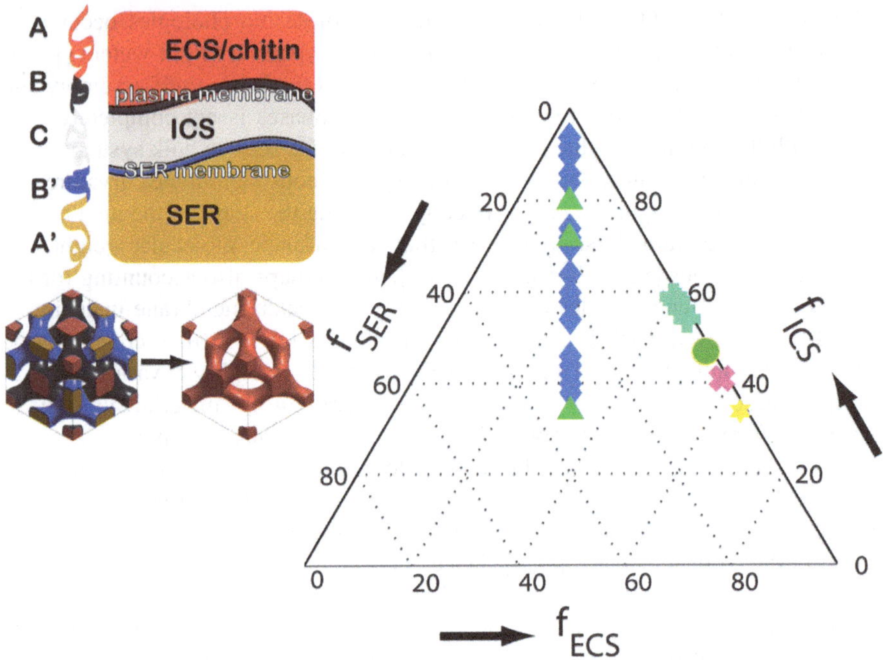

Fig. 11.4 Self-assembly of arthropod biophotonic nanostructures. As previously hypothesized in butterflies [35], single gyroid ($I4_132$) and single diamond (*Fd*-3*m* illustrated) PCs in weevils self-assemble via tandem infolding of plasma and SER membranes in a "balanced" parallel configuration, into a precursor core-shell double gyroid (*Ia*-3*d*) or double diamond (*Pn*-3*m*, illustrated) template within the interior lumen of the scale cell, by analogy to a hypothetical linear pentablock copolymer, ABCB'A' (although in practice, the bilayers B and B' have negligible volume fractions and can be ignored). In longhorn beetles, bees and spiders, the nanostructure likely develops by a severly "unbalanced" parallel membrane system to produce nanostructures with the observed ~50% or greater volume fractions. In either case, extra-cellular chitin is "backfilled", i.e., extruded and polymerized into only the volume in red, which is continuous with extra-cellular space, after which the cell dies to leave behind the chitin nanostructure in air. A hypothetical ternary phase diagram of the developing parallel membrane system within arthropod setae and scales is provided with chitin filling fractions taken from [1]. Key to shaded symbols: triangles—$I4_132$, diamonds—*Fd*-3*m*, pluses—*Pm*-3*m*, circle—quasi-ordered spheres, star—*bcc* sphere, X—*HEX*. Abbreviations: ECS (extra-cellular space), ICS (intra-cellular space), SER (intra-Smooth Endoplasmic Reticulum space)

morphology seen in triblock copolymer systems [118, 119], Saranathan et al. [35] have proposed that the plasma membrane of the developing photonic scales invaginate in tandem with smooth endoplasmic reticulum (SER) membranes to form a ("balanced") multi-continuous core-shell double gyroid (*Ia*-3*d*) precursor [35, 40, 79, 116]. This is analogous to the micro-phase separation of a linear ABC triblock copolymer that is compositionally asymmetrical about the mid-plane or in essence, a penta-block copolymer, ABCB′A′ (Fig. 11.4) [120]. It is unclear whether chitin, which is produced outside the scale cells and can therefore only be deposited into the extracellular space co-continuous with one of the two single gyroid cores, is extruded and polymerized inside the lumen simultaneously with the plasma membrane, once it starts invaginating or whether the precursor template is already in place within the scale cell before the chitin extrusion begins [38]. In any case, once the scale cell matures, the scale cell dies via apoptosis, leaving behind just the "back-filled" single gyroid ($I4_132$) PC network of chitin in air within its interior lumen [35]. Saranathan et al. [1] have hypothesized that single diamond PCs in weevil scales develop in a similar fashion to these single gyroid PCs in butterfly wing scales by virtue of some weevils also possessing photonic scales with single gyroid symmetry. However, based on the rather large chitin filling fractions of the nanostructures within the setae of longhorn beetles and bees, relative to those in weevil scales (Fig. 11.4), they argued for a distinction in the development of the former purely from physical space considerations that may well preclude the presence of a second (SER) parallel membrane system in addition to the plasma membrane within the confines of the developing setae [1]. The block copolymer analogy here then would be simply be a triblock copolymer (ABA′ or ABC) in order to produce nanostructures with nearly 50% dielectric fill fractions. An alternate, plausible developmental scenario in photonic longhorn beetle and bee setae could be that one or likely both bilayer membranes are highly displaced relative to the mid-plane of the unit cell in the developing cubic membrane template, leading to a severely "unbalanced" multi-continuous parallel membrane system in which the volumes enclosed on either side of the membranes are not equal (Fig. 11.3), unlike the "balanced" or zero-potential case. However, such "unbalanced" membranes need no longer be *minimal*, but can still be characterized by a constant mean curvature (the Schoen's I-WP is a notable exception) [92, 98, 121]. Interestingly, such "unbalanced" membranes have complex symmetries, as the volumes on either side of the mid-plane are no longer interchangeable and the complex topologies of such surfaces remain under-investigated [92, 98, 121].

11.4 Biomimetic Potential of Minimal Surface Biophotonic Nanostructures

Ordered bicontinuous cubic phases based on TPMS are increasingly being applied in diverse technological applications [31, 33, 49, 122–126], perhaps owing to their increasing ease of access and prevalence in self-assembled molecular and macromolecular systems [65, 66, 81]. In particular, the single diamond and single gyroid morphologies are of special relevance in opto-electronic applications given their large and complete bandgaps compared to their super-group double diamond and double gyroid and other competing structures [22, 127], the latter even more so given its intrinsic chirality [33, 128, 129]. To my knowledge, however, the direct experimental access of single gyroid and single diamond morphologies via self-assembly is not yet possible [22], without additional steps to manipulate a double or an alternating template [49, 50, 130, 131]. Only very recently, has a packing frustration induced epitaxial pathway for a double diamond to a single gyroid order-order phase transition been proposed using self-consistent field theory [132]. Even so, the direct synthetic self-assembly of single network minimal surface nanostructures at optical length scales will be far from facile [66, 81], while arthropods appear to have solved this particular problem over millions of years of evolutionary optimization to explore the lyotropic or amphiphilic phase space for a color production function [1, 35].

As such, these beetle minimal surface based nanoporous networks can serve as convenient positive templates for high-epsilon dielectric infiltration [133, 134] or as direct biotemplates for use in sensing [26, 30]. It would be more productive, however, to directly interrogate the self-assembly of these biological lipid bilayers into TPMS with large lattice constants and design biomimetic approaches to self-assemble tunable mesophases for multifunctional applications, but this requires considerable progress in comparative developmental studies of biophotonic nanostructure. In any case, membrane inclusions (binding proteins) with synthetic lipid bilayers may offer a novel bio-inspired paradigm to overcome both the intrinsic length scales of in vitro lipid self-assembly as well as facilely synthesize cubic TPMS phases with sub-group symmetries. Perhaps another tractable bio-inspired approach from these arthropod nanostructures in the short-term could be renewed attempts to stabilize the self-assembly of linear non-centrosymmetric pentablock copolymers to achieve optically-relevant lattice spacings [120].

11.5 Conclusion

We have seen how the diversity of arthropod cuticular biophotonic nanostructures within iridescent scales and setae [1] rival the phase behavior of amphiphilic or lyotropic macromolecules, such as block copolymers [64–67], surfactants [68], and lipids [69, 70], but are self-assembled at optical length scales. I have argued how these ordered and quasi-ordered triply periodic bicontinuous nanoporous networks, perforated lamellar, inverse hexagonal columnar and close-packed sphere morphologies found within the lumen of these photonic scales and setae can be topologically decomposed into saddle surfaces (*P*, *D*, and *G* surfaces), Riemann's surface, cylinder and sphere, all of which feature constant mean curvature [1]. However, these saddle and Riemann's surfaces also happen to be triply and singly periodic minimal surfaces. Biological lipid bilayer membranes are known to self-assemble into minimal surface and other morphologies reminiscent of lyotropic lipid phases but at much larger length scales, likely using the binding of proteins with intrinsic shapes to induce spontaneous curvature and control the final membrane topology [82, 100–102, 104, 106]. It appears that different lineages of arthropods have co-opted this for a color production function. From a materials perspective, the inclusions/binding of proteins to overcome intrinsic length scale limitations of bulk lipid self-assembly offers a new twist on a strictly topological or an elastic geometry perspective of self-assembly and phase transitions in lipid systems and perhaps offers a bio-inspired or in time, a biomimetic route for engineering multifunctional topologies at the increasingly relevant mesoscales.

Acknowledgements The author acknowledges support from Yale-NUS start-up funds, Simon Mochrie, Chinedum Osuji, Eric Dufresne, Rick Prum and Pietro De Camilli for stimulating and helpful discussions on membrane energetics and protein-membrane interactions, Heeso Noh and Hui Cao for help with recording light microscope images of scales, Ainsley Seago, Larry Gall, James Hogan, Ray Gabriel, Darren Mann, and Steve Lingafelter for access to specimens, Michael Rooks for help with EM, Suresh Narayanan and Alec Sandy at beamline 8-ID-I of the Advanced Photon Source (APS) at Argonne National Labs for help with collecting the SAXS data presented here with support by the U. S. Department of Energy, Office of Science, Office of Basic Energy Sciences, under Contract No. DE-AC02-06CH11357.

References

1. V. Saranathan, A.E. Seago, S. Narayanan, A. Sandy, S.G.J. Mochrie, E.R. Dufresne, H. Cao, C.O. Osuji, R.O. Prum, Nano Lett. **15**, 3735–3742 (2015)
2. D.L. Fox, *Animal Biochromes and Structural Colors* (Univ. California Press, Berkeley, CA, 1976)
3. E. Hecht, *Optics*, 2nd edn. (Addison-Wesley Publishing, Reading, MA, 1987)
4. H.M. Fox, G. Vevers, *The Nature of Animal Colors* (Macmillan, New York, 1960)
5. R.O. Prum, in *Bird Coloration: Mechanisms, vol. I: Mechanisms and Measurements* (Harvard University Press, Boston, 2006)
6. P. Vukusic, J.R. Sambles, Nature **424**, 852–855 (2003)
7. S. Kinoshita, S. Yoshioka, J. Miyazaki, Rep. Prog. Phys. **71**, 076401 (2008)
8. M. Srinivasarao, Chem. Rev. **99**, 1935–1961 (1999)
9. A.E. Seago, P. Brady, J.-P. Vigneron, T.D. Schultz, J. R. Soc. Interface **6**, S165–S184 (2009)
10. L.M. Mathger, E.J. Denton, N.J. Marshall, R.T. Hanlon, J. R. Soc. Interface **6**, S149–S163 (2009)
11. R.O. Prum, T. Quinn, R.H. Torres, J. Exp. Biol. **209**, 749–765 (2006)
12. A.E. Seago, V. Saranathan, in *Nature's Nanostructures* (Pan Stanford Publishing, Singapore, 2012)
13. G.E. Hill, K.J. McGraw, *Bird Coloration, Volume 2 Function and Evolution* (Harvard University Press, Cambridge, MA, 2006)
14. A. Sweeney, C. Jiggins, S. Johnsen, Nature **423**, 31–32 (2003)
15. R.D. Pope, H.E. Hinton, Biol. J. Linn. Soc. **9**, 331–348 (1977)
16. M.E. McNamara, V. Saranathan, E.R. Locatelli, H. Noh, D.E. Briggs, P.J. Orr, H. Cao, J. R. Soc. Interface **11**, 20140736 (2014)
17. K.L. Prudic, C. Jeon, H. Cao, A. Monteiro, Science **331**, 73–75 (2011)
18. H.L. Leertouwer, B.D. Wilts, D.G. Stavenga, Opt. Exp. **19**, 24061–24066 (2011)
19. B.T. Hallam, A.G. Hiorns, P. Vukusic, Appl. Optics **48**, 3243–3249 (2009)
20. J. Zi, X. Yu, Y. Li, X. Hu, C. Xu, X. Wang, X. Liu, R. Fu, Proc. Natl. Acad. Sci. USA **100**, 12576–12578 (2003)
21. H. Yin, B. Dong, X. Liu, T. Zhan, L. Shi, J. Zi, E. Yablonovitch, Proc. Natl. Acad. Sci. USA **109**, 10798–10801 (2012)
22. J.A. Dolan, B.D. Wilts, S. Vignolini, J.J. Baumberg, U. Steiner, T.D. Wilkinson, Adv. Opt. Mater. **3**, 12–32 (2015)
23. B.D. Wilts, I.J. N, D.G. Stavenga, BMC Evol. Biol. **14**, 160 (2014)
24. B.D. Wilts, X. Sheng, M. Holler, A. Diaz, M. Guizar-Sicairos, J. Raabe, R. Hoppe, S.H. Liu, R. Langford, O.D. Onelli, D. Chen, S. Torquato, U. Steiner, C.G. Schroer, S. Vignolini, A. Sepe, Adv. Mater. (2017). (In Press)

25. A.R. Parker, H.E. Townley, Nat. Nanotechnol. **2**, 347–353 (2007)
26. R.A. Potyrailo, H. Ghiradella, A. Vertiatchikh, K. Dovidenko, J.R. Cournoyer, E. Olson, Nat. Photon. **1**, 123–128 (2007)
27. M. Kolle, P.M. Salgard-Cunha, M.R.J. Scherer, F.M. Huang, P. Vukusic, S. Mahajan, J.J. Baumberg, U. Steiner, Nat. Nanotechnol. **5**, 511–515 (2010)
28. J.W. Galusha, L.R. Richey, M.R. Jorgensen, J.S. Gardner, M.H. Bartl, J. Mater. Chem. **20**, 1277–1284 (2010)
29. E. Kreit, L.M. Mathger, R.T. Hanlon, P.B. Dennis, R.R. Naik, E. Forsythe, J. Heikenfeld, J. R. Soc. Interface **10**, 20120601 (2012)
30. R.A. Potyrailo, T.A. Starkey, P. Vukusic, H. Ghiradella, M. Vasudev, T. Bunning, R.R. Naik, Z. Tang, M. Larsen, T. Deng, S. Zhong, M. Palacios, J.C. Grande, G. Zorn, G. Goddard, S. Zalubovsky, Proc. Natl. Acad. Sci. USA **110**, 15567–15572 (2013)
31. L. Lu, L. Fu, J.D. Joannopoulos, M. Soljacic, Nat. Photon. **7**, 294–299 (2013)
32. M. Xiao, Y. Li, M.C. Allen, D.D. Deheyn, X. Yue, J. Zhao, N.C. Gianneschi, M.D. Shawkey, A. Dhinojwala, ACS Nano **9**, 5454–5460 (2015)
33. M.D. Turner, M. Saba, Q. Zhang, B.P. Cumming, G.E. Schroder-Turk, M. Gu, Nat. Photon. **7**, 801–805 (2013)
34. V. Saranathan, J.D. Forster, H. Noh, S.F. Liew, S.G.J. Mochrie, H. Cao, E.R. Dufresne, R.O. Prum, J. R. Soc. Interface **9**, 2563–2580 (2012)
35. V. Saranathan, C.O. Osuji, S.G.J. Mochrie, H. Noh, S. Narayanan, A. Sandy, E.R. Dufresne, R.O. Prum, Proc. Natl. Acad. Sci. USA **107**, 11676–11681 (2010)
36. A. Singer, L. Boucheron, S.H. Dietze, K.E. Jensen, D. Vine, I. McNulty, E.R. Dufresne, R.O. Prum, S.G.J. Mochrie, O.G. Shpyrko, Sci. Adv. **2**, e1600149 (2016)
37. B. Winter, B. Butz, C. Dieker, G.E. Schroder-Turk, K. Mecke, E. Spiecker, Proc. Natl. Acad. Sci. USA **112**, 12911–12916 (2015)
38. B.D. Wilts, B. Apeleo Zubiri, M.A. Klatt, B. Butz, M.G. Fischer, S.T. Kelly, E. Spiecker, U. Steiner, G.E. Schroder-Turk, Sci Adv **3**, e1603119 (2017)
39. R.O. Prum, E.R. Dufresne, T. Quinn, K. Waters, J. R. Soc. Interface **6**, S253–S265 (2009)
40. H. Ghiradella, J. Morph. **202**, 69–88 (1989)
41. R. Maia, R.H. Macedo, M.D. Shawkey, J. R. Soc. Interface **9**, 734–743 (2012)
42. D.G. DeMartini, D.V. Krogstad, D.E. Morse, Proc. Natl. Acad. Sci. USA **110**, 2552–2556 (2013)
43. E.R. Dufresne, H. Noh, V. Saranathan, S.G.J. Mochrie, H. Cao, R.O. Prum, Soft Mater. **5**, 1792–1795 (2009)
44. M.D. Shawkey, L. D'Alba, M. Xiao, M. Schutte, R. Buchholz, J. Morphol. **276**, 378–384 (2015)
45. A.P. Hynninen, J.H. Thijssen, E.C. Vermolen, M. Dijkstra, A. van Blaaderen, Nat. Mater. **6**, 202–205 (2007)
46. S. Juodkazis, L. Rosa, S. Bauerdick, L. Peto, R. El-Ganainy, S. John, Opt. Express **19**, 5802–5810 (2011)
47. B.A.M. Urbas, M. Maldovan, P. Derege, E.L. Thomas, Adv. Mater. **14**, 1850–1853 (2002)
48. B. Hatton, L. Mishchenko, S. Davis, K.H. Sandhage, J. Aizenberg, Proc. Natl. Acad. Sci. USA **107**, 10354–10359 (2010)
49. S. Vignolini, N.A. Yufa, P.S. Cunha, S. Guldin, I. Rushkin, M. Stefik, K. Hur, U. Wiesner, J.J. Baumberg, U. Steiner, Adv. Mater. **24**, OP23-27 (2012)
50. M. Stefik, S. Guldin, S. Vignolini, U. Wiesner, U. Steiner, Chem. Soc. Rev. **44**, 5076–5091 (2015)
51. M.D. Shawkey, V. Saranathan, H. Palsdottir, J. Crum, M.H. Ellisman, M. Auer, R.O. Prum, J. R. Soc. Interface **6**, S213–S220 (2009)
52. L. D'Alba, V. Saranathan, J.A. Clarke, J.A. Vinther, R.O. Prum, M.D. Shawkey, Biol. Lett. **7**, 543–546 (2011)
53. H. Durrer, *Biology of the Integument 2: Vertebrates* (Springer, Berlin, 1986)
54. J. Dyck, Proc. Int. Ornith. Cong. **16**, 426–437 (1976)
55. S.T. Rohrlich, J. Cell Biol. **62**, 295–304 (1974)

56. P.J. Herring, Comp. Biochem. Physiol. A **109**, 513–546 (1994)
57. E.J. Denton, J.A.C. Nicol, J. Marine Biol. Assoc. UK **46**, 685–722 (1966)
58. S. Gupta, A. Saxena, MRS Bull. **39**, 265–279 (2014)
59. S. Hyde, *The Language of Shape: The Role of Curvature in Condensed Matter–Physics, Chemistry, and Biology* (Elsevier, Amsterdam, Netherlands, 1997)
60. V.N. Manoharan, Science **349**, 1253751 (2015)
61. J.D. Joannopoulos, S.G. Johnson, J.N. Winn, R.D. Meade, *Photonic Crystals: Molding the Flow of Light*, 2nd edn. (Princeton University Press, New Jersey, 2008)
62. R. Maia, D.R. Rubenstein, M.D. Shawkey, Proc. Natl. Acad. Sci. USA **110**, 10687–10692 (2013)
63. R. Maia, D.R. Rubenstein, M.D. Shawkey, Evolution **70**, 1064–1079 (2016)
64. I.W. Hamley, *The Physics of Block Copolymers* (Oxford University Press, Oxford, 1998)
65. F.S. Bates, G.H. Fredrickson, Phys. Today **52**, 32–38 (1999)
66. E.L. Thomas, Science **286**, 1307 (1999)
67. N. Hadjichristidis, S. Pispas, G. Floudas, in *Block Copolymers: Synthetic Strategies, Physical Properties, and Applications* (Wiley, Hoboken, NJ, 2003)
68. K. Fontell, Coll. Poly. Sci. **268**, 264–285 (1990)
69. V. Luzzati, P.A. Spegt, Nature **215**, 701–704 (1967)
70. S.M. Gruner, P.R. Cullis, M.J. Hope, C.P. Tilcock, Annu. Rev. Biophys. Biophys. Chem. **14**, 211–238 (1985)
71. T. Hahn, The 230 space groups, in *IUCr International Tables for Crystallography* (Springer, 2006)
72. A. Ingram, A. Parker, Phil. Trans. R. Soc. B: Biol. Sci. **363**, 2465–2480 (2008)
73. S. Förster, A. Timmann, C. Schellbach, A. Frömsdorf, A. Kornowski, H. Weller, S.V. Roth, P. Lindner. Nat. Mater. **6**, 888–893 (2007)
74. J.W. Galusha, L.R. Richey, J.S. Gardner, J.N. Cha, M.H. Bartl, Phys. Rev. E **77**, 2–5 (2008)
75. J. Jung, J. Lee, H.-W. Park, T. Chang, H. Sugimori, H. Jinnai, Macromolecules **47**, 8761–8767 (2014)
76. M.E. Vigild, K. Almdal, K. Mortensen, I.W. Hamley, J.P.A. Fairclough, A.J. Ryan, Macromolecules **31**, 5702–5716 (1998)
77. Y. Bouligand, Tiss. Cell **4**, 189–217 (1972)
78. B.D. Wilts, H.L. Leertouwer, D.G. Stavenga, J. R. Soc. Interface **6**, S185–S192 (2009)
79. H. Ghiradella, *Advances in Insect Physiology* (Academic Press, 2010)
80. Z. Bálint, K. Kertész, G. Piszter, Z. Vértesy, L.P. Biró, J. R. Soc. Interface **9**, 1745–1756 (2012)
81. G.C. Shearman, O. Ces, R.H. Templer, J.M. Seddon, J. Phys.-Condens. Mater. **18**, S1105–S1124 (2006)
82. W. Helfrich, *Z. Naturforsch. C* **28**, 693–703 (1973)
83. L. Golubovic, Phys. Rev. E **50**, 2419–2422 (1994)
84. E.A. Matsumoto, R.D. Kamien, C.D. Santangelo, Interf. Focus **2**, 617–622 (2012)
85. M. Terasaki, T. Shemesh, N. Kasthuri, R.W. Klemm, R. Schalek, K.J. Hayworth, A.R. Hand, M. Yankova, G. Huber, J.W. Lichtman, T.A. Rapoport, M.M. Kozlov, Cell **154**, 285–296 (2013)
86. C. Delaunay, *J. Math. Pures et Appl.* **6**, 309–320 (1841)
87. N. Kapouleas, Ann. Math. **131**, 239–330 (1990)
88. M. Wohlgemuth, N. Yufa, J. Hoffman, E.L. Thomas, Macromolecules **34**, 6083–6089 (2001)
89. A.H. Schoen, Interface Focus **2**, 658–668 (2012)
90. H.A. Schwarz, *Gesammelte Mathematische Abhandlungen* (Springer, Berlin, 1890)
91. A.H. Schoen, *Infinite Periodic Minimal Surfaces Without Self-intersections* (NASA Research Center, Cambridge, MA, 1970)
92. A. Fogden, S.T. Hyde, Eur. Phys. J. B **7**, 91–104 (1999)
93. J.M. Seddon, R.H. Templer, Phil. Trans. R. Soc. Lond. A **344**(377–401), 1 (1993)
94. V. Luzzati, Curr. Opin. Struct. Biol. **7**, 661–668 (1997)
95. H. Ishikawa, J. Cell Biol. **38**, 51–66 (1968)

96. S. Schiaffino, A. Margreth, J. Cell Biol. **41**, 855–875 (1969)
97. S. Schiaffino, M.G. Nunzi, P. Burighel, Tiss. Cell **8**, 101–110 (1976)
98. T. Landh, FEBS Lett. **369**, 13–17 (1995)
99. Y. Deng, M. Marko, K.F. Buttle, A. Leith, M. Mieczkowski, C.A. Mannella, J. Struct, Biol. **127**, 231–239 (1999)
100. E.L. Snapp, J. Cell Biol. **163**, 257–269 (2003)
101. N. Borgese, M. Francolini, E. Snapp, Curr. Opin. Cell Boil. **18**, 358–364 (2006)
102. Z.A. Almsherqi, T. Landh, S.D. Kohlwein, Y.R. Deng, Int. Rev. Cel. Mol. Bio. **274**, 275–342 (2009)
103. Z.A. Almsherqi, S.D. Kohlwein, Y. Deng, J. Cell Biol. **173**, 839–844 (2006)
104. J. Zimmerberg, M.M. Kozlov, Nat. Rev. Mol. Cell Biol. **7**, 9–19 (2006)
105. L. Norlen, A. Al-Amoudi, J. Investig. Dermatol. **123**, 715–732 (2004)
106. M. Fuhrmans, S.J. Marrink, J. Am. Chem. Soc. **134**, 1543–1552 (2012)
107. E.Y. Lee, M. Marcucci, L. Daniell, M. Pypaert, O.A. Weisz, G.C. Ochoa, K. Farsad, M.R. Wenk, P. De Camilli, Science **297**, 1193–1196 (2002)
108. A. Frost, V.M. Unger, P. De Camilli, Cell **137**, 191–196 (2009)
109. V.M. Unger, A. Frost, P. De Camilli, Structure **15**, 751–753 (2007)
110. V.M. Unger, A. Frost, R. Perera, A. Roux, K. Spasov, O. Destaing, E.H. Egelman, P. De Camilli, Cell **132**, 807–817 (2008)
111. A. Dinwiddie, R. Null, M. Pizzano, L. Chuong, A. Leigh, H. Krup, H. Ee Tan, H. Patel, Dev. Biol. **392**, 404–418 (2014)
112. O. Shalem, N.E. Sanjana, E. Hartenian, X. Shi, D.A. Scott, T. Mikkelson, D. Heckl, B.L. Ebert, D.E. Root, J.G. Doench, F. Zhang, Science **343**, 84–87 (2014)
113. A. Reynolds, D. Leake, Q. Boese, S. Scaringe, W.S. Marshall, A. Khvorova, Nat. Biotech. **22**, 326–330 (2004)
114. P.A. Guerette, S. Hoon, Y. Seow, M. Raida, A. Masic, F.T. Wong, V.H.B. Ho, K.W. Kong, M.C. Demirel, A. Pena-Francesch, S. Amini, G.Z. Tay, D. Ding, A. Miserez. Nat. Biotechnol. **31**, 908–915 (2013)
115. H. Ghiradella, Ann. Entomol. Soc. Am. **78**, 252–264 (1985)
116. H. Ghiradella, M. Butler, J. R. Soc. Interface **6**, S243–S251 (2009)
117. S.B. Carroll, R. Galant, J.B. Skeath, S. Paddock, D.L. Lewis, Curr. Biol. **8**, 807–813 (1998)
118. T.A. Shefelbine, M.E. Vigild, M.W. Matsen, D.A. Hajduk, M.A. Hillmyer, E.L. Cussler, F.S. Bates, J. Am. Chem. Soc. **121**, 8457–8465 (1999)
119. H. Hückstädt, A. Göpfert, V. Abetz, Polymer **41**, 9089–9094 (2000)
120. T. Goldacker, V. Abetz, R. Stadler, I. Erukhimovich, L. Leibler, Nature **398**, 137–139 (1999)
121. E.A. Lord, A.L. Mackay, Curr. Sci. **85**, 346–362 (2003)
122. L. Martin-Moreno, F.J. Garcia-Vidal, A.M. Somoza, Phys. Rev. Lett. **83**, 73–75 (1999)
123. S.C. Kapfer, S.T. Hyde, K. Mecke, C.H. Arns, G.E. Schroder-Turk, Biomaterials **32**, 6875–6882 (2011)
124. A.M. Urbas, E.L. Thomas, H. Kriegs, G. Fytas, R.S. Penciu, L.N. Economou, Phys. Rev. Lett. **90**, 108302 (2003)
125. A.M. Urbas, M. Maldovan, P. DeRege, E.L. Thomas, Adv. Mater. **14**, 1850–1853 (2002)
126. M. Maldovan, C.K. Ullal, W.C. Carter, E.L. Thomas, Nat. Mater. **2**, 664–667 (2003)
127. M. Maldovan, A.M. Urbas, N. Yufa, W.C. Carter, E.L. Thomas, Phys. Rev. B **65**, 165123 (2002)
128. M. Saba, M. Thiel, M.D. Turner, S.T. Hyde, M. Gu, K. Grosse-Brauckmann, D.N. Neshev, K. Mecke, G.E. Schroder-Turk, Phys. Rev. Lett. **106**, 103902 (2011)
129. J.A. Dolan, M. Saba, R. Dehmel, I. Gunkel, Y. Gu, U. Wiesner, O. Hess, T.D. Wilkinson, J.J. Baumberg, U. Steiner, B.D. Wilts, ACS Photon. **3**, 1888–1896 (2016)
130. H.-Y. Hsueh, Y.-C. Ling, H.-F. Wang, L.-Y.C. Chien, Y.-C. Hung, E.L. Thomas, R.-M. Ho, Adv. Mater. **26**, 3225–3229 (2014)
131. X. Cao, D. Xu, Y. Yao, L. Han, O. Terasaki, S. Che, Chem. Mater. **28**, 3691–3702 (2016)
132. T. Sun, P. Tang, F. Qiu, Y. Yang, A.-C. Shi, Macromol. Theor. Simul. (2017)
133. C. Mille, E.C. Tyrode, R.W. Corkery, RSC Adv. **3**, 3109–3117 (2013)
134. M.R. Weatherspoon, Y. Cai, M. Crne, M. Srinivasarao, K.H. Sandhage, Angew. Chem. **47**, 7921–7923 (2008)

Index

S. Gupta and A. Saxena (eds.), *The Role of Topology in Materials*,
Springer Series in Solid-State Sciences 189,
https://doi.org/10.1007/978-3-319-76596-9

Zeitfracht Medien GmbH
Ferdinand-Jühlke-Straße 7
99095 Erfurt, Deutschland
produktsicherheit@kolibri360.de